HEALTHCARE TECHNOLOGIES SERIES 4

Biomedical Nanomaterials

Other volumes in this series:

Volume 1 **Nanobiosensors for Personalized and Onsite Biomedical Diagnosis** Prof. David A. Clifton (Editor)

Volume 2 **Machine Learning for Healthcare Technologies** Prof. David A. Clifton (Editor)

Biomedical Nanomaterials

From Design to Implementation

Edited by Thomas J Webster
and Hilal Yazici

The Institution of Engineering and Technology

Published by The Institution of Engineering and Technology, London, United Kingdom

The Institution of Engineering and Technology is registered as a Charity in England & Wales (no. 211014) and Scotland (no. SC038698).

First published 2016

The Institution of Engineering and Technology
Michael Faraday House
Six Hills Way, Stevenage
Herts, SG1 2AY, United Kingdom

www.theiet.org

British Library Cataloguing in Publication Data
A catalogue record for this product is available from the British Library

ISBN 978-1-84919-964-3 (hardback)
ISBN 978-1-84919-965-0 (PDF)

Typeset in India by MPS Limited
Printed in the UK by CPI Group (UK) Ltd, Croydon

Contents

Biographies

Thomas J. Webster's (H index 75 according to Google Scholar) degrees are in Chemical Engineering from the University of Pittsburgh (BS, 1995) and in Biomedical Engineering from Rensselaer Polytechnic Institute (MS, 1997; PhD, 2000). He is currently the Department Chair and Professor of Chemical Engineering at Northeastern University in Boston. His research explores the use of nanotechnology in numerous applications. Specifically, his research addresses the design, synthesis, and evaluation of nanophase materials (i.e. materials with fundamental length scales less than 100 nm) as more effective biomedical devices. He has completed extensive studies on the use of nanophase materials to regenerate tissues and has graduated/supervised over 109 visiting faculty, clinical fellows, post-doctoral students, and thesis completing BS, MS, and PhD students. To date, his lab group has generated over 9 textbooks, 48 book chapters, 306 invited presentations, at least 403 peer-reviewed literature articles, at least 567 conference presentations, and 32 provisional or full patents. Some of these patents led to the formation of nine companies. His research on nanomedicine has received attention in recent media publications including MSNBC (October 10, 2005), NBC Nightly News (May 14, 2007), PBS DragonFly TV (covered across the United States during the winter, 2008), ABC Nightly News via the Ivanhoe Medical Breakthrough Segment (covered across the United States during the winters of 2008 and separate research segments in 2010 and 2011), Fox News (December 18, 2013), and the Weather Channel (March 18, 2014). His work has been on display at the London and Boston Science Museums. He is the founding Editor-in-Chief of the *International Journal of Nanomedicine* (the first international journal in nanomedicine which in 5 years has achieved an impact factor of 4.97), serves on the editorial board of 15 additional journals, has helped to organize 22 conferences emphasizing nanotechnology in medicine, and has organized over 53 symposia at numerous conferences emphasizing biological interactions with nanomaterials. He also recently chaired the 2011 Annual Biomedical Engineering Society (BMES) Conference and has organized numerous symposia for AIChE, IEEE, MRS and ASME Annual Meetings. He has received numerous honors including, but not limited to, 2002, Biomedical Engineering Society Rita Schaffer Young Investigator Award; 2003, Outstanding Young Investigator Award Purdue University College of Engineering; 2005, American Association of Nanomedicine Young Investigator Award Finalist; 2005, Coulter Foundation Young Investigator Award; 2006, Fellow, American Association of Nanomedicine; 2010, Distinguished Lecturer in Nanomedicine, University of South Florida; 2011, Outstanding Leadership Award for the

Biomedical Engineering Society (BMES); 2011, Fellow, American Institute for Medical and Biological Engineering (AIMBE, representing the top 2% of all medical and biological engineers); 2013, Fellow, Biomedical Engineering Society; 2014, Fellow, Ernst Strugmann Foundation; 2014, President-elect, U.S. Society For Biomaterials; and 2016, Fellow, FBSE.

Dr. Hilal Yazici joined TUBITAK (The Scientific and Technological Research Council of Turkey)-GEBI (Genetic Engineering and Biotechnology Institute) in 2016 as Senior Researcher. She was previously a postdoctoral research associate in Prof. Thomas J. Webster's Nanomedicine Group at Northeastern University. Prior to that, she was a postdoctoral research associate at the University of Potsdam in the Institute of Biology and Biochemistry and Technical University of Berlin in the Department of Biophysical Chemistry in between 2012 and 2014. She received her MSc and PhD from Istanbul Technical University in the Molecular Biology-Genetics and Biotechnology program in 2005 and 2012, respectively. During her PhD, she was also a graduate student and a research assistant in the Material Science and Engineering Department and Genetically Engineered Materials Science and Engineering Center at the University of Washington between 2006 and 2010. She was involved in various projects that were founded by the Turkish State Planning Organization, NSF, NIH, EU and The Scientific and Technological Research Council of Turkey (TUBITAK). She has authored over 50 peer-reviewed journal papers, book chapters, conference proceedings, and presentations. Dr. Yazici's research interests include molecular biomimetics, material binding peptides, biological coating of biomaterials, nanoparticle-based therapies and protein-enabled biosensors.

Preface

Nanotechnology has already begun to revolutionize numerous industries including aerospace, building construction, solar, electronics, and others. Over the last decade, the so-called "nanotechnology era" has also brought new and immense perspectives for the development of new biomaterials. This broad view has created enormous opportunities to improve the prevention, diagnosis, and treatment of numerous diseases. Thus, "nanotechnology-derived biomaterials" is an emerging field that intersects materials science and biology with an extraordinary impact on medicine. Specifically, nanotechnology-derived biomaterials refer to structures or components that exhibit novel and significant properties when their dimensions are at the nanoscale (i.e., $1-100$ nm or $10^{-9}-10^{-7}$ m particle or grain size diameters). They have been widely used in a broad spectrum of biological and biomedical applications from artificial implants to drug delivery to medical imaging. These nanobiomaterials include (but are not limited to) metals, ceramics, polymers, hydrogels, and novel self-assembled materials with structures from 0D (e.g., dots and particles) to 3D (e.g., tissue engineering scaffolds).

Undoubtedly, a new multidisciplinary area across biology, materials science, and nanotechnology has emerged through the rapid development of nanobiomaterials. Although there are various reasons behind this newly establishment area, nanotechnology has not only provided novel materials and tools for biological and medical purposes, but it is also reshaping our thinking, imagination, and creativity toward applying these materials to science and technology. Therefore, today, studies on nanostructured biomaterials have evolved into more comprehensive and systematic studies, resulting in the further understanding of the mechanisms behind biological responses to materials and the consequent better design of such materials. Thus, this book is organized first by discussing the rationale and application of nanobiomaterials in medicine, covering nanobiomaterials used in a wide range of tissue engineering applications (hard and soft tissues), preventing bacterial infection, and ending with an emphasis on current and future potential clinical applications of nanomaterials. In addition, the application of nanomaterials in the clinics concerning current regulatory policies from all over the world is also included. This book will serve as a critical resource for all graduate students, undergraduate students, and faculty at universities, hospitals, and industries in nanomedicine.

Preface

Part I

Nanomaterials for hard tissue engineering

Chapter 1

Biomimetic coatings of calcium phosphates on titanium alloys

Bengi Yilmaz[1] and Zafer Evis[1,2]

Abstract

Metallic orthopedic implants have been widely used to repair skeletal defects and poorly functioning joints for many years. Among them, titanium (Ti) and its alloys can be described as the most frequently used orthopedic metals, due to their unique properties, such as a good mechanical match with hard tissues, resistance to corrosion in body fluids, and being lightweight. However, Ti and its alloys fail to show bioactivity which is a very important issue for the integration of implants with bone tissue. Therefore, hydroxyapatite [HA, $Ca_{10}(PO_4)_6(OH)_2$], which is the main constituent of the inorganic part of natural bone, has been traditionally used as a coating to achieve the promoted bioactivity on the metallic implant surfaces. There are various techniques for coating HA on metallic implants such as plasma spraying, electrodeposition, pulsed-laser deposition, sputtering, sol–gel, and biomimetics. Simulated body fluid (SBF) used in biomimetic calcium phosphate (CaP) coatings basically mimics the inorganic composition, pH, and temperature of normal human blood plasma. Normally, a Ti alloy implant can be biomimetically coated in 14–28 days depending on the desired coating thickness if SBF solution is replenished every 2 days. Therefore, more concentrated formulations of SBF such as 1.5× SBF, 5× SBF, and 10× SBF were suggested for faster deposition rates with a drawback of decreased solution stability. To deposit CaPs on Ti alloy substrates, there are many different formulations of SBF with varying ionic compositions of certain ions, such as chloride (Cl^-), bicarbonate (HCO_3^-), and sulfate (SO_4^{2-}), and with different buffers such as tri(hydroxymethyl)aminomethane (TRIS) and Hank's balanced salt solution (HEPES). In recent studies, it was shown that it is also possible to coat CaPs doped with specific ions or loaded with antibacterial agents and bone stimulating factors via biomimetic routes.

[1]Middle East Technical University, Department of Biomedical Engineering, Ankara, Turkey
[2]Middle East Technical University, Department of Engineering Sciences, Ankara, Turkey

1.1 Biomimetics

The definition of biomimetics is the application of methods and systems that are already present in nature to technology and engineering [1]. The science of transfer between nature and man-made devices is a focus of interest for many different disciplines from materials science and architecture to computer science and robotics [2]. The rationale of mimicking biology of natural materials in biomaterials science is based on getting a superior biological response from the natural tissue. In particular, biomimetics is an area of great interest for calcified tissue engineering, since highly controlled structures and patterned arrangements are the basis of biomimetics such as that of bone tissue.

To mimic a natural tissue, it is essential to understand the normal structure and functioning of the tissue. The mineralized hard tissues are formed by a natural route in which the organic matrix is laid down first and the inorganic reinforcing phase grows within this organic matrix/template [3]. Specifically, bone has sophisticated macro-, micro-, and nanoscale hierarchical structures. Tissues comprising bone include osseous tissue, marrow, endosteum and periosteum, nerves, blood vessels, and cartilage [4]. It is very difficult to replicate the total complexity of the bone tissue for now. However, the mineral phase of bone is very well characterized and known to consist of HA in the form of thin (1.5–4 nm thick) plate-shaped crystals that contain carbonate in its lattice [5]. This inorganic structure can be produced artificially through different methods.

In bone tissue engineering, the term "biomimetic" can be used to refer 3D printed bone scaffold materials mostly of inorganic/organic hybrid compositions that may or may not contain any members of the CaP family. Biomimetics, as used herein, is intended to mean the route of depositing HA in the same way that it happens in the human body.

A biomimetic HA coating method is based on dipping metallic implants, such as Ti alloys, tantalum (Ta), and magnesium (Mg) alloys, in SBF at the physiological temperature (37°C) and pH (7.4). It has the following main advantages: (i) it is applicable to any heat-sensitive substrate (including polymers) because of it being a low temperature process, (ii) bone-like apatite crystals having high bioactivity and good resorption characteristics can be formed, (iii) it can be deposited on porous or complex implant geometries, (iv) it can incorporate bone-growth-stimulating factors [6], and (v) no costly investments are required. However, there is a concern about the relatively poor bonding between the coating and substrate still waiting to be solved.

This method yields poorly crystalline HA close to the inorganic component of bone, which is calcium deficient hydroxyapatite (CDHA) with small amounts of sodium (Na^+), potassium (K^+), magnesium (Mg^{2+}), chloride (Cl^-), and carbonate (CO_3^{2-}) ions present in the lattice structure [7]. Many elements such as Sr^{2+}, Pb^{4+}, Ba^{2+}, Fe^{3+}, Zn^{2+}, and Cu^{2+} may substitute at Ca^{2+} sites in human bone mineral, while various other substitutions, especially HPO_4^{2-}, are possible at PO_4^{3-} sites, both in trace amounts [8]. The biomimetic method also makes it possible to dope ions into the HA structure for different purposes.

As stated before, bone mineralization process occurs in situ by immersing implants in SBF that mimics the inorganic composition, pH, and temperature of human blood plasma [9]. SBF simulates just the inorganic part of human blood plasma and does not contain proteins, glucose, vitamins, hormones, etc.

1.2 Simulated body fluid (SBF) as a coating solution

Since the first formulation proposed by Kokubo for the SBF calcifying solution, a number of slightly different compositions were proposed [10]. A comparison of the concentrations of ions (mM/L) in human plasma and in several different SBF solutions is given in Table 1.1.

It can be noted that the original SBF used by Kokubo and Takadama [11] lacks the SO_4^{2-} ions contained in human blood plasma. This was corrected then as c-SBF (sometimes referred as conventional SBF [14]) and has been used as "SBF" by many researchers. However, the corrected c-SBF was still rich in Cl^- ions and poor in HCO_3^- ions when compared to human blood plasma [11]. Oyane *et al.* [14] tried to correct this difference. They prepared r-SBF and i-SBF which had ion concentrations equal to that of blood plasma and m-SBF whose only HCO_3^- ion concentration was different from that of blood plasma. However, r-SBF and i-SBF lacked long-term stability due to a strong tendency to precipitate both apatite and calcite from these SBF [9, 14]. In 2004, Takadama *et al.* [15] proposed n-SBF in which they decreased only the Cl^- ion concentration to the level of human blood plasma, leaving the HCO_3^- ion concentration equal to that of c-SBF. This improved SBF was compared with c-SBF in its stability and the reproducibility of apatite formation on synthetic materials via round robin testing. As a result, it was confirmed that the c-SBF does not differ from n-SBF in stability and reproducibility [11,15].

Table 1.1 Ionic concentrations (mM) of blood plasma and several SBF solutions in literature

Ion	Na^+	K^+	Mg^{2+}	Ca^{2+}	Cl^-	HCO_3^-	HPO_4^{2-}	SO_4^{2-}	Buffer
Blood plasma [10]	142.0	5.0	1.5	2.5	103.0	27.0	1.0	0.5	
Original SBF [11]	142.0	5.0	1.5	2.5	148.8	4.2	1.0	0	TRIS
Corrected (c-SBF) [12]	142.0	5.0	1.5	2.5	147.8	4.2	1.0	0.5	TRIS
Tas-SBF [13]	142.0	5.0	1.5	2.5	125.0	27.0	1.0	0.5	TRIS
Bigi-SBF [10]	141.5	5.0	1.5	2.5	124.5	27.0	1.0	0.5	HEPES
Revised SBF (r-SBF) [14]	142.0	5.0	1.5	2.5	103.0	27.0	1.0	0.5	HEPES
Modified (m-SBF) [14]	142.0	5.0	1.5	2.5	103.0	10.0	1.0	0.5	HEPES
Ionized (i-SBF) [14]	142.0	5.0	1.0	1.6	103.0	27.0	1.0	0.5	HEPES
Newly improved SBF (n-SBF) [15]	142.0	5.0	1.5	2.5	103.0	4.2	1.0	0.5	TRIS
5× SBF [6]	714.8	0	7.5	12.5	723.8	21.0	5.0	0	

SBF, simulated body fluid; TRIS, tri(hydroxymethyl)aminomethane; HEPES, Hank's balanced salt solution.

Other CaP solutions were reported continuously by the scientists preparing biomimetic coatings on metallic implant materials. Duan *et al.* [16] suggested an evaporation-induced surface crystallization process, which does not require surface etching, high supersaturation, or tight control of solution conditions. In this method, metallic samples were dipped into a supersaturated CaP solution consisting of 2.32 mM $NH_4H_2PO_4$, 3.87 mM $CaCl_2$, 150 mM NaCl, 40 mM HCl, and 50 mM TRIS, and it was termed as the accelerated calcification solution (ACS).

It is possible to use other types of artificially prepared simulated physiological solutions, such as HBSS [17], supersaturated calcification solution (SCS) [18], phosphate buffered saline (PBS) [19], as well as Dulbecco's modified Eagle medium (DMEM) [20], as the calcifying solution that allows deposition of a uniform apatite layer on metallic biomaterials. Theoretically, any solution that mimics the concentration of inorganic ions present in the human hard tissue environments can be used in biomimetic coatings at the proper temperature and pH. Most of the time, the solutions other than SBF are developed for different purposes, such as *in vitro* testing of synthetic biomaterials or growing cells at tissue culture conditions, and they are not widely used in the biomimetic synthesis of CaPs.

1.3 HA coating of titanium alloys

The titanium element (Ti) was discovered by William Gregor in 1790. However, it was named after Titans of Greek mythology by Martin Heinrich Klaproth in 1795. It is one of the transition elements with an atomic number of 22 and an atomic weight of 47.9 [21]. The studies about titanium as an implant material began in the 1950s at Cambridge University [22]. In 1952, Per Ingvar Brånemark implanted a titanium cylinder by screwing into a rabbit bone to observe its healing reactions. He fortuitously discovered that it was tightly integrated in the bone while he was trying to remove it from the bone after completion of the experiment [23]. Brånemark named this phenomenon osseointegration and later, he developed pure titanium screws for dental implant fixtures [22,23]. Nowadays, more than 1,000 tons of Ti are implanted into patients in various forms worldwide annually [24].

Ti, which is a low-density element (approximately 60% of the density of steel), can be strengthened by alloying and deformation processes [22]. For example, the addition of aluminum (Al) and vanadium (V) into pure Ti greatly increases its tensile strength [25]. Ti alloys have a Young's modulus about half that of stainless steel and Co–Cr alloys, and they are preferred materials for bone fixators because of their low Young's modulus [26].

Ti and its alloys are generally chemically durable and bioinert, since they are normally covered with a passive thin titania (TiO_2) layer [27] which does not support the precipitation of HA intrinsically. The bioactivity of Ti implants can be improved by using various chemical treatments, such as acid, hydrogen peroxide (H_2O_2), alkali treatments, and anodic oxidation (i.e. hydrothermal treatment). It was reported that only by heat treatment, the rutile TiO_2 surfaces were found to be bioactive: After 7 days of immersion into Dulbecco's PBS, a layer of HA several micrometers thick covered the Ti plates [28]. However, an alkali treatment provides a novel and simple technique [29] and it is widely applied. In an alkali treatment, the TiO_2 layer on the surface of Ti alloys can react with an alkali solution of sodium hydroxide (NaOH) to form a sodium titanate hydrogel layer. This treatment can be applied prior to a biomimetic HA coating as a way to enhance the bioactivity of Ti and its alloys.

The process of the apatite formation in SBF with the effects of the alkali treatment is demonstrated in Figure 1.1.

When Ti metal and its alloys are soaked in an alkaline solution, a hydrated titanium oxide gel layer containing alkali ions is formed on their surfaces [30]. During heat treatment, the hydrogel layer is dehydrated and densified to form a stable alkali titanate layer [31]. More specifically, when the metal is treated with 5 M NaOH aqueous solution at 60°C for 24 h, a sodium titanate hydrogel layer can be formed on its surface [32]. If the NaOH and heat-treated metal is soaked in SBF, it releases Na^+ ions via ion exchange with the H_3O^- ions in the SBF and forms Ti–OH groups on its surface [30]. The Ti–OH groups formed immediately combine with Ca^{2+} ions in the SBF to form calcium titanate. This calcium titanate later combines with phosphate ions in the SBF to form amorphous CaP during this complex process of apatite formation.

This can be interpreted in terms of the electrostatic interaction of the functional groups with the ions in the SBF. The Ti–OH groups formed on the surface of sodium

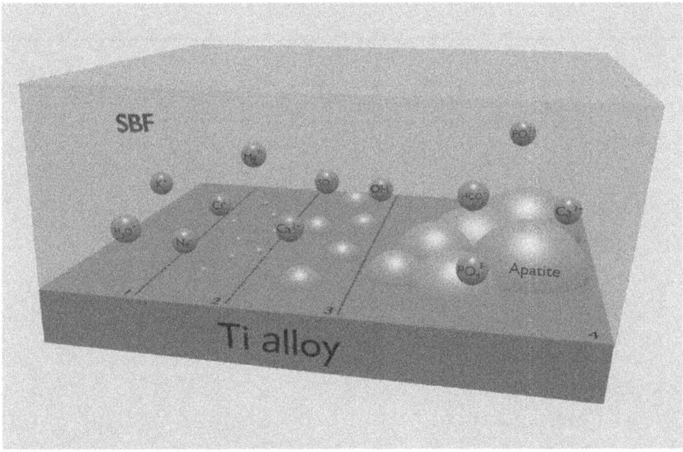

Figure 1.1 A schematic illustration of the apatite formation on the surface of an alkali-heat treated Ti based alloy soaked in SBF. Numbers (1–4) indicate the stages of apatite deposition: (1) alkali and heat-treated bioactive Ti alloy surface with a sodium titanate layer, (2) nucleation of CaP crystals, (3) deposition of CaP on the nuclei, and (4) formation of an apatite layer in the form of half spheres with nanotexture (sizes are arbitrary)

titanate after soaking in SBF are negatively charged and hence combine selectively with the positively charged Ca^{2+} ions in the fluid to form calcium titanate. As the calcium ions accumulate on the surface, the surface gradually gains an overall positive charge. As a result, the positively charged surface combines with negatively charged phosphate ions to form CaP [33]. Once the apatite nuclei are formed, they spontaneously grow by consuming the calcium and phosphate ions from the surrounding fluid, since the body fluid is already supersaturated with respect to the apatite [34]. This is also the reason of the necessity of SBF replenishment (e.g. in every 2 days) during the coating process with the purpose of supplying calcium and phosphate ions for the continuation of the apatite formation.

The equation of the precipitation of HA in SBF is as follows [35]:

$$5Ca^{2+} + 3PO_4{}^{3-} + OH^- \rightarrow Ca_5(PO_4)_3(OH)\downarrow \qquad (1)$$

The possible equilibrium chemical reactions involved in the process are listed in Table 1.2.

The concentrations of the ionic constituents of the human blood plasma in SBF are important, but the buffering agent also has an influence on the final coating. To coat CaP on Ti6Al4V alloy substrates, a SBF solution (Lac-SBF) was developed by Pasinli *et al.* [36], in which Ca^{2+}, Na^+, Mg^{2+}, K^+, $HPO_4{}^{2-}$, $HCO_3{}^-$, Cl^-, and $SO_2{}^-$

Table 1.2 Association/dissociation reactions in SBF and their equilibrium constants (K) [35]

Reaction	K
$H_2CO_3(aq.) \rightleftharpoons H^+ + HCO_3{}^-$	$10^{-6.31}$
$HCO_3{}^- \rightleftharpoons H^+ + CO_3{}^{2-}$	$10^{-10.25}$
$H_3PO_4(aq.) \rightleftharpoons H^+ + H_2PO_4{}^-$	$10^{-2.196}$
$H_2PO_4{}^- \rightleftharpoons H^+ + HPO_4{}^{2-}$	$10^{-7.185}$
$HPO_4{}^{2-} \rightleftharpoons H^+ + PO_4{}^{3-}$	$10^{-12.19}$
$Ca^{2+} + HCO_3{}^- \rightleftharpoons CaHCO_3{}^+$	$10^{1.16}$
$Ca^{2+} + CO_3{}^{2-} \rightleftharpoons CaCO_3(aq.)$	$10^{3.38}$
$Ca^{2+} + OH^- \rightleftharpoons CaOH^+(aq.)$	25.12
$Ca^{2+} + H_2PO_4{}^- \rightleftharpoons CaH_2PO_4{}^+$	31.9
$Ca^{2+} + HPO_4{}^{2-} \rightleftharpoons CaHPO_4(aq.)$	6.81×10^2
$Ca^{2+} + PO_4{}^{3-} \rightleftharpoons CaPO_4{}^-$	3.46×10^6
$Mg^{2+} + HCO_3{}^- \rightleftharpoons MgHCO_3{}^+$	$10^{0.62}$
$Mg^{2+} + CO_3{}^{2-} \rightleftharpoons MgCO_3(aq.)$	$10^{1.87}$
$Mg^{2+} + OH^- \rightleftharpoons MgOH^+$	$10^{2.19}$
$Mg^{2+} + H_2PO_4{}^- \rightleftharpoons MgH_2PO_4{}^+$	$10^{0.4}$
$Mg^{2+} + HPO_4{}^{2-} \rightleftharpoons MgHPO_4(aq.)$	$10^{1.8}$
$Mg^{2+} + PO_4{}^{3-} \rightleftharpoons MgPO_4{}^-$	$10^{3.3}$

ion concentrations are similar to those of human blood plasma. However, it contains appropriate amounts of sodium lactate (NaL) and lactic acid (HL). The coating thickness was reported as 4 μm after 48 h. The binding force of the surface coatings was measured by a scratch test and differed as 9 or 3 mN according to the processes applied to Ti6Al4V plates before the coating process. It was reported that Ca^{2+} and HPO_4^{2-} rich solutions (compared to traditional SBF) resulted in shortening of the coating time.

More recent studies focus on the feasibility to use a biomimetic method to coat HA on the metallic substrates in short soaking times. The geometry of the substrate, especially high porosity, and the type of the metallic biomaterial other than Ti alloys are still questions for scientists to determine the applicability of the method. Since the method is solution based, it is known to be applicable to porous implants provided that the implant has open porosity. Lindahl *et al.* [37] prepared partially crystalline, slightly CDHA coatings composed of plate-like crystallites forming roundish flowers in the size range of 300–500 nm on flat and porous CoCr substrates by using PBS with Ca^{2+} as the coating solution. The results also showed a coating with relatively similar morphology and thickness inside the pore structures as on the outer surfaces and on the flat CoCr substrates.

Human bone is comprised of homogeneous plate-like HA crystals with widths of 15–30 nm and lengths of 30–50 nm, and HA crystals in enamel are rod-like with diameters of 25–100 nm and lengths of 100 nm to microns [38]. Since the biomimetic deposition procedure simply mimics the natural precipitation process in human bone, it also allows the deposition of HA crystals with nanoscale dimensions [39].

The HA coating obtained by the biomimetic process is given in Figure 1.2. It generally consists of globular aggregates with a finer lamellar structure. Figure 1.2(a) shows the HA spheroids which generally exhibit a mean diameter of several

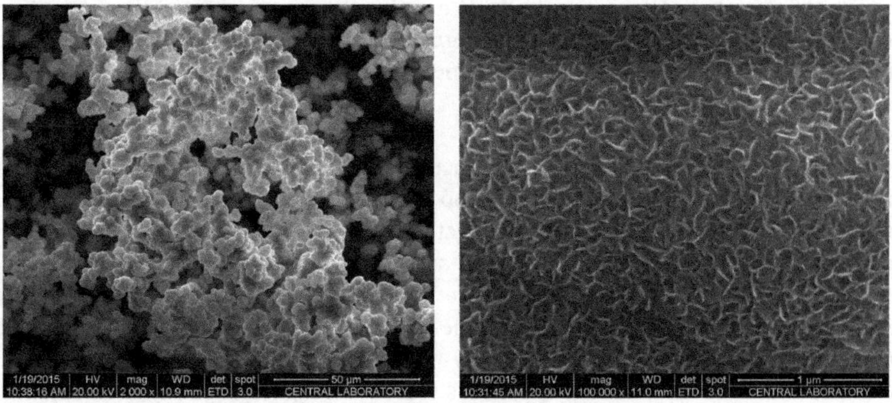

Figure 1.2 Scanning electron microscopy (SEM) micrographs of Ti6Al4V plates after soaking in 1.5× SBF for 14 days. Magnification: (a) 2,000× and (b) 100,000×

micrometers. The spheroids are made up of thin elongated plate-like crystals with a mean length of about several hundred nanometers (see Figure 1.2(b)).

Transmission electron microscopy (TEM) is a useful tool for determining the gradual development of HA with a morphological variation. With the use of TEM, it was previously shown that the biomimetic HA nanocrystals gradually grew from a crystallite size of 20 to 50 nm after immersing the substrates for a period of 6 days in SBF [40]. Bigi *et al.* [10] reported that the apatite layer grown in 1.5× SBF after 14 days exhibits HA spheroids with a mean diameter of 3–4 μm. These fine nanoapatite crystallites with half sphere shapes are considered as porous apatite structures [41].

1.4 Functionalization of HA coating via biomimetic method

When compared to the pure HA, ion doped HAs have more advantages in terms of enabling the production of orthopedic implants with functionalized coatings. Although human hard tissues are mainly composed of HA, they also have impurity ions other than Ca^{2+} and PO_4^{3-}.

HA is a member of inorganic crystalline compounds with the general formula $M_{10}(XO_4)_6Y_2$. M is usually a bivalent cation, such as Ca^{2+}, Sr^{2+}, Ba^{2+}, Cd^{2+}, Pb^{2+}, but monovalent and trivalent cations, such as Na^+, K^+, and Al^{3+}, can also be incorporated into this site as well; XO_4 is usually PO_4^{3-}, VO_4^{3-}, and AsO_4^{3-}, but also SiO_4^{3-}, CO_3^{2-}, and SO_4^{2-} are possible substitutions; Y is a monovalent anion, OH^-, F^-, Cl^-, Br^- [42]. Based on this, scientists attempted to study doping and co-doping of various ions into the structure of HA with different purposes via different methods. Biomimetic HA coating methods allow ion doping into the structure of HA by changing the formulation of the coating solution.

There are a limited number of studies to produce orthopedic implants with functionalized surfaces, especially which are designed to prevent cancer occurrence or the proliferation of cancerous cells. When this is achieved, one of the difficulties in systemic drug delivery, which is failure in sending the proper amount of a drug to the intended target, would be overcome. It makes sense to use the implant itself as a drug delivery system (local drug delivery), which has been problematic for currently used metals [43].

On the other hand, the studies that focus on doping ions into the structure of HA coating to obtain antibacterial, antiosteoporotic, and bone stimulant properties are more common. Ballo *et al.* [44] investigated the effects of biomimetic silicon (Si)-doped HA and strontium (Sr)-doped HA surface modifications for implant osseointegration at the early stages of implantation in an animal model. Ion-doped HA surfaces were prepared by immersing the pre-treated Ti screws into the PBS with Ca^{2+}, $H_2PO_4^-$, HPO_4^{2-}, and Sr^{2+}/SiO_3^{2-} ions. Si-doped HA yielded a higher amount of bone–implant contact, whereas significant higher bone area was obtained with the Sr-doped HA, when compared with the nondoped HA modification.

Similarly, modified PBS solutions, containing additions of Sr, Si, or fluoride (F), were previously used to obtain biomimetically ion-doped HA coatings [45]. The solubility of HA and therefore the bioactivity and osteoconductivity of the ion-doped material change according to the nature and amount of dopant element.

It was shown that the release of ions from the doped HA coatings was SrCaP > SiHA > FHA in TRIS–HCl, which could be also be related to the solubility for these coatings. In the phosphate buffer solution, a concurrent redeposition of new apatite crystals on the surface of all of the ion-doped coatings addressed both a dissolution and a remineralization process.

The biomimetic HA coating method was also used to take the advantage of the antibacterial property of silver nanoparticles. Ionita *et al.* [46] prepared a 5× SBF solution containing NaCl (733 mM/L), $CaCl_2{\cdot}2H_2O$ (12.5 mM), $MgCl_2{\cdot}6H_2O$ (7.5 mM), $Na_2HPO_4{\cdot}2H_2O$ (5.0 mM), and $NaHCO_3$ (21.0 mM). A thin amorphous calcium phosphate (ACP) layer was obtained on the TiAlZr substrates biomimetically. A CaP supersaturated solution (CPS) with same reagents in different concentrations and $AgNO_3$ (0.1 mM) and tobramycin was used to obtain an antibacterial HA coating [46].

1.5 Conclusions

Biomimetic HA with nano- and microcrystals can be synthesized to exhibit excellent properties like biocompatibility, bioactivity, osteoconductivity, and direct bonding to bone, exciting new applications in the fields of bone tissue engineering and in orthopedic and dental therapies [47]. This method can form nanostructured bone-like apatite layers on the surfaces of Ti and Ti alloy substrates even with complex shapes and also HA can incorporate bone growth stimulating factors or dopant ions biomimetically. The resulting coatings are beneficial as artificial bone substitutes due to their enhanced bioactivity, excellent adhesion to the natural tissue, uniform thickness, and low production costs.

References

[1] Pallua N., Suscheck C.V. (eds.). *Tissue Engineering.* Berlin Heidelberg: Springer; 2011. pp. 31–39.

[2] Lepora N.F., Verschure P., Prescott T.J. 'The state of the art in biomimetics'. *Bioinspiration & Biomimetics.* 2013;**8**(1):013001.

[3] Reis R.L. 'Biomimetics'. *Current Opinion in Solid State and Materials Science.* 2003;**7**(4):263–264.

[4] Porter J.R., Ruckh T.T., Popat K.C. 'Bone tissue engineering: a review in bone biomimetics and drug delivery strategies'. *Biotechnology Progress.* 2009;**25**(6):1539–1560.

[5] Reznikov N., Shahar R., Weiner S. 'Bone hierarchical structure in three dimensions'. *Acta Biomaterialia.* 2014;**10**(9):3815–3826.

[6] Habibovic P., Barrere F., Blitterswijk C.A., Groot K., Layrolle P. 'Biomimetic hydroxyapatite coating on metal implants'. *Journal of the American Ceramic Society.* 2002;**85**(3): 517–522.

[7] Zhou H., Lawrence J.G., Touny A.H., Bhaduri S.B. 'Biomimetic coating of bisphosphonate incorporated CDHA on Ti6Al4V'. *Journal of Materials Science: Materials in Medicine.* 2012;**23**(2):365–374.

[8] Pan H., Zhao X., Darvell B.W., Lu W.W. 'Apatite-formation ability – predictor of "bioactivity"?' *Acta Biomaterialia.* 2010;**6**(11):4181–4188.

[9] George A. (ed.). *Advances in Biomimetic.* Croatia: InTech; 2011. pp. 397–428.

[10] Bigi A., Boanini E., Bracci B., *et al.* 'Nanocrystalline hydroxyapatite coatings on titanium: a new fast biomimetic method'. *Biomaterials.* 2005;**26**(19): 4085–4089.

[11] Kokubo T., Takadama H. 'How useful is SBF in predicting in vivo bone bioactivity?' *Biomaterials.* 2006;**27**(15):2907–2915.

[12] Cui X., Kim H.M., Kawashita M., *et al.* 'Apatite formation on anodized Ti-6Al-4V alloy in simulated body fluid'. *Metals and Materials International.* 2010;**16**(3):407–412.

[13] Tas A.C. 'Synthesis of biomimetic Ca-hydroxyapatite powders at 37°C in synthetic body fluids'. *Biomaterials.* 2000;**21**(14):1429–1438.

[14] Oyane A., Onuma K., Ito A., Kim H.M., Kokubo T., Nakamura, T. 'Formation and growth of clusters in conventional and new kinds of simulated body fluids'. *Journal of Biomedical Materials Research Part A.* 2003;**64**(2):339–348.

[15] Takadama H., Hashimoto M., Mizuno M., Kokubo T. 'Round-robin test of SBF for in vitro measurement of apatite-forming ability of synthetic materials'. *Phosphorus Research Bulletin.* 2004;**17**(0):119–125.

[16] Duan K., Tang A., Wang R. 'A new evaporation-based method for the preparation of biomimetic calcium phosphate coatings on metals'. *Materials Science and Engineering: C.* 2009;**29**(4):1334–1337.

[17] Hiromoto S., Shishido T., Yamamoto A., Maruyama N., Somekawa H., Mukai T. 'Precipitation control of calcium phosphate on pure magnesium by anodization'. *Corrosion Science.* 2008;**50**(10):2906–2913.

[18] Choi J., Bogdanski D., Köller M., *et al.* 'Calcium phosphate coating of nickel–titanium shape-memory alloys. Coating procedure and adherence of leukocytes and platelets'. *Biomaterials.* 2003;**24**(21):3689–3696.

[19] Chen C., Kong X., Zhang S.M., Lee I.S. 'Characterization and in vitro biological evaluation of mineral/osteogenic growth peptide nanocomposites synthesized biomimetically on titanium'. *Applied Surface Science.* 2015; **334**:62–68.

[20] Tas A.C. 'Grade-1 titanium soaked in a DMEM solution at 37°C'. *Materials Science and Engineering: C.* 2014;**36**:84–94.

[21] Liu X., Chu P.K., Ding C. 'Surface modification of titanium, titanium alloys, and related materials for biomedical applications'. *Materials Science and Engineering: R: Reports.* 2004; **47**(3):49–121.

[22] Narayan R. (ed). *ASM Handbook, Volume 23, Materials for Medical Devices.* ASM International; 2012. pp. 223–249.

[23] Ratner B.D., Hoffman A.S., Schoen F.I., Lemons J.E. *Biomaterials Science: An Introduction to Materials in Medicine*. 3rd edn. Canada: Academic Press; 2013, pp. xxvii.

[24] Addison O., Davenport A.J., Newport R.J., *et al.* 'Do "passive" medical titanium surfaces deteriorate in service in the absence of wear?' *Journal of the Royal Society Interface.* 2012;**9**(76):3161–3164.

[25] Hermawan, H. *Biodegradable Metals: From Concept to Applications.* Berlin Heidelberg: Springer; 2012. p. 4.

[26] Niinomi M. (ed.). *Metals for Biomedical Device.* Elsevier; 2010. pp. 3–24.

[27] Somiya S., Aldinger F., Claussen N., Spriggs R.M., Uchino K., Kuomoto K. (eds.), *Handbook of Advanced Ceramics, Volume II, Processing and Their Applications.* Oxford: Elsevier Academic Press; 2003. pp. 385–416.

[28] Forsgren J., Svahn F., Jarmar T., Engqvist H. 'Formation and adhesion of biomimetic hydroxyapatite deposited on titanium substrates'. *Acta Biomaterialia.* 2007;**3**(6):980–984.

[29] Mukherjee A. (ed.). *Biomimetics, Learning From Nature.* Croatia: Intech; 2010. pp. 415–450.

[30] Kokubo T., Miyaji F., Kim H.M., Nakamura T. 'Spontaneous formation of bonelike apatite layer on chemically treated titanium metals'. *Journal of the American Ceramic Society.* 1996;**79**(4):1127–1129.

[31] Kim H.M., Miyaji F., Kokubo T., Nakamura T. 'Preparation of bioactive Ti and its alloys via simple chemical surface treatment'. *Journal of Biomedical Materials Research.* 1996;**32**(3):409–417.

[32] Kokubo T. 'Apatite formation on surfaces of ceramics, metals and polymers in body environment'. *Acta Materialia.* 1998;**46**(7):2519–2527.

[33] Kokubo T., Kim H.M., Kawashita M. 'Novel bioactive materials with different mechanical properties'. *Biomaterials.* 2003;**24**(13):2161–2175.

[34] Wei M., Uchida M., Kim H.M., Kokubo T., Nakamura T. 'Apatite-forming ability of CaO-containing titania'. *Biomaterials.* 2002;**23**(1):167–172.

[35] Lu X., Leng Y. 'Theoretical analysis of calcium phosphate precipitation in simulated body fluid'. *Biomaterials.* 2005;**26**(10):1097–1108.

[36] Pasinli A., Yuksel M., Celik E., Sener S., Tas A.C. 'A new approach in biomimetic synthesis of calcium phosphate coatings using lactic acid–Na lactate buffered body fluid solution'. *Acta Biomaterialia.* 2010;**6**(6):2282–2288.

[37] Lindahl C., Xia W., Engqvist H., Snis A., Lausmaa J., Palmquist A. 'Biomimetic calcium phosphate coating of additively manufactured porous CoCr implants'. *Applied Surface Science.* 2015;**353**:40–47.

[38] Cai Y., Tang R. 'Calcium phosphate nanoparticles in biomineralization and biomaterials'. *Journal of Materials Chemistry.* 2008;**18**(32):3775–3787.

[39] Christenson E.M., Anseth K.S., van den Beucken J.J.J.P., *et al.* 'Nanobiomaterial applications in orthopedics'. *Journal of Orthopaedic Research.* 2007;**25**(1):11–22.

[40] Chakraborty J., Daneu N., Recnik A., *et al.* 'Stepwise formation of crystalline apatite in the biomimetic coating of surgical grade SS 316L substrate: a TEM analysis'. *Journal of the Taiwan Institute of Chemical Engineers.* 2011;**42**(4):682–687.

[41] Ma J., Wong H., Kong L.B., Peng K.W. 'Biomimetic processing of nanocrystallite bioactive apatite coating on titanium'. *Nanotechnology.* 2003;**14**(6):619–623.

[42] Boanini E, Gazzano M, Bigi A. 'Ionic substitutions in calcium phosphates synthesized at low temperature'. *Acta Biomateriala.* 2010;**6**:1882–1894.

[43] Tran P.A., Sarin L., Hurt R.H., Webster T.J. 'Opportunities for nanotechnology-enabled bioactive bone implants'. *Journal of Materials Chemistry.* 2009;**19**(18):2653–2659.

[44] Ballo A.M., Xia W., Palmquist A., *et al.* 'Bone tissue reactions to biomimetic ion-substituted apatite surfaces on titanium implants'. *Journal of the Royal Society Interface.* 2012;**9**:1615–1624.

[45] Lindahl C., Xia W., Lausmaa J., Engqvist H. 'Incorporation of active ions into calcium phosphate coatings, their release behavior and mechanism'. *Biomedical Materials.* 2012;**7**(4):045018.

[46] Ionita D., Grecu M., Ungureanu C., Demetrescu I. 'Antimicrobial activity of the surface coatings on TiAlZr implant biomaterial'. *Journal of Bioscience and Bioengineering.* 2011;**112**(6):630–634.

[47] Roveri N., Iafisco M. 'Evolving application of biomimetic nanostructured hydroxyapatite'. *Nanotechnology Science and Applications.* 2010;**3**:107–125.

Chapter 2

Surface modifications of orthopedic implant materials using an electroplating process

Yardnapar Parcharoen[1] and Sirinrath Sirivisoot[2]

Abstract

The failure incidents of orthopedic implants originate from many reasons, for example, the wear corrosion, fibroblast encapsulation, allergic reaction to metal materials, and infection. To minimize these failures, surface modifications of biomaterials have been widely used. Among those, electroplating is one of the methods that are the most convenient, cheap, and environmental friendly. Currently, the electroplating method is used to coat several types of nanomaterials, such as hydroxyapatite (HA), gold, carbon nanotubes, and graphene oxide, on orthopedic implants. This method shows a high capability to replace patient bone fractures with new bone for long-term use. This chapter reviews the current solutions and examples of practical ideas in the area of surface coating using electrodeposition for improving bone implants.

2.1 Introduction

Bone is a hard material, made by specialized bone cells, which is composed of calcium, phosphate, and a protein called collagen. Bone is composed of both trabecular and cortical bones, which make bone very strong but lightweight. From the reason that bones have high strength-to-weight ratios and complex structure, biomedical treatment of bone defects is required [1]. In 1 year, more than a million of people need bone replacements, but they still have main problems associated with suitable bone implants for the long-term use.

Recently, there are many strategies which are used to treat bone defects including autografts, allografts, synthetic implants, and cell-based therapies. However, each of these strategies has both advantages and disadvantages. For example, autograft

[1]Chulabhorn International College of Medicine at Thammasat University 95 Phahonyothin Rd Klonglung, Pathum Thani, Thailand

[2]Biological Engineering Program, Faculty of Engineering, King Mongkut's University of Technology Thonburi (KMUTT), Bangkok, Thailand

therapy can cause pain, infection, or injury in the neurovascular system [2]. To use allografts, which is donated by another within the same species, a risk of infection to the donor occurs. Moreover, the allograft source is limited, expensive, and its storage is very complicated [3]. The implant materials in the form of ceramics, metals, polymers, gels, and other composite materials are wildly used in dentistry and orthopedics [4]. However, such synthetic materials can have a fixation problem with surrounding bone or cells, infection from microorganisms, or mechanical loosening, causing implant failures [5]. Tissue engineering and cell-based orthopedic therapies are currently being developed to solve the problems of implant failures and are in stages of in vitro study and animal tests [6] as well as early stage clinical trials [7]. The advances in biomaterial engineering can extend the life of implants for long-term use in patients. Therefore, advance and simple techniques are necessary to improve synthetic implant materials for use in clinical orthopedic practice.

2.1.1 Current synthetic implant materials

Multifunctional implant materials made recently are from alternative synthetic materials. Some synthetic implant materials have low fibroblast growth around them (less fibrous encapsulation) and are degraded, then allowing new bone growth juxtaposed to the implant. Current implant materials contain only one or two of these ideal properties. For example, the implant materials should have a mechanical strength similar to that of the bone which it will replace. Such comparable strength of orthopedic implant materials to the bone is necessary, because this provides a stress shielding effect and also continuing appropriate toughness avoiding a fatigue fracture in surrounding bone. Examples of implant materials which exhibit good mechanical properties for orthopedic applications are silicon, aluminum, titanium, and its alloys.

Ceramic scaffolds have been used to replace bone for more than 35 years, especially the ceramics that are made from calcium phosphate being more popular than others [8]. Calcium phosphate scaffolds are synthesized in the laboratory into many forms, but tricalcium phosphate (TCP) and HA are the most used clinically. A chemical stoichiometry of TCP (calcium phosphate) is similar to the precursor of bone which is a noncrystal composition, while HA is a crystal composition which has a similar chemical stoichiometry to natural bone. Although the formation of such ceramics through a natural process has not been found, the results of using synthetic ceramics as bone replacement materials show as good biological responses as found in natural bone scaffolds. Most synthetic ceramic materials did not possess osteoconduction or osteoinduction properties. They showed only a slight direct supporting of bone structure. However, calcium phosphate ceramics, after connecting to the surrounding bone, result in osteoid produced directly onto the surface of the ceramic. Moreover, the juxtaposed bone to the implant occurs without a fibroblast or soft tissue encapsulation. After the osteoid is mineralized with calcium phosphate and gradually causes bone formation, bone will go through a remodeling process as found in natural bone. Under biological conditions, porous TCP-ceramic has a fast remodeling by process bone cells which have migrated into the material, while HA crystals have a slower process but more permanent than the TCP-ceramic [9]. However, HA has a low stiffness and slow degradation. Therefore, to implant HA ceramic materials into patient bodies, it is necessary to focus on the mechanical property of the materials.

Licon-based compounds are materials made with a silicon-based component. One of most widespread licon-based compound is bioactive glass. Bioactive glass is a hard solid without porous, and it is composed of calcium, phosphorus, and silicon dioxide (silicate as the main component). This material expresses both osseointegrative (structural and functional connection between the implant material and the surrounding natural bone) and osteoconductive (bone growth on the implant surface) properties [10]. The bioactive glass shows a higher mechanical strength when compared to the ceramic HA. Moreover, it is resistant to drilling and shaping, but some might crack via the processes. Unfortunately, bioactive glasses are hard to stand in the skeletal system but they are effectively used as a bone graft expander [11]. Other licon-based compounds are composed of calcium/aluminum/fluorosilicate glass powder mixed with polycarboxylic acid which shows mechanical properties similar to cortical bone. This licon compound is called a glass ionomer which is produced in the form of porous semen paste. The glass ionomer has been applied to avoid an exothermic reaction from polymethyl-methacrylate (PMMA) used as an implant material. Furthermore, the glass ionomer filled with antibiotics, proteins, or molecules shows slower release of additives than using PMMA, and thus less damage to heat-labile proteins [12].

Various well-known polymers are the products of the polymerization reaction of monomers (polystyrene, polyethylene, polypropylene, etc.). Many types of polymers are used as orthopedic implant materials [polyaryletherketone (PAEK), polyglycolic acid (PGA), polylactide (PLLA), etc.]. A natural tissue can be continuously repaired and remodeled according to changes in its physiologic environment. Ideally, synthetic biopolymers should possess a degradable function and replaced by natural bone. An example of a degradable polymer is a polylactic polymer or its copolymers which are currently used as bone screws and as sutures. However, to use such polylactic acid (PLA) in a highly loaded position, polymer engineering needs to be improved.

Bone is the main structure of the body which has high strength and fracture toughness to support and protect various organs. Therefore, when comparing to other synthetic implant materials, metals have been used for a long time to replace bone. Stainless steel was the first metal which was successfully implanted in the human body with aseptic surgery practice, as shown in Figure 2.1 [13]. After that, vitallium and cobalt-based alloys were used in dentistry and artificial joints. Later, titanium (Ti) and its alloys are the newest metals which are widely used, and currently have become the most popular choice of metals in orthopedic applications.

2.1.2 Metals as implant materials

Metals are widely used as orthopedic implant materials when compared to other synthetic implant materials including polymers and ceramics. Many types of metals

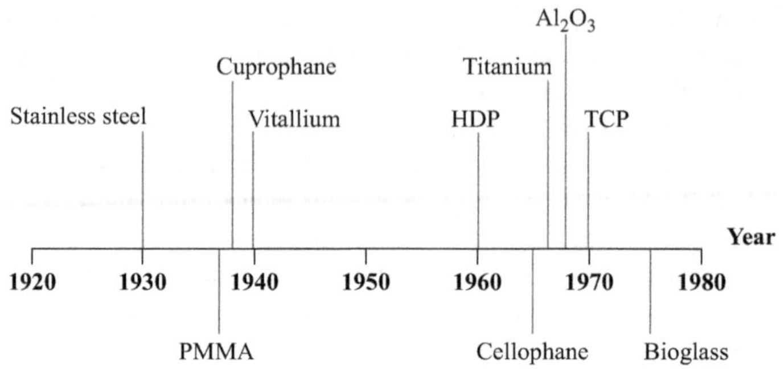

Figure 2.1 Timeline of synthetic implant materials

are manufactured but only a few types can be used in this area such as stainless steel, cobalt-chrome (CoCr) alloys, and Ti alloys. Firstly, stainless steel is tougher than steel and highly resistant to corrosion which is contributed from chromium (Cr) and nickel (Ni). Moreover, molybdenum (Mo) is added into stainless steel, called as "316 stainless steel," to improve overall corrosion resistance. To increase chloride corrosion resistance, the composition percentage of solid state carbon (C) in stainless steel is reduced from 0.08 wt% to 0.03 wt%, which is called as "316L stainless steel." Ti is widely popular also because it has low weight and density. The densities of Ti, 316 stainless steel, and CoCrMo alloy are 4.5 g/cm^3, 7.9 g/cm^3, and 8.3 g/cm^3, respectively [14]. Currently, there are two new materials commercially pure tantalum (Ta) and commercially pure niobium (Nb), which show better electrochemical properties and biocompatibility than Ti and its alloys, which researchers expect will be new candidate orthopedic implant materials [15].

2.1.2.1 Stainless steels

Since early 1900s, surgical practices have used stainless steel and it remains the popular medical device chemistry. As previously discussed 316L stainless steel has a corrosion resistance greater than other stainless steels, therefore it is used the most in the human body. As is well-known, stainless steel is a metal alloy that is corrosion resistant and has other properties that can be adjusted by adding other elements. For example, adding chromium increases corrosion resistance of the stainless steel alloy. Carbon and nickel are added into stainless steels to stabilize the austenite (gamma-phase iron) structure of stainless steel under heat treatment. The 316L stainless steel contains 17–19% of Cr and 14% of Ni. Specifically, an additive of Mo in stainless steels can lead to the formation of a protective layer against an acidic environment [16]. The addition of the ferrite (magnet) element in stainless steel is not widely used, because this affects bioimaging process from Magnetic Resonance Imaging (MRI) equipment. A ferrite additive can increase a magnetic property of stainless steels which can heat up under MRI and affect the shape structure of orthopedic implant materials [17].

2.1.2.2 CoCr alloys

CoCr alloys have been used as implant materials, especially, in artificial joints. They are commonly famous for outstanding wear resistance. For example, the cobalt-nickel-chromium-molybdenum (CoNiCrMo) alloy is used in manufacturing of high-loaded artificial joints. Superior hardness and resistance to wear of CoCr alloys make them highly appropriate for moving parts, such as femoral heads. In comparison, Ti alloys are about 15% softer than CoCr alloys, which are more appropriate to use as bearing parts such as lightweight surface parts (femoral necks, stems, and porous coatings). However, Ti alloys are high in corrosion resistance and mechanical properties are nearer to that natural bone than other alloys, reducing the stress shielding effect on the bone more than other alloys [18]. Most alloy compositions of CoCr are based on Ni which is at a high risk to irritation to the tissue around it. Among the CoCr alloys, following the ASTM F799 standard will develop Ni free to replace the traditional one [16].

2.1.2.3 Ti and its alloys

Ti is one of the most interesting metals applied in hard tissue replacement because it has a high strength, fracture toughness, rigidity, and reliability. Another important property which supports the popular use of titanium in orthopedic applications is biocompatibility. A comparison of biocompatibility for various implant materials is shown in Table 2.1 [16]. As given in Table 2.1, Ti and its alloys show superior biocompatibility than other metallic implant materials. It is because Ti-based materials has a naturally oxide layer (inert layer) on the surface that makes them nontoxic to the host cells [19].

In orthopedic applications, an important mechanical property that researchers are concerned is Young's modulus. Ti and Ti alloys have a Young's modulus similar to that of cortical bone (or called "compact bone"), which is the main and most important part of the human skeleton. There are many benefits and drawbacks in using Ti and Ti alloys. They have been shown to effectively prevent atrophy and bone strengthening. However, fatigue and low strength make them not suitable to be used as implant materials especially in long-term use. The results from tensile tests of commercially pure Ti when compared with aluminum (Al) and vanadium (V) are inferior. However, the biocompatibility of commercially pure Ti when compared with Al and V with cells is better [20]. However, only osseointegrative and biological features of implants are often not enough to introduce good adhesion between the implant and bone. When the implant is used

Table 2.1 Biocompatibility of various biomaterials [16]

Biomaterials	
Stainless steel, vitallium, PMMA	Biotolerant materials
Titanium, titanium alloys, carbon, alumina, zirconia, titania, titanium nitride, silicon nitride	Bioinert materials
Bioglass, ceravital, TCP, HA, glass ceramic	Bioactive materials

long term in the human body, it may cause a mechanical failure, and required surgery to remove the implant [21].

The metals, ceramics, polymer, and carbons listed in Table 2.1 are often used in medical applications because of their biotolerant, bioinert, or bioactive properties. In general, Ti is often used as a prosthetic for replacing bones that have been damaged, such as hip implants, knee implants, bone plates, and dental implants. Ti alloys are also very attractive for both medical and dental caries due to its biocompatibility, lightweight, appropriate mechanical property, and good corrosion resistance.

2.1.2.4 Other metals

Examples of other metals used as a bone substitute material include tantalum (Ta), amorphous and biodegradable metal alloys. First, Ta is a metal which does not result in a radiographic image which was established in a standard specification for unalloyed Ta for surgical implant applications (ASTM: F560). However, the mechanical strength of Ta is lower than commercially pure Ti, but its strength can increase when surrounding temperature is low [15]. Second, amorphous alloys have high strength and resistance to corrosion, as well as have a lower modulus compared to a polycrystalline metal with the same composition, currently in development stage. For example, zirconium (Zr)-based amorphous metals, specifically "Zr–7.6Ni–12.3Cu–3.5Al" have the strength and Young's modulus equal to 1,300 MPa and 75 GPa, respectively [22]. Finally, metals proposed for biodegradable implants, named as biodegradable metals, are either iron-based or magnesium (Mg)-based alloys. The Mg-based alloys include magnesium aluminate (MgAl), magnesium-rare earth (MgRE) [23], and magnesium-calcium (MgCa) [24] alloys. The popular biodegradable metals that have been implanted in the body include iron and magnesium alloys.

2.1.3 Failure problems of metallic implant materials

For more than a decade, implant materials have been developed continuously by material engineers. However, an implant often fails after 10–20 years of use. The failure problems commonly come from bone fracture around the implanted materials, wear or corrosion of the implant, inflammation, and infection [25]. Such problems cause a high volume of revision surgeries when the age of the population is higher. Once implants have been used in patients under the age of 65, an estimate of the longevity of the implants in patients is about 15 years [4].

Metallic orthopedic implants are often fixed into the skeletal system of the human body as constituents of reconstructive devices. The design of these implants is following the anatomy and physiology of the skeletal structure of the human body. The mechanical and chemical stabilities and biocompatibilities of the implant materials in host tissue are important for the successful treatment of bone replacements. Unfortunately, metals have a tendency to fail after long-term use due to mechanical, electrochemical and/or biological failures. The mechanical failure due to a material mechanical modulus being higher in comparison with bones causes fatigue and wear. Moreover, electrochemical failure is related to low

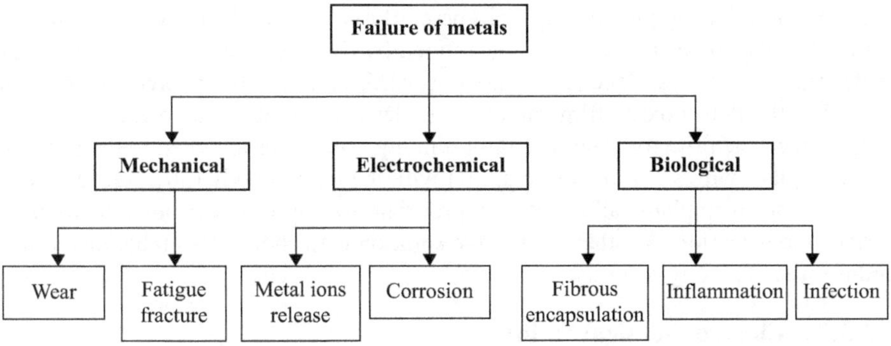

Figure 2.2 Failures of metal implants from mechanical, electrochemical, and biological causes that lead to a revision surgery in patient

corrosion resistance under corrosive medium at the implantation site. The patient condition from the failures will be worse if the metal releases toxic metal ions from implanted metals in the patient's body. Biological failures are from infection, inflammation, or adverse reactions in the host such as fibrous encapsulation.

Another surgery to fix the failure of bone material implants can increase a patient's risk, increasing complications after surgery and bone regeneration at different speeds. Moreover, the cost of surgery is very high. Key points in recovery efficiency after revision surgery are osseointegration and bioactivity. Once biomaterials allow bone-binding between the implant and the surrounding bone failure rates can decrease due to biological reasons. Various causes for revision surgery in metals are summarized in Figure 2.2. Failures may also result from the combinations of these processes [26]. Importantly, the success of implants depends on the company or bonding between bone and prosthetic material, which shows new bone ingrowth and its lastingness.

2.1.3.1 Mechanical failure

Once implanted in the body, most metallic bone substitutes will be under pressure all the time inside the human body. This can lead to fracture of metal from fatigue. The major cause of the fatigue behavior of artificial materials contributes from its microstructure (internal phase boundaries) characteristic. For example, the tensile strength of an Ti-6Al-4V α-equiaxed structure was increased rapidly in low-cycle-fatigue [27]. In contrast, the elongation of sample fatigue increased by a rapid decline in the low-cycle-fatigue. An atomic structure of metals is crystalline which has a regular atomic arrangement, and thus mechanical properties are controlled by dislocation motion and twinning closely involving with the crystal structure. Importantly, their mechanical properties can be improved by control of the crystal lattice defect or phase transformation at the atomic level.

Wear is a type of surface degradation and generated particles are typically very small with its size of about 1–2 μm. These small particles may or may not affect surrounding tissue which is usually associated with inflammation and/or fibroblast encapsulation. The application of commercially pure Ti and Ti-6Al-4V alloy as a material replacement for bones is constrained by the low resistance of the material to degradation at its surface [28]. The Ti surface is normally covered by an oxide layer, but the dense oxide film can occur under the simultaneous reaction of wear [29]. In fact, the primary causes of failure in hip prostheses are wear to degradation of the implant with a complex interaction with the human body [30]. Because wear is one cause of implant failure, it is appropriate to improve resistance to wear by surface modification. Another alternative solution is to choose a suitable material to endure wear for long-term use.

2.1.3.2 Electrochemical failure

When material made of a metal is implanted into the human body, the metal surface is continually exposed to liquids which causes an electrochemical reaction. Decomposition of the material but at a limited rate always occurs. This is because of the fact that a contact surface with the surrounding environment forms. The tissues in the human body contained water, organic, and inorganic compounds, dissolved oxygen/carbon dioxide, amino acids/proteins, blood or other liquid compositions, saliva, etc. The beginning of corrosion can be due to several factors surrounding implant surfaces. Generally, it is formed from a small electrochemical reaction. Then, it becomes a result of corrosion in pitting attacks or crevice corrosion. The difference is caused by the types of the interface and a metal ion, or may be both. There may be other forms of corrosion that can occur (uniform corrosion, intergranular corrosion, selective corrosion, etc.) [31] following the corrosion process, and thus metal ions are released.

CoCrMo alloys are used as artificial bone materials because their corrosion resistance and mechanical properties are superior to other metals. However, when CoCrMo alloys are placed in human body environment, they can release alloying ions to human tissues, which can increase the risk to the human health over time of implantation [32]. The important causes of metal-ion release are coming from many reasons, such as the protective oxide layers which naturally forming on metal surfaces. If the oxide layers are destroyed or broken, metal ions will be possibly released from the bulk metal. Although in vivo the release profile of metal ions can change due to the low concentration of dissolved oxygen, inorganic ions, proteins, or cells surrounding the material, which are the catalyst for release of metal ions. [33]. Another reason of ionic release is the pH of the host environment where those ion releases are affected by the pH of the body [34].

2.1.3.3 Biological failure

Each company that produces commercial metal implants has their own different theories on implant designs that are related to different orthopedic applications. The success of implants is due to both biomechanical and biological factors. Even if those implant materials have benefits and successful uses, it must be without any risk of adverse effects when placed in the host body. The body may respond to unsolicited metals including fibroblast encapsulation and inflammation which can affect long-term performance of implants [35].

Millions of orthopedic implant materials are used each year; although many advances in biomaterial sciences, there are some devices that become colonized by bacteria [36]. The problem of infection is mainly focused of implant materials [37]. Additionally, after bacteria contact, grow, and form biofilms on the implant surface, it brings three major problems. First, after bacterial presence on the surface, this pool of bacteria can spread into the body system, leading to a chronic infection. Second, bacterial biofilms can cause an antibiotic resistance, because the biofilms protect bacteria which are really hard to get rid of using conventional antimicrobials for even a long time. If the host response and antibiotic treatment are not work, to get rid of the bacteria in a biofilm, one needs to do it from the site of biofilm. That is to produce an antibacterial implant surface by surface modification [38]. Symptoms of infection can show immediately after surgery, or may take longer to develop the infection such as a months or even years.

The knowledge of bacterial or cell surfaces is very important because it is related to the pathway of flocculation and adhesion to a solid surface of the bacterial cells, which are based on the features of bacterial cell membrane in each species, especially the hydrophobic and electrical properties of the cell surface. For this reason, various techniques have been achieved for surface modification to change surface energy, electrochemical properties, and biocompatibility of implant materials.

2.1.4 Surface modification on metal surfaces

Surface modifications can increase the success of implant materials once used in the human body for a lifetime. The possible surface modification methods of materials include:

1. To clean material surfaces from possible contaminations before the materials are implanted,
2. To add bioactive properties on material surfaces to support growth of cells and tissue integration,
3. To strengthen artificial materials and also to reduce wear rate, especially in the case of materials used as artificial bone joints,
4. To increase the inert layer of material surfaces that can prevent ion release into the patient's body,
5. To increase antibacterial properties of implants, and
6. To increase fatigue strength of implants.

In orthopedic applications, many metallic materials fail due to lacking of inertia of the surface and biofunctionalities (compatibility of blood and bone conduction). Therefore, surface modifications of metallic implants through any physical or chemical process are needed to increase bone healing and implant integration with healthy tissues [39]. Improving the biological response on the implant surface can be achieved by coating it with bioactive ceramics such as HA [40], or with bioplastics such as PLA to increase compatibility of blood [41]. Recently, researchers have developed new composites or alloys from metal elements that are nontoxic and allergy free as biomaterials. For example, new forms of biodegradable metallic glasses (or amorphous metals) such as Mg and zinc (Zn) have been suggested to use as impermanent implant materials [42].

Physical methods are mostly used to modify surface morphology or topography of material implants such as machining, etching, plasma spraying, grit-blasting, and electrochemistry to change the surface topography into the micro- or nano-scale. Importantly, these processes can change surface roughness and energy of artificial materials. The high surface roughness may lead to higher surface energy, which affects absorbing proteins once implanted, adhesion of bone-forming cells (osteoblasts), osteoblast functions, and osseointegration in the end [43].

Chemical methods, which are mainly used to increase bioactivity of materials, are various coatings with biochemical and physicochemical techniques. A biochemical technique is a coating technique using organic molecules (such as growth factors, peptides, or enzymes) to alter responses of a specific cell type [44]. These biochemical surface coatings can provide direct biochemical signals, which can regulate and control the assembly and differentiation of cells to create the environment for the formation of matrix proteins and promote cell–extracellular matrix (ECM) interactions on surface implants. A physicochemical technique is a combination of an inorganic phase such as calcium phosphate, which enables synchronization of biochemical interlocking of the bone–protein matrix, and existing surface materials with mechanically strengthening bonds between them [43]. Several modifications of material surfaces in practice are subject to both physical and chemical engineering methods.

Among all surface modifications, an electrochemical method is a good method to produce different surface coatings on metals. It can be both physical and/or chemical modification of materials. The main advantage of an electrochemical process is an easiness to adjust the thickness of coating or porosity, increase the corrosion protection, and reduce the possible releases of metallic ions. Structural and chemical properties of the coating can be changed over a relatively broad range of important parameters in the electrochemical process such as electrical potential, electrolyte composition, temperature, and current density.

2.2 Surface modification by electrochemical process

An electrochemical process is a chemical reaction which involves an electron transfer to or from an electrode surface in an electrolyte. Typically, a charge transfer takes place across an electrode and electrolyte interface. Metals are electronic conductors which are usually used as electrodes of an electrochemical process. The net current flow is unidirectional and in a different direction with electron flow along the electrode. As shown in Table 2.2, anodic- and cathodic-electrode reactions are involved together in various reaction types regarding the nature of phase changes at the electrode surface. Each specific type of electrode reaction can have several application areas. For example, in the case of cathodic electrodeposition of metals, the product can be a compact deposit (called electroplating), colloidal particles suspended in a liquid medium migrate under the influence of an

Table 2.2 Various electrode reactions of important surface processing of metals [45]

Type of electrode reaction	Direction of change	Nature of phase change	Application area
Electrodeposition	Cathodic/anodic	Liquid → solid	Electroplating, electrophoretic deposition (EPD), electroforming, electrorefining, and electrowinning metal ion removal
Dissolution of metals	Anodic	Solid → liquid	Soluble anodes in electroplating, electroforming, etching of printed circuit boards, corrosion of metals
Gas evolution	Cathodic/anodic	Liquid → gas	Hydrogen evolution at cathode and oxygen evolution at inert anodes
Oxide formation	Anodic	Solid → solid	Anodizing of metal, conversion coating

electric field (electrophoresis) and are deposited onto an electrode, a detachable foil or self-standing work piece (called electroforming), a pure metal (called electro-refining), a roughened deposit needed further processing (called electrowinning), or a toxic/precious metal for a metal-ion removal in environmental control and recycling [45]. However, electrodeposition processes also have some drawback reactions such as a hydrogen evolution or reduction of soluble ionic species which may occur as a secondary reaction after the primary coating layer formed which could interfere the primary layer. In other case, some reaction process at electrode can entirely be occurred by a chance, for example, a classical corrosion of metals.

2.2.1 Electrodeposition

The term "electrodeposition" is commonly called for electrolytic, electroplating, or EPD. A composition of electrolyte in EPD is different from ELD. The EPD comprises solid particles suspended in a solvent, while in ELD the electrolyte is a mixture of ions [47]. Table 2.3 shows a distinction between EPD and ELD. Both EPD and ELD under a large number of novel applications are highly interested in both research and industrial communities. The reason for that is because they are highly flexible, which can be applied with different materials and shapes, and also cost-effective techniques, which usually require simple equipment. Moreover, EPD and ELD have a high potential for scalability in a large scale production industries.

EPD is a process which has a motion of charged particles dissolved in an electrolyte moving toward an electrode under the influence of an electric field. After charged particles arrive at an electrode, the deposition occurs by coagulation of those charged particles. ELD is often used for manufacturing a thin ceramic coating from a solution of salt ions which produces colloidal particles by the reaction at the electrode. Therefore, both the reaction at electrodes in ELD and the motion of charged particles in EPD lead to a deposition of ceramic particles and

Table 2.3 Characteristics of electrodeposition techniques [46]

	Electrophoretic deposition (EPD)	**Electrolytic deposition (ELD)**
Medium	Suspension	Solution
Moving species	Particles	Ions or complexes
Electrode reactions	None	Electrogeneration of OH^- and neutralization of cationic species
Preferred liquid	Organic solvent	Mixed solvent (water-organic)
Required conductivity of liquid	Low	High
Deposition rate	$1–10^3$ m/min	10^{-3}–1 m/min
Deposit thickness*	$1–10^3$ m	10^{-3} 10 m
Deposit uniformity†	Limited by size of particles	On nanometer scale
Deposit stoichiometry	Controlled by stoichiometry of powders used for deposition	Can be controlled by use of precursors

*Controlled by variation of deposition time, voltage, or current density.
†Controlled by electric field.

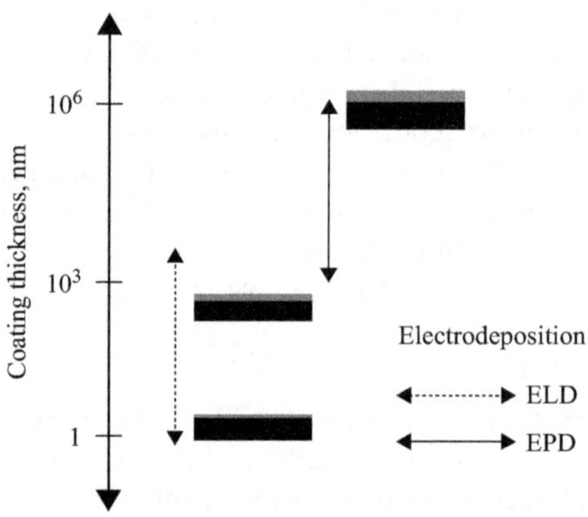

Figure 2.3 Thickness in nanometers of the coatings which are formed using ELD and EPD. Gray color represents the coating on substrates

formation of ceramic films at the electrodes. A comparison of coating thickness formed by both processes is shown in Figure 2.3. EPD is a tool to prepare thick ceramic films $1–10^4$ μm, while ELD is used to form nanostructured thin ceramic films 1–1,000 nm [48]. As charged particles deposit under the electric field the coating surface (EPD coating) depends on particles size, the bigger the particle, the thicker the coating formed.

Understanding the mechanisms of particle aggregation is required to succeed in coating materials by the electrochemical method. Currently, the fundamental in understanding EPD [49] and ELD [50] comes from the theory of the stability of colloidal, named Derjaguin-Landau-Verwey-Overbeek (DLVO) [51]. The DLVO theory discusses two main forces which are double-layer repulsion and van der Waals' attraction. These forces result in the forming of colloidal particles, influence stability and internal structure of the coatings.

2.2.1.1 The DLVO theory

The DLVO theory states that particles touch each other by forces, which are coulombic double-layer repulsion and van der Waals' attraction [48]. The total energy V_T of interaction of two isolated, identically charged particles maybe defined as:

$$V_T = V_A + V_R \tag{2.1}$$

The attractive energy V_A of the London–van der Waals' interaction between two spherical particles can be expressed by:

$$V_A = \frac{A}{6}\left(\frac{2}{s^2 - 4} + \frac{2}{s^2} + ln\frac{s^2 - 4}{s^2}\right) \tag{2.2}$$

where A is the Hamaker constant and $s = 2 + H/a$, with H is the shortest distance between the two spheres and a is the particle radius. If $H = a$, (2.2) can be simplified to:

$$V_A = -A\frac{a}{12H} \tag{2.3}$$

The repulsive energy V_R is:

$$V_R = 2\pi\varepsilon\varepsilon_0 a\psi^2 \ln\left[1 + e^{-\kappa H}\right] \tag{2.4}$$

where ε is the dielectric constant of the solvent, ε_0 is the vacuum dielectric permittivity, ψ is the surface potential, $1/\kappa$ is the Debye length:

$$\kappa = \left(\frac{e_0^2 \sum n_i z_i^2}{\varepsilon\varepsilon_0 \kappa T}\right)^{1/2} \tag{2.5}$$

where e_0 is the electron charge, k is the Boltzmann constant, T is the absolute temperature, and n_i is the concentration of ions with valence z_i. The repulsion between colloidal particles is directly related to the diffused layer charge on the particles. The equation of interaction energy (2.1)–(2.5) describe the force of interaction of two charged particles. Two forces of attraction and repulsion mainly subtract each other. The DLVO theory shows that the coagulation of particles or ions depends on many factors such as particle radius, distance between two particles, dielectric constant, temperature, etc.

The coating system contains two particle species or the single shape of the electrode. Therefore, the following studies report multiple theories to describe particle interactions and kinetics of deposition [52], and further theoretical and computer modeling work is carried out. For example, a study of the electrodynamic properties of the combination of particles under the EPD process under both steady and alternating electric fields has been conducted recently [53]. This study shows the time involved in the process for a possible layer separation of particles into the coating under certain different conditions. The other study reports that colloidal particles in the electrolyte deposited near the electrode in DC by convection flow and osmotic pressure of particles [54]. Another very important recent fundamental study on different parameters to form colloidal films during EPD has been provided by Sarkar *et al.* [55]. They studied the single crystal deposition of silica particles in which the yield of production was dependent on deposition time.

2.2.1.2 EPD

EPD is one of colloidal processes used in the production of ceramics that has been known since the 19th century. Its applications in ceramic production can be found in the last 40 years [56]. The advantages of EPD include a simple apparatus, short coating time, and flexibility of substrate shape. Moreover, this process is a green coating method which contains only a few or no toxins in the process [57]. This technique can prepare multi-layered composites which is a combination of various coating layers [58].

EPD is usually composed of two electrodes. The samples can be placed on either cathode (cathodic-EPD) or anode (anodic-EPD) depending on the charge of interested particle species to be deposited on the electrode as the coating, as shown in Figure 2.4. The movement of charged particles in the electrolyte toward the working electrode is completely different under the effect of an electrical field. This process can also be called as electrophoresis. In the beginning of the process,

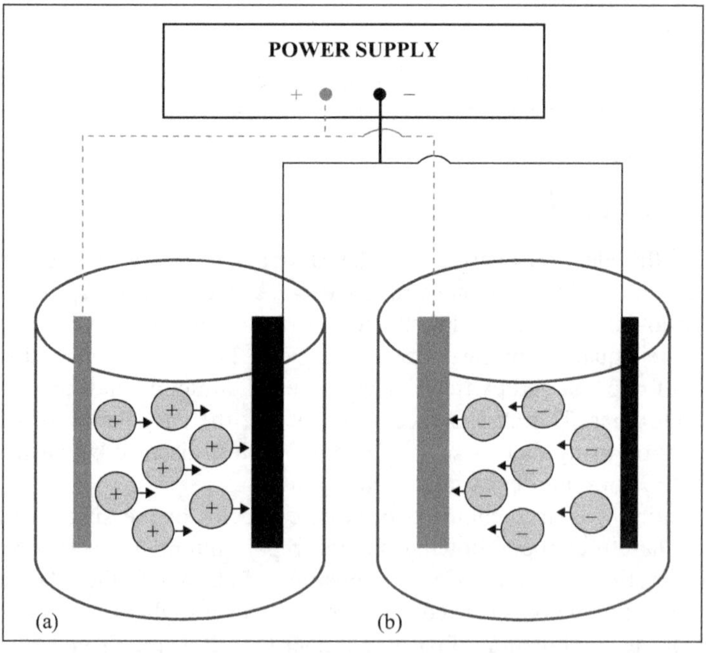

Figure 2.4 A schematic of EPD processes: (a) cathodic-EPD and (b) anodic-EPD

particle coagulation leads to the deposition and growth of a coating layer on the electrode [47, 49].

Particles charging mechanisms

When ceramic particles are in an electrolyte solution, the particles will be charged under the influence of an electrical field before they deposit onto the electrode. The mechanism of EPD involved with these charged particles can be separated into four consecutive mechanisms:

a. The specifically adsorbed ions onto the solid ceramic particles from electrolyte liquids.
b. The changing of ions between liquid and solid.
c. Absorbing or arranging of dipolar molecules on the surface of the solid ceramic particles.
d. The transfer of electrons between solid and liquid.

The explanation of particle-charged mechanisms and interactions has been discussed in many literature articles [59]. In summary, EPD is mainly composed of two stages of mechanisms which are electrophoresis and deposition:

1. Electrophoresis is the kinetics of the EPD process which shows a relationship of systematic parameters with time.
2. After solid ceramic particles arrive at the electrode surface, the particles deposit on the electrode surface which will reproduce the shape of the original electrode surface.

The coating and arrangement of particles on the electrode surface depend on the chemical nature of the particle surface and the adhesion or repulsion forces between particle agglomeration and particles in solution as well as between particles in solution and the electrode substrate [60]. The basic mechanisms of EPD have been discussed in the previous section, which includes a principal basis of DLVO theory and other multiple theories.

Parameters influencing EPD

The parameters that determine the characteristics of EPD process are divided into two groups: the first group which is related to the electrolyte solution (suspension); and the second group which is related to physical parameters such as electrical properties of electrodes and process/electrical parameters (voltage, deposition time, etc.) [57].

The influencing parameters in EPD are simply described by Hamaker's law [61] which relates the deposit yield (w) to the electric field strength (E), the electrophoretic mobility (μ), the surface area of the electrode (A), and the particle mass concentration in the electrolyte (C), as concluded in the following equation:

$$w = \int_{t_1}^{t_2} \mu \cdot E \cdot A \cdot C \cdot \mathrm{d}t \tag{2.6}$$

Consequently, the weight of the deposited particles (w) in the EPD method is a function of E, t, and C. Therefore, the production yield of the deposited particles, specifically the thickness of the coating, can be controlled by the concentration of the particles in electrolyte, electrical voltage, and deposition time in the EPD method. There are other theories that explain the deposition yield that depends on the shape of electrode (cylindrical, sphere, or coaxial), phase of deposition system (liquid, suspension, or mixing phase), or no free-ion interruption in the electrolyte as mentioned in the initial state of EPD reaction [46].

Parameters on suspension

Many parameters on suspension influencing the yield of EPD coating include: (i) particle sizes, (ii) dielectric constant of suspension, (iii) conductivity of suspension, (iv) suspension viscosity, (v) zeta potential of the particles, and (vi) stability of suspension. First, particle sizes between about 1 and 20 μm [62] have very high potential to move and deposit on the electrode. However, the particle size outside of this range is still suitable for EPD. However, the particles must be able to move along the electrical field, rather than settling down followed the gravity force. If the gravity has high influence on coating thickness, then the vertical cylinder electrode coating does not have the same thickness throughout the electrode. The lower part must be thicker than the upper part due to gravity. The nano-size particles have a tendency to deposit on the electrode better than the micro-size particles. The cracking of the coating also affects by the particle size as shown by Sato *et al.* [63], which the larger crack area was found when using 3-μm particles in EPD. They found that the smaller crack was found in the coating with 0.06-μm particles. Second, the dielectric constant is the capacity of a solution to store the electrical energy. The comparatively low level of dielectric constant can lead to the failure of the coating because the dissociation of particles from liquid phase is inadequate. Third, electrical conductivity of suspension affects on the motility and stability of particles. When there is too low electrical conductivity, particles slowly move in the electrolyte. However, when the electrical conductivity is too high, it may result in a loss of particle charge, and thus immobility of particles. Fourth, suspension viscosity should be between low to optimum for EPD coating. Fifth, zeta potential of the particles can determine direction, migration, and stabilization of the particles in the electrophoretic system. Sixth, for stability of the suspension, when the electrolyte is stabilized, the particles are not likely to slow flocculating precipitation patterns, and then the adhesion of the coating can be slowly dense at the bottom of the container. However, when the solution is too stable, electrical repulsion between the particles will not be dominated by the electric field and accumulation will not occur.

Parameters on process

The main parameters that have an influence on electrophoretic formation are the process parameters including: (i) deposition time, (ii) applied voltage, (iii) concentration of solid in solution, and (iv) electrical conductivity of substrate. First, for deposition time, when a voltage is kept constant, the electric field induced electrophoresis will be reduced when coating time increases. It is because the time will increase the thickness of the coating, which is an insulating layer that can reduce the influence of the electrical field during the electrophoretic process [64]. However, in the initial stage of the EPD process, it has commonly a linear relationship between deposition yield and time. Second, the thickness of the coating is proportional to the applied electrical voltage. However, too high electrical potential can be lowering the quality of the coating. It is because in a high voltage, particles move too fast that lead to the deposition at a nonsuitable site of particles. Third, a concentration of solid particles in solution is a very important parameter in the case of multi-component coatings. If the concentration of each solid in suspension is high, the particles will deposit at the same rate. Nevertheless, if the solid concentration in the suspension is low, the particles will be coated at a rate proportional to electrophoretic motility of each particle. Fourth, when electrical conductivity of the substrate is too low, it can lead to a slow coating process and nonuniform coating.

From the discussion above mentioned in terms of kinematics of electrophoretic process and the quality of the coating, ones need to carefully control each parameter during the electrophoretic process to obtain a desirable coating. Moreover, the parameters of the process need to be controlled carefully and the practical technique is also important. For example, a good cleansing of equipment and substrates to avoid contamination might affect the electrokinetic property. The uniform shape and surface of electrode also influence the uniformity of coating on the electrodes.

Applications of EPD

EPD is a widely used method to prepare cathode- and anode-supported coating materials. Solid oxide fuel cells (SOFC) is one of the EPD applications in various complex shapes such as planar and tubular geometry. Numerous publications have explained the application of EPD in this area [65]. The reason of high success in fuel production is because of the fact that it is possible to adapt and scale up the production line, to obtain a short production time, and to use a simple device. Moreover, fibrous and porous substrate coatings can also be produced by the EPD method. For example, Zhitomirsky *et al.* [66] used the EPD method to deposit carbon fibers with HA and applied the coating on biomaterials. Other examples include membranes and porous materials such as alumina membranes [67], or hollow zeolite fiber coatings obtained by coating nanozeolites onto carbon fibers and subsequent burnout of the carbon core [68]. Furthermore, the EPD process is one of the processes used to protect wear and oxidation of biomedical implants by coating with bioactive substances such as HA and related calcium phosphate minerals. In the case of biomedical coatings obtained using the EPD method, such as HA/TiO_2 composites, Kumar *et al.* [69] showed that the functional composite of HA, TiO_2, and Ti can be completely achieved by EPD alone. The EPD method has been used to coat various materials on electrodes such as carbon nanotubes [70], ceramic composites [71], oxide nanorods [72], and layered ceramics [73].

Thick film coatings can also be produced using EPD such as silica coatings [74], nanoscaled zeolite membranes [75], luminal materials [76], various coatings for electrode and sensor productions [77], multi-layer composites [78], infiltrations of ceramic particles onto fiber fabrics to produce glass/ceramic composites [79], oxide nanorods [72], carbon nanotube films [70], superconductors [80], high-temperature superconductors superconducting films [81], piezoelectric materials [82], HA coatings on metals [83], etc. The EPD process is also used to fabricate fibrous reinforced composites with high quality production [84]. Moreover, EPD is used to grow ceramic nanotubes, nanorods, or other nanosizes to obtain an efficient deposition of these nanoceramic structures on various substrates. For example, Limmer *et al.* studied the production of TiO_2 nanorods in polycarbonate membranes 200 nm in diameter. Smeets *et al.* coated nano-sized particles (titania (TiO_2), barium titanate ($BaTiO_3$), silica (SiO_2), strontium niobate ($Sr_2Nb_2O_7$), and lead zirconate titanate ($Pb(Zr_{0.52}Ti_{0.48})O_3$, PZT) nanorods) into nano-porous glass using a low temperature process [85].

The disadvantage of using EPD to immobilize materials in colloidal form when compared with other methods (such as dip and slurry coatings) is that the EPD process cannot use water as an electrolyte. Water in EPD can cause a gas evolution (such as hydrogen or oxygen) at the electrode surface which can affect physical, electrical, and chemical characteristics of the coating. Nevertheless, nonaqueous electrolytes such as methanol, ethanol, and isopropanol can be used to replace water in EPD. The EPD is a great method to produce desirable coatings of various materials, and thus it is popularly used in research and industries for various applications including modification of biomaterials for orthopedic implants.

2.2.1.3 ELD

ELD can be used to produce thin ceramic films on substrates from solutions of ion salts. ELD is a quite new technique in ceramic processing; however, it is a good approach to prepare nanostructured or nano-biostructured coatings for implant materials. There are three main features that make ELD a suitable method for nano-, bio- and microtechnologies. First, ELD can be used to grow functional materials having complex material shapes. Second, it can be performed at room or near room temperature from aqueous-based solutions. Third, this method can be applied in a thin coating of a few atoms, or can scale up the production into a large scale dimension.

Mechanisms and influence of deposition parameters on ELD

The basic mechanisms of ELD for ceramic coatings include cathodic- and anodic-ELD processes. Cathodic ELD is suitable to coat various oxide materials on the surface of electrodes. The cathodic reactions produce the OH^- groups at the electrode surface, and then increasing the pH of the electrolyte solution around the electrode. Then, metal or complexes ions, which are normally stable in the bulk of solutions at low pH, are hydrolyzed at the electrode surface under the influence of an electrical field, and then forming colloidal particles on the electrode surface. These particles become coagulated and form cathodic ceramic deposits [54]. If the electrolyte concentrations are above the flocculation values for stabilizing ions, the colloidal particles would be unstable and deposit at the cathodic electrode.

Recently, understanding the mechanism of cathodic electrodeposition has come from the application of the classical DLVO theory of colloidal stability as described in the previous section [48]. The important point is that deposition of layers formed during the ELD process occurs by flocculation which is initialized by the electrolyte. In the cathodic electrodeposition process, the sample which is in need of a coating must be immersed in a chamber containing the electrolyte, and connected to another electrode and an external power supply, which feeds the flow of current to the whole system. The coated sample is attached to the cathode of the electrochemical system. From this process, the dissolved ions in the electrolyte are reduced into atoms of the substance of the coating on the cathode. Later on, a thicker coating is found on the surface of the material. Electrolytes must be the ion conductor, which consists of the interesting charged ions dissolved in a suitable solvent or in a liquid state in the form of a molten salt. The solvent is mainly composed of water, but in the recent research, there are electrolytes of various organic liquids and ionic others (fused-salt) that can bring to the process of coating with ELD. In an aqueous solution, metal ions M^{Z+} are represented by

$$M^{z+}_{solution} + ze \rightarrow M_{lattice} \tag{2.7}$$

The ELD coating process can be achieved by two main different processes: (1) deposition by ELD process with electron z coming from an external power source and (2) an autocatalytic coating process caused by a chemical reducing agent in the electrolyte. Without the power of origin, these reducing agents are the only source of electrons in reaction and the coating process is called electroless ELD. Equation (2.7) is the interaction between charged particles at the interface between solid metal electrodes and a liquid solution. Both types of charged particles, metal ions or electrons can cross this interface between solution and the metal.

There are many parameters and process conditions such as current density, electrode potential, temperature of the solution, pH, concentration of the reactants, type of surface, and interference reaction (generation of hydrogen gas at the surface) that these factors could influence plated with electrochemical processes [86].

Current density ($A{\cdot}m^{-2}$) has a direct impact on coating of the coating substances. The current density is measured using ampmeter and then it is calculated comparing with the reaction area. In general, current density is assumed to be uniform across the entire surface area. However, surface electrodes have different properties in each area. The current density while the reaction is very uneven is likely to change in various ranges the process with increasing reaction time.

Applied voltage is one more physical parameter, beside the current density, which will control the deposition method. The voltage electrode φ decreases during the distance of the solution and the metal of the electrode. This is also known as the Galvani potential difference.

Temperature is one of the most important parameters. High temperature of the reaction has several advantages, including a good solubility, enhanced conductivity of the solution, and reduced inertia of the anode. In contrast, high temperature has a tendency to rush the evaporation of the solution and accelerate corrosion. Therefore, in practice to limit the temperature is essential. Low reaction temperature is low current density or reduces the rate of growth of the coating layer. Low temperature causes the slow kinetics of electrons, and the coating is likely to become brittle. The reaction rate of the process is also influenced by temperature, i.e. higher temperatures affect the increasing coating rate of the electrochemical process.

pH of the solution is an important coating parameter, which influence the performance of the coating process, hydrogen evolutions and hydroxide formation at the electrode surface. The pH of the solution balances the whole electrochemical system and controls the concentration of the process.

Ion concentration of the reactants (electroactive) is coated from the solution which has low ion concentration under a specific potential and the coating surface will have a fine-grained structure and coated at a slow rate. Positive and negative ions dissolved in the solution which do not participate in the reaction at the electrode can influence the process which will change the strength of the ionic conductivity of the solution.

Substrates influence the deposition process, mainly in the initial coating stage of the clean substrate. In some cases, especially when the surface is free of oxide crystals. If crystal structure of the surface is similar to that of the coating, the coating may be up to about 10 microns. The smooth and mechanically polished surface has little influence on the growth of the coating while etching with chemicals that affect the coating as well.

Hydrogen evolution, from the reaction, hydrogen may occur in parallel with the surface coating. Under the solution, with an agitated solution, the reaction will affect the formation of hydrogen bubbles in a nearby electrode. This formed hydrogen can be diffused into the surface coating and cause embrittlement and other unwanted side effects.

Cathodic electrodeposition has assumed major commercial significance due to its use to prevent electrochemical dissolution of the metal, and the better corrosion protection. Several cathodic reactions can cause an increase in the pH from the reaction at the cathode electrode surface, because that chemical reaction consumes H^+ ions. These reactions cause a slight increase of the pH at electrode surface. By creating OH^- (pH increase) due to the reduction reaction of water, oxygen dissolves in the electrolyte, nitrate, and perchlorate ions. There are some examples of cathodic coatings in the orthopedic field which are HA coatings. The changing of local pH and HA at electrode has been mentioned in the HA coating process. Figure 2.5 shows the formation of HA from the electrochemical plating method. By this mechanism, when voltage is sufficient, the various exchanges between a group of electrochemical reactions, chemical reactions, and an increase in the pH of the surface area of the electrode could occur. Under neutral pH, the sequence of reactions may be presented as following:

$$2H_2O + 2e^- \rightarrow H_2 + 2(OH)_{ads}^- \tag{2.8}$$

$$HPO_4^{2-} + (OH)_{ads}^- \rightarrow H_2O + PO_4^{3-} \tag{2.9}$$

$$10Ca^{2+} + 6PO_4^{3-} + 2(OH)_{ads}^- \rightarrow Ca_{10}(PO_4)_6(OH)_2 \tag{2.10}$$

The formation of OH^- groups on sample surfaces ($(OH)_{ads}^-$) enhance the formation of HA crystals. Electrodeposited HA, prepared under high pH, has a crystal structure and morphology similar to those of natural bone [87].

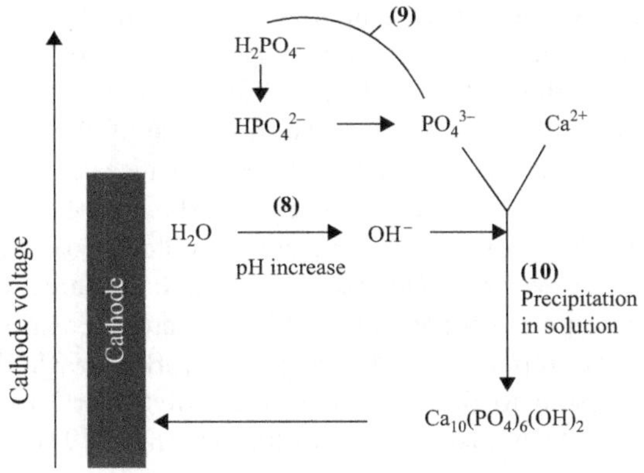

Figure 2.5 The processes that allow formation of HA on cathode in electrolytes. The increase in pH first facilitates the reaction (8). Since the cathodic potential is high enough to allow a direct production of sufficient phosphate ions, which is the reactions (9), then the HA forming on the cathode is being more dominant, as shown in the reaction (10) [88]

Anodic process

For the anodic process, the anode electrode connects to the positive pole of the power supply and is dipped in an electrolyte. Both soluble and insoluble anodes are used as electrodes. Examples of the soluble electrode include copper, zinc, nickel, and silver. Other metals such as chrome, gold, and palladium are used as the insoluble electrode. Soluble anodes have an important role in completing the electrical circuit of a coating system: they are the sources of metal ions to the cathode coating of that metal, such as nickel, where nickel anodes should be soluble anodes and were melted and coated on the cathode under the active system. The nickel ions are dissolved when a potential difference is more positive (anode) than the voltage equilibrium of the nickel metal. The anode inactivity can occur in some cases of input high voltage to the anode. This phenomenon consists of the formation of a layer with low electrical conductivity on the surface; this will cause current density of many areas to increase, which resulted in a crisis of electric potential of that passive surface. Under the terms of these inert thin films that are formed on the surface, the current density will rapidly decline in this area. This phenomenon is not desirable in the process of electrochemical plating. Therefore, in some cases dividing chloride or fluoride ions in the electrolyte can be used to treat the surface of anodes for active states, by making oxygen to dissolve in water or other oxidation agents.

Applications of ELD

An ELD coating has more applications, especially for the development of thin film coatings of titanium dioxide and more complicated titanates in electronics [89], TiO_2 nanowires used in photoelectrochemical applications [90], ZnO films in the development of electrical film deposition [91], etc. What is very important about this recent work is that a relatively low temperature is required for film crystallization, i.e. there are many important developments of coatings to use in high temperature applications, such as coating stainless steel against corrosion at high temperature with thin films of yttria stabilized zirconia and $Ce_{1-x}Gd_xO_{2-y}$.

Recently, there has been strong interest in the use of electrochemical methods for ceramic coatings for medical applications, because the electrolytic process could achieve a high purity of coating. Moreover, the complex shapes of materials which are necessary in biological systems are possible to coat with this process as same as in the EPD process. Dinamani and Kamath prepared various phosphate materials using cathodic electrodeposition [92]. Electrodeposition of HA coatings in basic conditions was reported by Manso and co-workers [93]. These HA coatings exhibited good adhesion to the substrates. Electrolytic ZrO coatings have been deposited on Co–Cr–Mo implant alloys for hip prosthesis [94].

2.2.2 Oxidation

Oxidation is a common chemical reaction between metals and oxygen. However, oxidation can start with an external force that is stimulating the atoms by providing the electrical potential. By the way, oxidation is a chemical reaction between oxygen and a metal by creating an oxide layer. Oxidation of metals based on multiple factors and mechanisms of reaction which are often quite complex. The phenomenon began with the absorption of oxygen molecules from the atmosphere or from the solution, followed by nucleation of the oxide, formation of a thin oxide layer, and an increase in the oxide layer thickness eventually.

2.2.2.1 Mechanisms of oxidation reaction

Similar to EPD or ELD, many materials such as metals, semiconductors, and conductive polymers, a thin layer of oxide can be direct coated onto their surface directly from the solution by electrochemical plating. The junction between the electrodes (and the surface coating) and the electrolyte, metal ions are oxidized and that metallic oxide solid state is coated onto the sample (the anode) through the following three reactions [95].

$$M \rightarrow M^{Z+} + ze^- \tag{2.11}$$

$$M + \frac{z}{2}H_2O \rightarrow MO_{z/2} + zH^+ + ze^- \tag{2.12.1}$$

$$M^{z+} + zH_2O \rightarrow M(OH)_z + zH^+ \tag{2.12.2}$$

$$M(OH)_z \rightarrow MO_{z/2} + \frac{z}{2}H_2O \tag{2.13}$$

The reaction of hydrogen gas at the cathode follows this (2.14):

$$zH_2O + ze^- \rightarrow \frac{z}{2}H_2\uparrow + zOH^- \tag{2.14}$$

Typically, the oxide layer positioned between the oxide growth and the electrolytes, that is a less dense oxide layer [96], whereas the growing layer positioned between the metal and the oxide that will consists of a dense and stable TiO_2. Compared to other oxides of transition metals, TiO_2 is widely used and extensively researched. Especially in the last 10 years, TiO_2 is the only substance which has many studies and research results in materials science. It is known that the TiO_2 compound is very useful because of its nontoxic nature, corrosion protection, and production processes that are environmental-friendly. TiO_2 is often used as a blending component in white color or UV protection. With key features of good biocompatibility of TiO_2, it is suitable for applications in medical science (implant materials).

The anodic oxidation process has been successfully used to change the flat surface of titanium to nanotubular structures of TiO_2 in a fluoride containing electrolyte. Adding fluoride ions into electrolytes has deeply impacted anodization, the fluorides ions and metal ions could form water soluble $[TiF_6]^{2-}$ species. $[TiF_6]^{2-}$ complexes were formed from Ti^{4+} ions dissolved out at the interface between the oxide and electrolyte [after the diffusion through the oxide film; (2.15)] and from the chemical dissolution of TiO_2 (2.16). From (2.15) and (2.16) reactions, competition between formation of oxides and the dissolved Ti^{4+} causes pores or nanotubes on surfaces.

$$Ti^{4+} + 6F^- \rightarrow [TiF_6]^{2-} \tag{2.15}$$

$$TiO_2 + 6F^- + H^+ \rightarrow [TiF_6]^{2-} + 2H_2O \tag{2.16}$$

Adjusting parameters such as the chemistry of the electrolyte, voltage, and current density, one can control physicochemical properties of oxide surfaces, and the diameter and the spacing between nanotubes can be precisely modulated.

2.2.2.2 Applications

An oxide surface is commonly used due to some unique feature and to optimize the reaction rates or the rate of diffusion of ions, thus the surface has to be made in a small scale to obtain a large surface area or over a short distance for ionic diffusion. The use of Ti/TiO_2 is approximately 40% among all other artificial materials used in medicine. This is the material that is mostly used worldwide. Today, it is more relevant to modify titanium surfaces with biomolecules on the surface, to be linked with organic materials (mostly enzymes, proteins, or DNA) which are important biosignals on the materials surface and may be coated with biofunctionals such as APTES [95,97] that transport a terminal NH_2 group. Another method which is similar to this APTES coating can be used for forming a covalent bond for many other organic materials [98].

Several studies [99] have shown that the formation of HA can be induced on the surface of the TiO_2 nanotubes which is better when compared with formations on the surface of flat TiO_2. It was also found that the size and surface roughness obviously influence the formation of HA [100]. The trial production of titanium tube nanoparticles from compounds of two metals to be used as implant materials (include TiNb, TiZr, TiTa, TiW, TiMo, TiAl) have yielded successful results [95].

The shape of the tube TiO_2 nanoparticle demonstrates trends for their use as capsules which may be used for storage and delivery of drugs, especially if the layer of nano-sized tubes can be separated by a single pipe. The experiments of

Shrestha *et al.* [101] shows that TiO_2 nanotubes can contain magnetic particles of Fe_3O_4, and a magnetic field can be used to take the tube for other specific requirements. The nano-sized tubes can be coated easily with drug molecules attached to the appropriate link. The drug release can be induced by UV light reaction, electrical activation (voltage induced catalysis) [88], or X-rays [102], which can be used directly to treat tissues in the body at the point of use, such as killing cancer cells specifically inside the patient's body [101].

Other applications and aspects of TiO_2 are gas sensing which could be sensing CO, H_2, or NO_x gases. The nanotube has been used as the nano-test tube to observe small volume reactions. TiO_2 is a material that has high photocatalytic properties and is used for the decomposition of organic materials (in case of organic pollution). To increase this activity, photocatalytic researchers have synthesized nano-sized tubes of mixed oxides including TiMo or TiW [103].

2.3 Future directions of surface modification by electrochemical processes

Various electrochemical surface modification processes have been used for improving surface properties of implant materials as explained in the previous sections. It has been commonly noted that the trend of modifying the surface seems to have moved away from the traditional methods (heating and gas) to more applications of advanced technology to improve materials (room temperature and multi-layer coatings) because traditional methods have low efficiency of the manufacturing process that requires more time, larger systems, and high energy. Research in the field of implant materials seems to focus more on cellular compatibility, for example, cell growth, cell attachment, antibacterial effects, as well as on mechanical issues (wear resistance, corrosion resistance, and hardness). The further research needs to focus on modifying implant surfaces are as follows:

- The development of the coating on implant material surfaces which includes biomolecules
- Techniques for expanding production of surface coatings which is appropriate for orthopedic implant materials
- Investigate and apply nanotubes on implant material surfaces
- Research on the high efficiency of animal tests to determine the toxicity of bone substitute materials
- Methods to prevent implant related infections
- Degradation measurements to evaluate more accurate lifetimes of materials
- Development of the real time indicators of implant material functionality

When moving to novel coatings on implants, there are many problems waiting to be fixed, such as bonding strength between the coating and the surface that is not enough to be used in the human body or manufacturing processes that use too high temperatures. The EPD process has shown the potential to produce coatings at low temperatures which can fix this problem. Moreover, EPD has significant importance in industrial and commercial applications with advantages over other fabrication methods, as it can be possibly scaled up using inexpensive instruments. Many researchers show a tendency to use the EPD process to improve orthopedic implant surfaces in the future and EPD may be used in some other advanced fields. ELD is also increasingly used for the preparation of nanostructured thin films for various applications. This technique provides opportunities in the design of novel materials and devices.

From new materials to new techniques and coating types which are used to improve bone substitute materials, many research groups require further development to create materials with higher biocompatibility and integration with patient cells. From this, the ability to control, promote tissue growth, and differentiation of cells efficiently by using these novel materials may also increase the ability to work and increase the lifespan of materials as well. Moreover, the future is more likely to pay attention to implants which are smart materials, a material that can have various features for many purposes in one type of material, for example, a transplanted material that can provide anti-infection properties in the material, or, nanoscale magnetic particles used in the treatment of bone cancer. The development of these techniques improves the treatment system, reduces the discomfort of the patient, improves implant materials which include anti-infection, in situ sensor, and high osseointegration on one surface of the implant material, etc. Reproducible, accurate, and scalable methods for surface modification with nanoscale properties are needed for the production of new bone implant materials.

References

[1] Rey, C., Combes, C., Drouet, C., and Glimcher, M.J., 2009, "Bone mineral: update on chemical composition and structure", *Osteoporosis International*, Vol. 20, No. 6, pp. 1013–1021.

[2] LaPrade, R.F. and Botker, J.C., 2004, "Donor-site morbidity after osteochondral autograft transfer procedures", *Arthroscopy*, Vol. 20, No. 7, pp. e69–e73; Devine, J.G., 2013, "Bone grafting techniques in idiopathic scoliosis: a confirmation that allograft is as good as autograft but dispels the purported pain associated with the iliac crest bone graft harvest", *Spine Journal*, Vol. 13, No. 5, pp. 530–531.

[3] Cartwright, E.J., Prabhu, R.M., Zinderman, C.E., *et al.*, 2010, "Transmission of *Elizabethkingia meningoseptica* (formerly *Chryseobacterium meningosepticum*) to tissue-allograft recipients: a report of two cases", *Journal of Bone & Joint Surgery American*, Vol. 92, No. 6, pp. 1501–1506.

[4] Geetha, M., Singh, A.K., Asokamani, R., and Gogia, A.K., 2009, "Ti based biomaterials, the ultimate choice for orthopaedic implants – a review", *Progress in Materials Science*, Vol. 54, No. 3, pp. 397–425.

[5] (a) Hacking, S.A., Bobyn, J.D., Toh, K., Tanzer, M., and Krygier, J.J., 2000, "Fibrous tissue ingrowth and attachment to porous tantalum", *Journal of Biomedical Materials Research*, Vol. 52, No. 4, pp. 631–638;

(b) Kremers, H.M., Howard, J.L., Loechler, Y., *et al.*, 2012, "Comparative long-term survivorship of uncemented acetabular components in revision total hip arthroplasty", *Journal of Bone & Joint Surgery American*, Vol. 94, No. 12, p. e82.

[6] Horie, M., Driscoll, M.D., Sampson, H.W., *et al.*, 2012, "Implantation of allogenic synovial stem cells promotes meniscal regeneration in a rabbit meniscal defect model", *Journal of Bone & Joint Surgery American*, Vol. 94, No. 8, pp. 701–712.

[7] (a) Dawson, J.I., Kanczler, J., Tare, R., Kassem, M., and Oreffo, R.O.C., 2014, "Concise review: bridging the gap: bone regeneration using skeletal stem cell-based strategies—where are we now?" *Stem Cells*, Vol. 32, No. 1, pp. 35–44;
(b) Vangsness, C.T., Jr., Farr, J., 2nd, Boyd, J., Dellaero, D.T., Mills, C.R., and LeRoux-Williams, M., 2014, "Adult human mesenchymal stem cells delivered via intra-articular injection to the knee following partial medial meniscectomy: a randomized, double-blind, controlled study", *Journal of Bone & Joint Surgery American*, Vol. 96, No. 2, pp. 90–98.

[8] (a) Bohner, M., 2000, "Calcium orthophosphates in medicine: from ceramics to calcium phosphate cements", *Injury*, Vol. 31, No. Suppl 4, pp. 37–47;
(b) Hollinger, J.O., Brekke, J., Gruskin, E., and Lee, D., 1996, "Role of bone substitutes", *Clinical Orthopaedics and Related Research*, Vol., No. 324, pp. 55–65.

[9] McAndrew, M.P., Gorman, P.W., and Lange, T.A., 1988, "Tricalcium phosphate as a bone graft substitute in trauma: preliminary report", *Journal of Orthopaedic Trauma*, Vol. 2, No. 4, pp. 333–339.

[10] Gross, U., Brandes, J., Strunz, V., Bab, I., and Sela, J., 1981, "The ultra-structure of the interface between a glass ceramic and bone", *Journal of Biomedical Materials Research*, Vol. 15, No. 3, pp. 291–305.

[11] (a) Kinnunen, I., Aitasalo, K., Pollonen, M., and Varpula, M., 2000, "Reconstruction of orbital floor fractures using bioactive glass", *Journal of Cranio-Maxillofacial Surgery*, Vol. 28, No. 4, pp. 229–234;
(b) Peltola, M., Suonpaa, J., Aitasalo, K., *et al.*, 2000, "Experimental follow-up model for clinical frontal sinus obliteration with bioactive glass (S53p4)", *Acta Oto-Laryngologica. Supplementum*, Vol. 543, pp. 167–169.

[12] Brook, I.M. and Hatton, P.V., 1998, "Glass-ionomers: bioactive implant materials", *Biomaterials*, Vol. 19, No. 6, pp. 565–571.

[13] Park, J.B., 2000, *"The Biomedical Engineering Handbook"*, CRC Press LLC, Boca Raton.

[14] Dayal, V. and Totemeier, T.C., 2004, "21 – Mechanical testing", In *Smithells Metals Reference Book* (8th ed.). Totemeier, W.F.G.C., Butterworth-Heinemann, Oxford, pp. 1–23.

[15] Papakyriacou, M., Mayer, H., Pypen, C., Plenk Jr, H., and Stanzl-Tschegg, S., 2000, "Effects of surface treatments on high cycle corrosion fatigue of metallic implant materials", *International Journal of Fatigue*, Vol. 22, No. 10, pp. 873–886.

[16] Niinomi, M., 2002, "Recent metallic materials for biomedical applications", *Metallurgical and Materials Transactions A*, Vol. 33, No. 3, pp. 477–486.

[17] Disegi, J.A., 1992, "Magnetic resonance imaging of AO/ASIF stainless steel and titanium implants", *Injury*, Vol. 23, No., Suppl 2, pp. S1–S4.

[18] (a) Kummer, F.J. and Rose, R.M., 1983, "Corrosion of titanium/cobalt-chromium alloy couples", *Journal of Bone & Joint Surgery American*, Vol. 65, No. 8, pp. 1125–1126;
(b) Niinomi, M., 2008, "Mechanical biocompatibilities of titanium alloys for biomedical applications", *Journal of the Mechanical Behavior of Biomedical Materials*, Vol. 1, No. 1, pp. 30–42.

[19] Lausmaa, J. and Linder, L., 1988, "Surface spectroscopic characterization of titanium implants after separation from plastic-embedded tissue", *Biomaterials*, Vol. 9, No. 3, pp. 277–280.

[20] Wang, W. and Poh, C.K., 2013, *Titanium Alloys in Orthopaedics.*

[21] Puleo, D.A. and Nanci, A., 1999, "Understanding and controlling the bone–implant interface", *Biomaterials*, Vol. 20, No. 23–24, pp. 2311–2321.

[22] Gilbert, C.J., Schroeder, V., and Ritchie, R.O., 1999, "Mechanisms for fracture and fatigue-crack propagation in a bulk metallic glass", *Metallurgical and Materials Transactions A*, Vol. 30, No. 7, pp. 1739–1753.

[23] Witte, F., Kaese, V., Haferkamp, H., *et al.*, 2005, "In vivo corrosion of four magnesium alloys and the associated bone response", *Biomaterials*, Vol. 26, No. 17, pp. 3557–3563.

[24] Li, Z., Gu, X., Lou, S., and Zheng, Y., 2008, "The development of binary Mg–Ca alloys for use as biodegradable materials within bone", *Biomaterials*, Vol. 29, No. 10, pp. 1329–1344.

[25] Park J.B. and Lakes, R.S., 1992, *Biomaterials: An Introduction*, Plenum Press, New York.

[26] (a) Hanawa, T., 2010, "1 – Overview of metals and applications", In *Metals for Biomedical Devices*. Niinomi, M., Woodhead Publishing, pp. 3–24;
(b) Okazaki, Y., 2010, "2 – Material selection", In *Metals for Biomedical Devices*. Niinomi, M., Woodhead Publishing, pp. 25–67.

[27] Akahori, T. and Niinomi, M., 1998, "Fracture characteristics of fatigued Ti–6Al–4V ELI as an implant material", *Materials Science and Engineering: A*, Vol. 243, No. 1–2, pp. 237–243.

[28] Dearnley, P.A., Dahm, K.L., and Çimenoğlu, H., 2004, "The corrosion–wear behaviour of thermally oxidised CP-Ti and Ti–6Al–4V", *Wear*, Vol. 256, No. 5, pp. 469–479.

[29] Komotori, J., Hisamori, N., and Ohmori, Y., 2007, "The corrosion/wear mechanisms of Ti–6Al–4V alloy for different scratching rates", *Wear*, Vol. 263, No. 1–6, pp. 412–418.

[30] Buford, A. and Goswami, T., 2004, "Review of wear mechanisms in hip implants: paper I – general", *Materials & Design*, Vol. 25, No. 5, pp. 385–393.

[31] Hansen, D.C., 2008, "Metal corrosion in the human body: the ultimate bio-corrosion scenario", *The Electrochemical Society Interface*.

[32] Öztürk, O., Türkan, U.U., and Eroğlu, A.E., 2006, "Metal ion release from nitrogen ion implanted CoCrMo orthopedic implant material", *Surface and Coatings Technology*, Vol. 200, No. 20–21, pp. 5687–5697.

[33] Hanawa, T., 2004, "Metal ion release from metal implants", *Materials Science and Engineering: C*, Vol. 24, No. 6–8, pp. 745–752.

[34] Okazaki, Y. and Gotoh, E., 2008, "Metal release from stainless steel, Co–Cr–Mo–Ni–Fe and Ni–Ti alloys in vascular implants", *Corrosion Science*, Vol. 50, No. 12, pp. 3429–3438.

[35] Branemark, R., Branemark, P.I., Rydevik, B., and Myers, R.R., 2001, "Osseointegration in skeletal reconstruction and rehabilitation: a review", *Journal of Rehabilitation Research and Development*, Vol. 38, No. 2, pp. 175–181.

[36] Khardori, N. and Yassien, M., 1995, "Biofilms in device-related infections", *Journal of Industrial Microbiology*, Vol. 15, No. 3, pp. 141–147.

[37] Whitehouse, J.D., Friedman, N.D., Kirkland, K.B., Richardson, W.J., and Sexton, D.J., 2002, "The impact of surgical-site infections following orthopedic surgery at a community hospital and a university hospital: adverse quality of life, excess length of stay, and extra cost", *Infection Control & Hospital Epidemiology*, Vol. 23, No. 4, pp. 183–189.

[38] Ribeiro, M., Monteiro, F.J., and Ferraz, M.P., 2012, "Infection of orthopedic implants with emphasis on bacterial adhesion process and techniques used in studying bacterial–material interactions", *Biomatter*, Vol. 2, No. 4, pp. 176–194.

[39] Castellani, C., Lindtner, R.A., Hausbrandt, P., *et al.*, 2011, "Bone–implant interface strength and osseointegration: biodegradable magnesium alloy versus standard titanium control", *Acta Biomaterialia*, Vol. 7, No. 1, pp. 432–440.

[40] Habibovic, P., Barrere, F., Blitterswijk, C.A.v., Groot, K.d., and Layrolle, P., 2002, "Biomimetic hydroxyapatite coating on metal implants", *Journal of the American Ceramic Society*, Vol. 85, No. 3, pp. 517–522.

[41] Seeger, J.M., Ingegno, M.D., Bigatan, E., *et al.*, 1995, "Hydrophilic surface modification of metallic endoluminal stents", *Journal of Vascular Surgery*, Vol. 22, No. 3, pp. 327–336.

[42] Yusop, A.H., Bakir, A.A., Shaharom, N.A., Abdul Kadir, M.R., and Hermawan, H., 2012, "Porous biodegradable metals for hard tissue scaffolds: a review", *International Journal of Biomaterials*, Vol. 2012, p. 641430.

[43] Coelho, P.G., Granjeiro, J.M., Romanos, G.E., *et al.*, 2009, "Basic research methods and current trends of dental implant surfaces", *Journal of Biomedical Materials Research B Applied Biomaterials*, Vol. 88, No. 2, pp. 579–596.

[44] Morra, M., 2007, "Biomolecular modification of implant surfaces", *Expert Review of Medical Devices*, Vol. 4, No. 3, pp. 361–372.

[45] Anon., 2002, "What is electrometallurgy", In *Fundamental Aspects of Electrometallurgy*, Springer US, pp. 1–3.

[46] Besra, L. and Liu, M., 2007, "A review on fundamentals and applications of electrophoretic deposition (EPD)", *Progress in Materials Science*, Vol. 52, No. 1, pp. 1–61.

[47] Heavens, N., 1990, "Advanced ceramic processing and technology". In *Electrophoretic Deposition as a Processing Route for Ceramics.* Binner, J.G.P. Noyes Publications, USA, Vol. 1, pp. 255–283.

[48] Zhitomirsky, I., 2002, "Cathodic electrodeposition of ceramic and organoceramic materials. Fundamental aspects", *Advances in Colloid and Interface Science*, Vol. 97, No. 1–3, pp. 279–317.

[49] Sarkar, P. and Nicholson, P.S., 1996, "Electrophoretic deposition (EPD): mechanisms, kinetics, and application to ceramics", *Journal of the American Ceramic Society*, Vol. 79, No. 8, pp. 1987–2002.

[50] Zhitomirsky, I., 2000, "Ceramic films using cathodic electrodeposition", *The Minerals, Metals & Materials Society*, Vol. 52, No. 1, pp. 1–11.

[51] (a) Derjaguin, B. and Landau, L., 1993, "Theory of the stability of strongly charged lyophobic sols and of the adhesion of strongly charged particles in solutions of electrolytes", *Progress in Surface Science*, Vol. 43, No. 1–4, pp. 30–59;
(b) Marshall, C.E., 1949, "Theory of the stability of lyophobic colloids. The interaction of particles having an electric double layer." E. J. W. Verwey and J. T. G. Overbeek, with the Collaboration of K. Van Ness. Elsevier, New York-Amsterdam, 1948, 216 pp., $4.50, *Journal of Polymer Science*, Vol. 4, No. 3, pp. 413–414.

[52] (a) Raggi, G., Stace, A.J., and Bichoutskaia, E., 2013, "Surface–charge distribution on a dielectric sphere due to an external point charge: examples of C60 and C240 fullerenes", *Physical Chemistry Chemical Physics*, Vol. 15, No. 46, pp. 20115–20119;
(b) Larsen, A.E. and Grier, D.G., 1997, "Like-charge attractions in metastable colloidal crystallites", *Nature*, Vol. 385, No. 6613, pp. 230–233;
(c) Qian, Y. and Bowen, W.R., 1999, "Long-range electrostatic interaction between a charged wall and two similarly charged colloidal spheres at low surface potentials", *Journal of Colloid and Interface Science*, Vol. 213, No. 2, pp. 316–321;
(d) Trau, M., Saville, D.A., and Aksay, I.A., 1997, "Assembly of colloidal crystals at electrode interfaces", *Langmuir*, Vol. 13, No. 24, pp. 6375–6381.

[53] B. Ferrari, R.M., 1996, "The conductivity of aqueous Al_2O_3 slips for electrophoretic deposition", *Materials Letters*, Vol. 28, No. 4–6, pp. 353–355.

[54] Boccaccini, A.R. and Zhitomirsky, I., 2002, "Application of electrophoretic and electrolytic deposition techniques in ceramics processing", *Current Opinion in Solid State and Materials Science*, Vol. 6, No. 3, pp. 251–260.

[55] Sarkar, P., De, D., Yamashita, K., Nicholson, P.S., and Umegaki, T., 2000, "Mimicking nanometer atomic processes on a micrometer scale via electrophoretic deposition", *Journal of the American Ceramic Society*, Vol. 83, No. 6, pp. 1399–1401.

[56] Boccaccini, A.R., Keim, S., Ma, R., Li, Y., and Zhitomirsky, I., 2010, "Electrophoretic deposition of biomaterials", *Journal of The Royal Society Interface*, Vol. 7, No. Suppl 5, pp. S581–S613.

[57] Zhitomirsky, I., 2002, "Cathodic electrodeposition of ceramic and organoceramic materials. Fundamental aspects", *Advances in Colloid and Interface Science*, Vol. 97, No. 1–3, pp. 279–317.

[58] Kaya, C., 2003, "Al_2O_3–Y-TZP/Al_2O_3 functionally graded composites of tubular shape from nano-sols using double-step electrophoretic deposition", *Journal of the European Ceramic Society*, Vol. 23, No. 10, pp. 1655–1660.

[59] Lewis, J.A., 2000, "Colloidal processing of ceramics", *Journal of the American Ceramic Society*, Vol. 83, No. 10, pp. 2341–2359.

[60] Riccardis, M.F.D., 2012, "Ceramic coatings obtained by electrophoretic deposition: fundamentals, models, post-deposition processes and applications", *Ceramic Coatings – Applications in Engineering, Prof. Feng Shi (Edition).*

[61] Hamaker, H.C., 1940, "Formation of a deposit by electrophoresis", *Transactions of the Faraday Society*, Vol. 35, pp. 279–287.

[62] Heavens, S., 1990, "Electrophoretic deposition as a processing route for ceramics", *Noyes Publications, Advanced Ceramic Processing and Technology*, Vol. 1, pp. 255–283.

[63] Sato, N., Kawachi, M., Noto, K., Yoshimoto, N., and Yoshizawa, M., 2001, "Effect of particle size reduction on crack formation in electrophoretically deposited YbCo films", *Physica C: Superconductivity*, Vol. 357, pp. 1019–1022.

[64] Zhitomirsky, I. and Gal-Or, L., 1997, "Electrophoretic deposition of hydroxyapatite", *Journal of Materials Science: Materials in Medicine*, Vol. 8, No. 4, pp. 213–219.

[65] (a) Ishihara, T., Shimose, K., Kudo, T., Nishiguchi, H., Akbay, T., and Takita, Y., 2000, "Preparation of yttria-stabilized zirconia thin films on strontium-doped Lamno3 cathode substrates via electrophoretic deposition for solid oxide fuel cells", *Journal of the American Ceramic Society*, Vol. 83, No. 8, pp. 1921–1927;
(b) Basu, R.N., Randall, C.A., and Mayo, M.J., 2001, "Fabrication of dense zirconia electrolyte films for tubular solid oxide fuel cells by electrophoretic deposition", *Journal of the American Ceramic Society*, Vol. 84, No. 1, pp. 33–40;
(c) Negishi, H., Sakai, N., Yamaji, K., Horita, T., and Yokokawa, H., 2000, "Application of electrophoretic deposition technique to solid oxide fuel cells", *Journal of The Electrochemical Society*, Vol. 147, No. 5, pp. 1682–1687.

[66] Zhitomirsky, I., 2000, "Electrophoretic hydroxyapatite coatings and fibers", *Materials Letters*, Vol. 42, No. 4, pp. 262–271.

[67] Chen, C.Y., Chen, S.Y., and Liu, D.M., 1999, "Electrophoretic deposition forming of porous alumina membranes", *Acta Materialia*, Vol. 47, No. 9, pp. 2717–2726.

[68] Ke, C., Yang, W.L., Ni, Z., *et al.*, 2001, "Electrophoretic assembly of nanozeolites: zeolite coated fibers and hollow zeolite fibers", *Chemical Communications*, No. 8, pp. 783–784.

[69] Roop Kumar, R. and Wang, M., 2002, "Functionally graded bioactive coatings of hydroxyapatite/titanium oxide composite system", *Materials Letters*, Vol. 55, No. 3, p. 133.

[70] Du, C., Heldbrant, D., and Pan, N., 2002, "Preparation and preliminary property study of carbon nanotubes films by electrophoretic deposition", *Materials Letters*, Vol. 57, No. 2, pp. 434–438.

[71] Put, S., Vleugels, J., Anné, G., and Van der Biest, O., 2003, "Functionally graded ceramic and ceramic–metal composites shaped by electrophoretic deposition", *Colloids and Surfaces A: Physicochemical and Engineering Aspects*, Vol. 222, No. 1–3, pp. 223–232.

[72] Limmer, S.J. and Cao, G., 2003, "Sol–gel electrophoretic deposition for the growth of oxide nanorods", *Advanced Materials*, Vol. 15, No. 5, pp. 427–431.

[73] Ferrari, B., Sánchez-Herencia, A.J., and Moreno, R., 1998, "Electrophoretic forming of Al_2O_3/Y-TZP layered ceramics from aqueous suspensions", *Materials Research Bulletin*, Vol. 33, No. 3, pp. 487–499.

[74] Hasegawa, K., Kunugi, S., Tatsumisago, M., and Minami, T., 1999, "Preparation of thick films by electrophoretic deposition using surface modified silica particles derived from sol–gel method", *Journal of Sol-Gel Science and Technology*, Vol. 15, No. 3, pp. 243–249.

[75] Wei, S., Yahong, Z., Wuli, Y., *et al.*, 2004, "Electrophoretic deposition of nano-size zeolites in non-aqueous medium and its application in fabricating thin zeolite membranes", *Microporous and Mesoporous Materials*, Vol. 69.

[76] Jun-ho, Y., Soo-Yeon, S., Seonghoon, L., and Yung-Eun, S., 2003, "$Y_3Al_5O_{12}$: $Ce_{0.05}$ phosphor coating on gallium nitride for white light emitting diodes", *Journal of The Electrochemical Society*, Vol. 150, No. 2.

[77] Hayashi, K. and Furuya, N., 2004, "Preparation of gas diffusion electrodes by electrophoretic deposition", *Journal of The Electrochemical Society*, Vol. 151, No. 3, pp. A354–A357.

[78] Yamashita, K., Yonehara, E., Ding, X., Nagai, M., Umegaki, T., and Matsuda, M., 1998, "Electrophoretic coating of multilayered apatite composite on alumina ceramics", *Journal of Biomedical Materials Research*, Vol. 43, No. 1, pp. 46–53.

[79] Boccaccini, A.R., Kern, H., Krueger, H.G., Trusty, P.A., and Taplin, D.M. R., 1998, "Electrophoretic deposition of nanoceramic particles onto electrically conducting fibre fabrics", In *Proceedings of the 43rd International Scientific Colloquium. Technical University of Ilmenau.*

[80] Maiti, H.S., Datta, S., and Basu, R.N., 1989, "High-Tc superconductor coating on metal substrates by an electrophoretic technique", *Journal of the American Ceramic Society*, Vol. 72, No. 9, pp. 1733–1735.

[81] Ochsenkühn-Petropoulou, M., Altzoumailis, A., Argyropoulou, R., and Ochsenkühn, K.-M., 2004, "Superconducting coatings of MgB_2 prepared by electrophoretic deposition", *Analytical and Bioanalytical Chemistry*, Vol. 379, No. 5–6, pp. 792–795.

[82] Van Tassel, J. and Randall, C.A., 1999, "Electrophoretic deposition and sintering of thin/thick PZT films", *Journal of the European Ceramic Society*, Vol. 19, No. 6–7, pp. 955–958.

[83] (a) Wei, M., Ruys, A.J., Milthorpe, B.K., Sorrell, C.C., and Evans, J.H., 2001, "Electrophoretic deposition of hydroxyapatite coatings on metal

substrates: a nanoparticulate dual-coating approach", *Journal of Sol-Gel Science and Technology*, Vol. 21, No. 1–2, pp. 39–48;
(b) Sridhar, T.M., and Mudali, U.K, 2003, "Development of bioactive hydroxyapatite coatings on type 316l stainless steel by electrophoretic deposition for orthopaedic application", *Transactions of the Indian Institute of Metals*, Vol. 56.

[84] Timms, L.A., Westby, W., Prentice, C., Jaglin, D., Shatwell, R.A., and Binner, J.G., 2001, "Reducing chemical vapour infiltration time for ceramic matrix composites", *Journal of Microscopy*, Vol. 201, No. 2, pp. 316–323.

[85] Limmer, S.J., Seraji, S., Wu, Y., Chou, T.P., Nguyen, C., and Cao, G.Z., 2002, "Template-based growth of various oxide nanorods by sol–gel electrophoresis", *Advanced Functional Materials*, Vol. 12, No. 1, pp. 59–64.

[86] Gamburg, Y. and Zangari, G., 2011, "Introduction to electrodeposition: basic terms and fundamental concepts", In *Theory and Practice of Metal Electrodeposition*, Springer New York, pp. 1–25.

[87] Xie, Y., Liu, X., Chu, P.K., and Ding, C., 2006, "Nucleation and growth of calcium–phosphate on Ca-implanted titanium surface", *Surface Science*, Vol. 600, No. 3, pp. 651–656.

[88] Song, Y.-Y., Roy, P., Paramasivam, I., and Schmuki, P., 2010, "Voltage-induced payload release and wettability control on TiO_2 and TiO_2 nanotubes", *Angewandte Chemie International Edition*, Vol. 49, No. 2, pp. 351–354.

[89] Boccaccini, A.R. and Zhitomirsky, I., 2002, "Application of electrophoretic and electrolytic deposition techniques in ceramics processing", *Current Opinion in Solid State & Materials Science*, Vol. 6, No. 3.

[90] Zhang, X., Yao, B., Zhao, L., Liang, C., Zhang, L., and Mao, Y., 2001, "Electrochemical fabrication of single-crystalline anatase TiO_2 nanowire arrays", *Journal of The Electrochemical Society*, Vol. 148, No. 7, pp. G398–G400.

[91] Pauporté, T. and Lincot, D., 2001, "Hydrogen peroxide oxygen precursor for zinc oxide electrodeposition II—mechanistic aspects", *Journal of Electroanalytical Chemistry*, Vol. 517, No. 1–2, pp. 54–62.

[92] Dinamani, M. and Kamath, P.V., 2001, "Electrochemical synthesis of metal phosphates by cathodic reduction", *Materials Research Bulletin*, Vol. 36, No. 11, pp. 2043–2050.

[93] Manso, M., Jiménez, C., Morant, C., Herrero, P., and Martínez-Duart, J.M., 2000, "Electrodeposition of hydroxyapatite coatings in basic conditions", *Biomaterials*, Vol. 21, No. 17, pp. 1755–1761.

[94] Yen, S.K., Guo, M.J., and Zan, H.Z., 2001, "Characterization of electrolytic ZrO_2 coating on Co–Cr–Mo implant alloys of hip prosthesis", *Biomaterials*, Vol. 22, No. 2, pp. 125–133.

[95] Roy, P., Berger, S., and Schmuki, P., 2011, "TiO_2 nanotubes: synthesis and applications", *Angewandte Chemie International Edition*, Vol. 50, No. 13, pp. 2904–2939.

[96] Taveira, L.V., Macák, J.M., Tsuchiya, H., Dick, L.F.P., and Schmuki, P., 2005, "Initiation and growth of self-organized TiO_2 nanotubes anodically formed in $NH_4F/(NH_4)_2SO_4$ electrolytes", *Journal of The Electrochemical Society*, Vol. 152, No. 10, pp. B405–B410.

[97] (a) Song, Y.-Y. and Schmuki, P., 2010, "Modulated TiO_2 nanotube stacks and their use in interference sensors", *Electrochemistry Communications*, Vol. 12, No. 4, pp. 579–582;
(b) Chappanda, K.N., Smith, Y.R., Mohanty, S.K., Rieth, L.W., Tathireddy, P., and Misra, M., 2012, "Growth and characterization of TiO(2) nanotubes from sputtered Ti film on Si substrate", *Nanoscale Research Letters*, Vol. 7, No. 1, p. 388.

[98] (a) Nanci, A., Wuest, J.D., Peru, L., *et al.*, 1998, "Chemical modification of titanium surfaces for covalent attachment of biological molecules", *Journal of Biomedical Materials Research*, Vol. 40, No. 2, pp. 324–335;
(b) Kulkarni, M., Mazare, A., Gongadze, E., *et al.*, 2015, "Titanium nanostructures for biomedical applications", *Nanotechnology*, Vol. 26, No. 6, p. 062002;
(c) Morra, M., 2006, "Biochemical modification of titanium surfaces: peptides and ECM proteins", *European Cells & Materials*, Vol. 12, pp. 1–15.

[99] (a) Kar, A., Raja, K.S., and Misra, M., 2006, "Electrodeposition of hydroxyapatite onto nanotubular TiO_2 for implant applications", *Surface and Coatings Technology*, Vol. 201, No. 6, pp. 3723–3731;
(b) Parcharoen, Y., Kajitvichyanukul, P., Sirivisoot, S., and Termsuksawad, P., 2014, "Hydroxyapatite electrodeposition on anodized titanium nanotubes for orthopedic applications", *Applied Surface Science*, Vol. 311, pp. 54–61.

[100] Tsuchiya, H., Macak, J.M., Müller, L., *et al.*, 2006, "Hydroxyapatite growth on anodic TiO_2 nanotubes", *Journal of Biomedical Materials Research Part A*, Vol. 77A, No. 3, pp. 534–541.

[101] Shrestha, N.K., Macak, J.M., Schmidt-Stein, F., *et al.*, 2009, "Magnetically guided titania nanotubes for site-selective photocatalysis and drug release", *Angewandte Chemie International Edition*, Vol. 48, No. 5, pp. 969–972.

[102] Schmidt-Stein, F., Hahn, R., Gnichwitz, J.-F., *et al.*, 2009, "X-ray induced photocatalysis on TiO_2 and TiO_2 nanotubes: degradation of organics and drug release", *Electrochemistry Communications*, Vol. 11, No. 11, pp. 2077–2080.

[103] (a) Agarwal, P., Paramasivam, I., Shrestha, N.K., and Schmuki, P., 2010, "MoO_3 in self-organized TiO_2 nanotubes for enhanced photocatalytic activity", *Chemistry – An Asian Journal*, Vol. 5, No. 1, pp. 66–69;
(b) Paramasivam, I., Nah, Y.-C., Das, C., Shrestha, N.K., and Schmuki, P., 2010, "WO_3/TiO_2 nanotubes with strongly enhanced photocatalytic activity", *Chemistry – A European Journal*, Vol. 16, No. 30, pp. 8993–8997.

Chapter 3

Carbon-based nano biomaterials: design, fabrication and application

Xiao Lin[1,2], *Aaron Clasky*[2,3], *Kalyn Lai*[2,4] *and Lei Yang*[1,2,*]

Abstract

The emergence of carbon-based nanomaterials has changed many aspects of industry and academia, often in revolutionary ways. Carbon-based nanomaterials have found an increasingly wide utilization in numerous and diverse fields since the 1990s such as physics, electronics, optics, mechanics, biology and medicine. The carbon-based nanomaterials are of great promise for biomedical applications due to their unique characteristics. For example, carbon-based nanomaterials are the same scale as some biological components, such as collagen, allowing them to be used to form biomimetic structures. The ultrahigh mechanical strength and conductivity are also attractive in some biomedical applications. In this chapter, the use of carbon-based nanomaterials in the biomedical field is covered from fabrication and design methods to applications and toxicity analysis. First, the common existing forms of carbon on the nanoscale will be summarized, and the unique properties and fabrication methods of different carbon-based nanomaterials will be introduced. Second, how to modify or design carbon-based nanomaterials when one intends to use these materials for biomedical applications will be investigated, as well as common design and modification methods. Third, some potential applications of carbon-based nanomaterials and the performance of these materials in those applications will be shown, in addition to their toxicity concerns. Lastly, a summary and outlook of the carbon-based nanobiomaterials will be given based on the data presented in the chapter.

[1]Orthopaedic Institute and Department of Orthopaedics, The First Affiliated Hospital, Soochow University, China
[2]International Research Center for Translational Orthopaedics (IRCTO), Soochow University, China
[3]Departments of Chemistry, Chemical Engineering, and Electrical and Computer Engineering, University of Waterloo, Ontario, Canada
[4]Department of Chemical Engineering, Northeastern University, Boston, MA, USA
[*]Corresponding author.

3.1 Introduction

3.1.1 Overview of carbon-based nanomaterials

Carbon-based nanomaterials are important members of the nanomaterial family with various entities, each having unique properties. The carbon-based nanomaterials can be divided into zero-dimensional materials (e.g. fullerenes, particulate diamonds and carbon blacks), one-dimensional materials [e.g. carbon nanotubes (CNTs), carbon nanofibers (CNFs) and diamond nanorods], two-dimensional (2D) materials (e.g. graphene, graphite sheets and diamond nanoplatelets) and three-dimensional (3D) materials [e.g. nanocrystalline diamond (NCD) coatings and, nanostructured diamond-like carbon (DLC) scaffolds] according to their structures (as shown in Figure 3.1) [1]. These materials have been studied extensively for

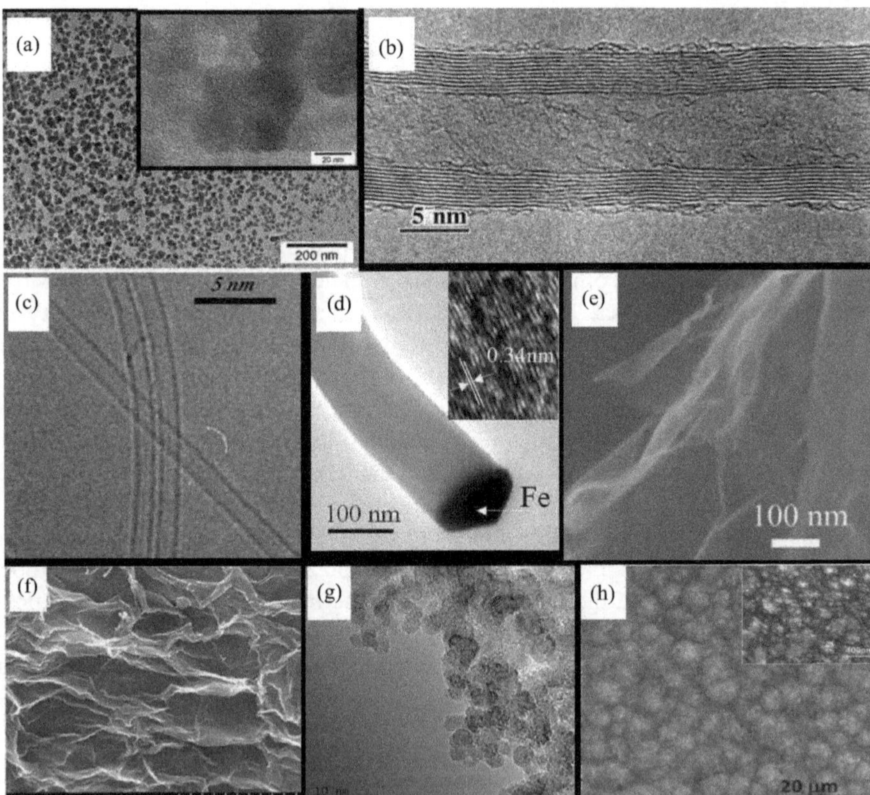

Figure 3.1 Typical microstructures of several carbon-based nanomaterials. TEM images of (a) fullerenes [188]; (b) multi-walled CNTs [197]; (c) single-walled CNTs [198]; (d) TEM image of a CNF fabricated with Fe nanoparticles as catalysts [199]; SEM image of (e) graphene [149] and (f) gaphene oxides [200]; (g) TEM image of particulated nanodiamonds [201]; and (h) SEM image of the surface topography of a NCD film [106]

medical applications based on their structures and properties during the past couple decades. Among all of the above materials, fullerene, CNF and CNTs, graphene, diamond and diamond-like structures have gained the most attention.

3.1.1.1 Fullerenes

Since their first discovery [2] and bulk production, fullerenes have played important roles in numerous fields. Fullerenes are in the form of a hollow sphere, ellipsoid or tube. At present, the main fabrication method of fullerenes includes arc-discharge deposition, thermal evaporation and chemical vapor deposition (CVD).

First produced in 1990 [3], C_{60} is the most abundant member of the fullerene family and is at the forefront of investigations and applications of fullerenes. The diameter of a C_{60} molecule is 0.7 nm. The C_{60} surface contains 20 hexagons and 12 pentagons with all the rings being fused and all the double bonds being conjugated. C_{60} has high symmetry, with 120 symmetrical operations that can map the molecule onto itself. The fullerenes behave chemically and physically as electron-deficient alkenes. C_{60} is insoluble and easily forms aggregates in aqueous media, which is a potential problem for biomedical applications. Chemical modification of fullerenes is the most versatile methodology to overcome this problem, which leads to a wide variety of fullerene derivatives with different physical and chemical properties [4]. The fullerene derivatives have been exploited for various biomedical applications including drug and gene delivery, antiviral activity, DNA cleavage, enzyme inhibition, antimicrobial activity, antioxidant activity and diagnostic applications for (metallofullerenes) photodynamic therapy.

3.1.1.2 Graphene and graphene oxide (GO)

Graphene was isolated for the first time in 2004 by Novoselov and Geim [5], who were awarded the 2010 Nobel Prize in Physics for this finding. Presently, graphene is mainly synthesized by the CVD method. The graphene is actually a 2D honeycomb-like crystal structure made of single atomic layers of graphite, composed almost entirely of sp^2 hybridized carbon atoms arranged in a hexagonal lattice. Similar to CNTs, graphene has shown its extraordinary mechanical, electrical, optical and thermal properties [6]. Besides, graphene is expected to be biocompatible in mild concentrations, which makes it an attractive material for biomedical applications. However, the apolar nature of the carbonaceous material makes graphene highly hydrophobic [7] with an extremely low solubility in aqueous solution, which limits its biomedical applications. GO is a graphene sheet derivatized with oxygen-containing functional groups. Today, the Hummers method [8] is the most popular method to fabricate GO. Compared to graphene, GO is more hydrophilic and displays colloidal dispersibility under an appropriate pH. In addition to the excellent aqueous processability, the chemically versatile characteristics, high surface-to-volume ratio, surface functionalizability, surface enhanced Raman scattering property, and fluorescence quenching ability make GO a promising material in the biomedical field. The versatile biomedical applications of graphene and GO include drug and gene delivery, bio-sensing, bio-imaging, photodynamic therapy, tissue engineering (TE) and antibacterial materials.

3.1.1.3 CNTs and CNFs

Since their discovery by Iijima in 1991 [9], CNTs have become a top class of materials both in academic research and industrial fields. CNTs are well-ordered, hollow nanostructures consisting of carbon atoms bonded to each other via sp^2 bonds which renders them excellent mechanical strength and high electrical and thermal conductivity. CNTs can be composed of a single tube (single-walled CNTs – SWCNTs) or concentric cylinders of carbon (multi-walled CNTs – MWCNTs). CNTs possess very large aspect ratios. For instance, the outer diameter of MWCNTs ranges from 2 nm to more than 100 nm, and the inner diameter is in the range of 1–3 nm, while their length can range from 0.2 to several micrometers [10]. The CNTs are usually prepared by CVD, arc-discharge deposition and laser vaporization deposition methods.

CNTs can be imagined as a graphene sheet rolling into a cylinder, while CNFs form when graphene sheets are curved at some angle, to create stacks of nanocones [11]. Sizes of CNFs can range from a few to several hundred nanometers in diameter and up to several micrometers in length [11]. CVD and arc-discharge deposition method, are usually used to fabricate CNFs. With its high aspect ratio, mechanical, electrical thermal and optical properties, there has been growing interest in using CNTs and CNFs for biomedical applications such as drug and gene delivery for cancer and infection treatments, biosensor vehicles for diagnosis and detection, and reinforcing components for biomaterials.

3.1.1.4 Nanostructured Diamond and DLC

Diamond nanostructures are another class of carbon allotrope, which mainly includes particulate nanodiamonds (PNDs), NCD films and nanostructured DLC. PNDs refer to diamond particles with a diameter of 2–10 nm. The PNDs were first fabricated by detonation in Soviet Russia in 1963 by V. Danilenko and two others [12]. Detonation is still the main method to fabricate PNDs nowadays. The unique properties of PNDs include superior hardness and stiffness, high thermal conductivity and electrical resistivity, large specific surface area, ultrahigh chemical stability, and unique optical properties. Most importantly, the PNDs were found more biocompatible than other carbon-based nanomaterials [13,14] and can be easily functionalized by versatile surface modifications [15]. Presently, the biomedical applications of PNDs include drug delivery, bio-labelling, bio-imaging, protein mimics, nanocomposites fabrication, TE scaffolds and surgical implants.

NCD film is a thin continuous layer of diamond crystallites with diameters smaller than 100 nm, which is usually fabricated by CVD methods. The NCD films with crystallite sizes less than 10 nm are also termed as ultra-nanocrystalline diamond (UNCD) films to distinguish them from the common NCD films (cystallite sizes between 10 and 100 nm). The NCD contains both sp^3 and sp^2 carbon bonds and in general NCD films contain less sp^2 hybridization and are thus more transparent than UNCD films. The NCD films possess some unique mechanical, thermal conductive, tribological, electrical and optical properties which could be adjusted by controlling the crystallite size and quality of the films [16,17]. In addition, the NCD films have excellent surface smoothness, high corrosion resistance, chemical inertness, good biocompatibility and anti-bacterial property [18,19]. All the above properties have made NCD films attractive for both therapeutic and diagnostic applications in the biomedical fields including coatings for orthopaedic and dental implants, neural repair prostheses, cardiovascular and ophthalmologic devices, biosensors and carriers for drug delivery.

The DLC film is a class of amorphous carbon materials first discovered in 1971 by Aisenberg and Chabot [20] and prepared mainly by DC plasma activated CVD, filtered arc deposition and DC discharge deposition. The mechanical properties of DLC films lie between those of graphite and diamond, featuring a low friction coefficient, high wear resistance, extremely high hardness, as well as high chemical stability, high electrical resistivity, and high optical transparency. The DLC films have also been proven to be bio-, haemo- and neuro-compatible [21,22]. The DLC films were used as coatings for prosthetic joints to prolong the service life, on vascular devices such as stents, heart waves and blood pumps, to suppress metal ion release and thrombus formation, as well as for contact lenses, dental implants and nanoscale instruments used in the nervous system.

3.2 Modification of carbon-based nanomaterials for biomedical applications

The fabrication methods of the carbon-based nanomaterials have been well documented in literature [23–26]. However, for biomedical applications, carbon-based nanomaterials need to be further modified and processed. For example, surface modification is usually needed to endow these materials with better functionality and biocompatibility. For some medical applications, the carbon-based nanomaterials need to be processed into large scale devices and structures without sacrificing their intrinsic advantages. In other situations, the carbon-based nanomaterials are used to fabricate solid or porous composites to enhance the mechanical and/or biological properties of the matrix, or achieve multi-functionality.

3.2.1 Surface modification

Surface modification of the carbon-based nanomaterials is very important for their biomedical applications. First, most carbon-based nanomaterials are not soluble in aqueous solutions. However, many biomedical applications of these materials could be realized only by making them water soluble and easier for interaction with biological system. Proper surface modification has the potential to increase the solubility of these materials. Second, surface treatment can functionalize the carbon-based nanomaterials with some properties which can be used in certain applications. Thirdly, surface treatment is effective in decreasing the toxicity of the carbon-based nanomaterials. The main surface modification strategies of different carbon-based nanomaterials are summarized in Table 3.1 and are further explained as follows.

Water-soluble malonic acid derivatives of C_{60} with a defined 3D structure were firstly synthesized by Hirsch and Lamparth [27]. They also synthesized supramolecular organizations of dendritic amphiphiles in which the fullerenes acted as core

Table 3.1 Surface modifications of carbon-based nanomaterials

Nanomaterials	Modification or its methods	Purposes/applications	Ref.
Gd@C_{60} species	Hydrolyzed to form Gd@C_{60} $[C(COOH)_2]_{10}$	Increased solubility, and used as MRI contrast agent	[31]
C_{60}	Addition of dendrimer containing 18 carboxylic acid groups	Improved antioxidation properties	[32]
	Cycloaddition of azomethane ylides to synthesize regioisomeric bis-fulleropyrrolidines	Adding anti-HIV activity	[33]
	Polyhydroxylated (Fullerenols) PEG-modified Adsorption of ascorbic acid	Getting a strong affinity for hydroxyapatite and used as, free-radical trapper	[34,35]
Graphene	Functionalized with nucleic acids, aptamers, carbohydrates	Monitoring of target-specific changes of electrical signals	[26,36]
	Complexed with therapeutic molecules (e.g. DOX)	For highly efficient drug delivery	[37]
	Coated with positively charged PEI	For highly efficient drug delivery and gene transfer	[36]
GO	Conjugated with fluorescein	Molecular imaging probe	[36,37]
CNTs	Oxidation and carboxyl-based couplings	Increased solubility, dispersity, and the possibility of conjugating with bioactive and therapeutic agents	[40,41]
	Attachment of organic groups (e.g. cycloaddition of azomethine ylides)	Control of solubility	[11]
	Attachment of amino groups	Ability to link with a series of amino acids, fluorescent probes and bioactive peptides	[43]
NCD	Oxygen, hydrogen or amine terminated	Enhanced cellular adhesion	[47,48]
	Conjugated with biomolecules or drugs	Bio-sensing and drug delivery	[50,51]
DLC	Doped or alloyed with elements such as Si, F and P; treated by O_2, CF_4 or N_2 glow discharge	Improved haemocompatibility	[53,54]

units and these compounds contained both hydrophobic and hydrophilic moieties and self-assembled into water-soluble supramolecular structures [28,29]. Presently, combining with amphiphilic polymers, amino acids, carboxylic acids, and polyhydroxyl groups is a widely used method for increasing the solubility of fullerenes [30]. The potential applications of Gd@C_{60} have been largely overlooked because of its mediocre solubility. Solubilized Gd@C_{60} species were synthesized by forming Gd@$C_{60}[C(COOCH_2CH_3)_2]_{10}$, and its hydrolyzed water-soluble form, Gd@C_{60} $[C(COOH)_2]_{10}$. The water-soluble derivatives can be used as a magnetic resonance imaging (MRI) contrast agent [31]. A dendrimer containing 18 carboxylic acid groups was added to C_{60} to form a dendro C_{60} which showed remarkable antioxidant properties due to its ability to scavenge reactive oxygen species often created by oxidation [32]. Regioisomeric bis-fulleropyrrolidines bearing two ammonium groups were synthesized using the [3+2] dipolar cycloaddition of azomethane ylides to C_{60} and were considered to have anti-HIV activity [33]. Fullerenes have also been functionalized by a variety of other methods to make them suitable for biomedical applications, such as 'fullerenols' (polyhydroxylated fullerenes) [34], poly-ethylene glycol (PEG) modification and surface adsorption of ascorbic acid [35].

Graphene and its derivatives such as GO have been functionalized with nucleic acids, aptamers and carbohydrates for monitoring the target specific changes of electrical signal with high signal-to-noise ratios in bio-sensing applications [26,36]. When conjugated with fluorescein, GO can be used for molecular imaging. However a linker, such as PEG, is always used to reduce direct interactions between fluorescein and GO [36,37]. When the graphene is modified with therapeutic molecules, it serves as an effective drug delivery system. The majority of drugs that have been complexed with graphene surfaces feature planar aromatic domains, such as doxorubicin (DOX), since the stable π–π stacking between their aromatic rings and the graphene carbon surface leads to stable complex formation, while avoiding chemical conjugation [36]. Graphene can be also covalently coated with polymers to improve the efficiency of delivering drugs [37]. Positively charged polyethylenimine (PEI) can be coated on negatively charged GO via electrostatic interactions, which allows condensation of negatively charged RNA or DNA molecules onto the GO for gene transfer purposes [36]. Decoration of graphene with metal nanoparticles and metal oxides has also been realized [38].

Similarly, pristine CNTs are not soluble in common solvents, and some of them were shown to be cytotoxic [39]. Surface modification is required to solubilize CNTs and lead to better compatibility. Oxidation and carboxyl (–COOH)-based couplings are effective surface treatment strategies for CNTs. In this method, the tube cap openings are created and holes in the sidewalls are formed by an oxidation process using strong acids. In addition to this, the oxidation conjugates the caps and sidewalls with carboxylic groups, enhancing the solubility of CNTs in aqueous solutions [40]. On one hand, carboxylic groups on the sidewalls of CNTs reduce Van der Waals interactions between the tubes, which are beneficial to separate nanotube bundles [41]. On the other hand, CNTs can be conjugated with various functional groups such as bioactive agents and therapeutic agents through carboxyl [11]. Besides this method, additional reactions have been employed to attach some organic groups to the sidewalls and/or tips of the CNTs. For example, the 1,3-dipolar cycloaddition of azomethine ylides is suitable for the preparation of soluble SWCNTs and MWCNTs alike [42]. This reaction occurs in a suspension of pristine CNTs in dimethylformamide (DMF) followed by the addition of an N-substituted glycine and an aldehyde. In addition, when free amino groups are present on the CNT surface, they can be covalently linked with a series of amino acids, fluorescent probes and bioactive peptides [43]. Besides the covalent approaches, the large aromatic (π-electrons) hydrophobic surface of CNTs makes them ideal partners for noncovalent interactions with suitable complementary molecules and macro-biomolecules like DNAs [44,45].

Many functional groups can be attached to the surface of PNDs, allowing quite sophisticated surface functionalization without compromising the useful properties of the diamond core [46]. It is convenient to start modification with carboxylated PNDs produced by air or ozone purification, so as to take advantage of the rich chemistry of the –COOH groups. PNDs can also be modified through the chemistry of graphitic carbon which is either present intrinsically or created through surface graphitization. The surface functionalization outcome of PNDs strongly depends on the purity and uniformity of the surface chemistry of the starting material [15]. The NCD film can also be further modified to gain better cellular response or special capabilities. First, the NCD can be made hydrophilic with oxygen or hydroxyl terminations, which supports cellular adhesion [47]. The terminations can be formed by UV irradiation [48], oxygen-contained plasma treatment [49] or oxidation in concentrated HNO_3 [50]. Biomolecules and drugs can also be conjugated on NCD films for bio-sensing or drug delivery purposes. Before linking the functional molecules, the NCD surface usually undergoes some treatment like oxidation [50] and electrochemical reduction of aryl diazonium salt [51] to produce groups serving as anchoring sites. Sometimes, linkers are needed for attachment of biomolecules like DNA, antibodies and enzymes [52]. Differently, DLC is specified for its chemical inertness, and few work reports the modification of DLC covalently bound with biomolecules. However, the amorphous structure of DLC is favorable for doping or alloying elements, such as Si, F, P, Ag, Cu and N, which imparts improved or new properties to DLC [53,54].

3.2.2 Large-scale modification of carbon-based nanomaterials

The carbon-based nanomaterials need to be built up into large-scale structures and even advanced devices for some applications. In the biomedical field, e.g. large porous structures made of carbon-based nanomaterials gained the most attention, particularly for tissue regeneration. Recently, porous graphene materials including graphene nanomesh (GNM, in-plane pores in graphene sheets), crumpled graphene (CG) and graphene foam (GF, the construction of macroscopic architectures using graphene sheets as building blocks) have been studied extensively [55]. Synthesis methods of GNM include plasma, electron, photo, template and chemical etching methods, while the CG can be fabricated by thermal expansion, cooling contraction, substrate regulation and capillary compression methods. Compared to GNM and CG, GF gained more attention recently due to its versatile potential applications. The template directed CVD method has been developed to make GF with unique network structure and outstanding electrical and mechanical properties [55,56]. Free-standing, porous and flexible 3D GO-based scaffolds (Figure 3.2(a)) have also been produced by a biocompatible freeze casting procedure [57].

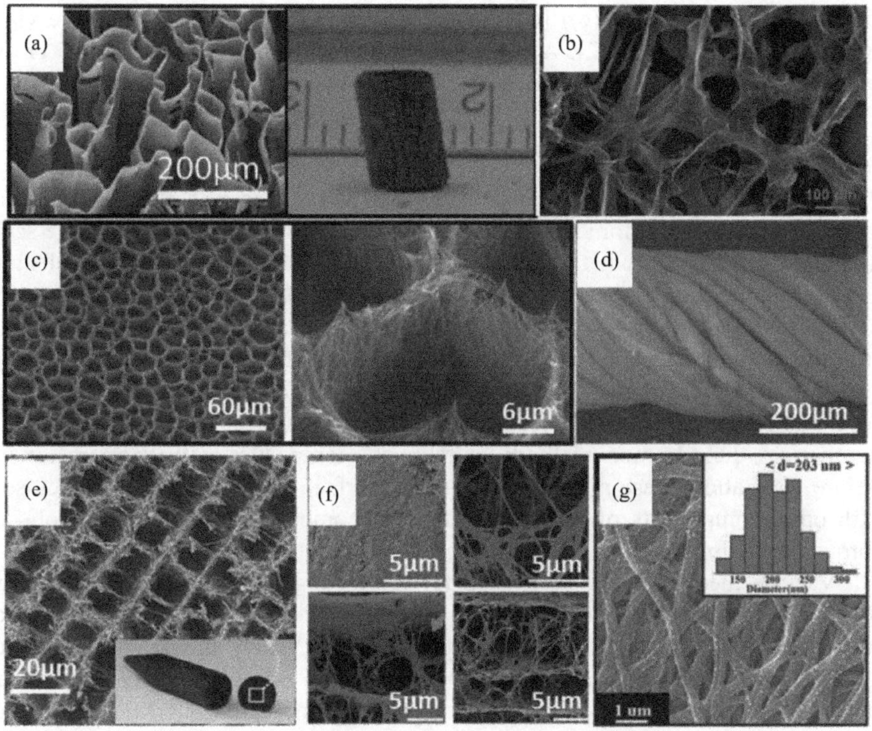

Figure 3.2 (a) 3D GOx scaffolds obtained by ice segregation-induced self-assembly method and thermal treatment [57]; (b) SEM images of 3D graphene foams formed by depositing graphene on a nickel template and followed by etching [90]; (c) Network of crosslinked CNT walls forming cavities [61]; (d) SEM image of the CNT rope surface. A spiral topography propagating throughout the surface of the CNT rope is observed. The ridge-like pattern originating from the spiral structure is similar to that of an actual rope [63]; (e) SEM micrograph of a cross-sectioned MWCNT/chitosan scaffold exhibiting a chamber-like structure. Inset shows the monolithic scaffolds resulting from the ice segregation-induced self-assembly process [80]; (f) Top (upper row) and cross-sectional (lower row) view of CNT-coated bacterial cellulose (left column) and amphiphilic comb-like polymer adsorbed CNT-coated bacterial cellulose hybrid scaffolds (right column) [82]; (g) SEM image of nanoporous structured CNF-bioactive glass (BG) composites. The inset shows the fiber diameter distribution [83]

Other larger scale structures have also been fabricated. For example, the roll-to-roll CVD synthesis is most successful in producing high quality graphene films up to the meter scale [58,59]. Organized CNT architectures have been fabricated using the CVD method [60], by exerting chemically induced capillary forces upon the nanotubes (Figure 3.2(c)) [61], or compacting CNTs in a steel-tool die via a uniaxial pressing cycle [62]. A rope-like structure (1 mm in diameter, 1.5 cm in length, Figure 3.2(d)) merely made of SWCNTs has also been fabricated [63].

3.2.3 Composites based on carbon nanostructures

Two major drawbacks exist regarding the use of large-scale carbon-based nanomaterial structures for biomedical applications. One is the fragility of the large-scale structures, and the other is the toxicity responses by the direct contact of cells and tissues with the carbon-based nanomaterials at the early stage after implantation [64]. Composites based on carbon nanostructures have arisen as novel materials for biomedical applications. CNTs, CNFs, graphene (and its derivatives) and PNDs are very promising for fabricating composites for biomedical applications due to their good combinations of mechanical, electrical and biological properties. Table 3.2 summarizes the common methods to fabricate composites with these materials. Before fabrication, these materials are usually surface-treated to bond more closely with other components of the composite. The composites can be either solid or porous depending on different applications.

Various methods have been used for fabricating solid CNTs/hydroxyapatite (HA) [65], CNTs/ultrahigh molecular weight polyethylene (UHMWPE) nanocomposites [66] and CNF/polyetheretherketone (PEEK) composites [67] for orthopaedic implants. In addition, the electrostatic spraying method has been established for synthesizing large-scale graphene nanoplatelet/UHMWPE composite films [68]. Hot-pressing sintering [69] and spark plasma sintering [70] have been used to fabricate CNF/HA composites and graphene/HA composites, respectively. In addition, melt mixing [71], solvent casting [72] and in situ precipitation method [73] have been used to prepare carbon-based nanomaterials/biodegradable polymers composites. CNTs are also added to injectable bone cements [74–76] and injectable nanocomposites [77]. Generally, carbon-based nanomaterials need to be dispersed homogeneously in the composite. To achieve this, thorough mixing of the carbon-based nanomaterials with different particles or evenly dispersion of carbon-based nanomaterials in the polymer melt (or solution) is critical. In addition, functional groups on the surfaces of carbon-based nanomaterials also assist their dispersion in a polymer melt or solution by preventing formation of agglomerated bundles and increasing interactions [71,73,78].

Porous composites containing carbon-based nanomaterials are gaining increasing attention with the rapid development of tissue engineering (TE) [11,79]. Several methods have been developed for the fabrication of porous carbon-based nanomaterials containing composites, including solvent casting/particulate leaching [78], freeze-drying (Figure 3.2(e)) [80], thermal sintering of microspheres [81], in situ hybridizing (Figure 3.2(f)) [82], as well as electrospinning (Figure 3.2(g)) [83]. Recently, a MWCNT reinforced porous magnesium composite was produced by a powder metallurgical method [84]. Other approaches to produce CNT-containing 3D structures involve coating the surface of the preformed 3D scaffolds of diverse composition, layer-by-layer deposition [85], in situ growth [86] or electrophoretic deposition [87].

Table 3.2 Fabrication techniques for composites based on carbon nanomaterials

Composites		Fabrication techniques	Details	Ref.
Solid composites	Graphene nanoplatelet/ UHMWPE films	Electrostatic spraying	An electrodeposition technique involving a liquid flow emerging from a small tip or tube under the influence of a strong electric field.	[68]
	CNF/HA	Hot-pressing sintering	Sintering powders by applying a compressive stress at high temperatures. The compressive stress drives the sintering process forward and decreases the energy required to densify the sample.	[69]
	Graphene/HA	Spark-plasma sintering	Sintering powders using DC pulses by generating high temperature plasma in gaps between powder granules. Performed in a few minutes as compared to hours or days for other sintering methods. Very effective for sintering of nanopowders.	[70]
	MWCNT/PCL	Melt mixing	Melting components down and mixing thoroughly by varying the processing parameters like temperature, mixing time and speed.	[71]
Porous composites	MWCNT/ chitosan	Freeze-drying	Rapidly freezing down the slurry and then transferring it to a high vacuum to remove all (frozen) moisture as a gas.	[80]
	MWCNT/Mg	Powder metallurgical method	The mixture of Mg powder, MWCNT, and carbamide ($CO(NH_2)_2$) followed a processing route including mixing, compacting, and sintering. The removal of carbamide leaves pores in the final products.	[84]
	PND/poly (L-lactide-co-ε-caprolactone)	Solvent casting and particulate leaching	Using a polymer solution with salt embedded throughout. The solvent evaporates and leaves behind a polymer matrix with the salt within. Then immersing the matrix in water and the salt leaches out of the matrix, forming porous structure.	[78]

(Continues)

Table 3.2 (Continued)

Composites		Fabrication techniques	Details	Ref.
	MWCNT/PLGA	Thermal sintering of microspheres	Heating the microspheres above their glass transition temperature so that the mixture becomes more rubbery and forms bonds with neighbouring microspheres.	[81]
	CNF/BG	Electrospinning and carbonization	A charged polymer jet is collected on a grounded collector that can either spin (to make fibers) or remain stationary (to make randomly oriented fibrous mats). Below the critical entanglement concentration of the polymer, electrostatic spraying can be achieved. The unwanted composition can be removed by carbonization.	[83]
	CNT/bacterial cellulose	In situ hybridization	Growing bacterial cellulose-producing bacteria in a medium containing CNTs.	[82]

3.3 Applications of carbon-based nanomaterials

3.3.1 Tissue repair and regeneration

Tissue repair and regeneration has become an active research area in the biomedical engineering field for treating tissue damage that cannot self-heal. In tissue repair and regeneration, substrates supporting or promoting tissue regeneration, facilitating the growth and function of tissue forming cells and/or inducing the differentiation of stem cells into tissue forming cells are very important. In recent years, studies on the application of carbon-based nanomaterials as scaffolds or implant coatings for tissue repair and regeneration experienced explosive growth, which revealed the great promise of carbon-based nanomaterials in bone, nerve, cardiac, vascular and cartilage applications.

3.3.1.1 Bone

Trauma, tumor excision, infection or congenital malformation are major causes of bone defects. When the defect size exceeds a critical value, bone is not able to self-heal, which is denoted as a critical-sized bone defect. TE is promising for critical-sized bone defect repair, so the osteogenic capability of scaffold materials is very important. Graphene and its derivatives exhibit osteogenic potential. For example, cellular proliferation of mesenchymal stem cells (MSCs) on graphene and GO substrates was higher than that on PDMS substrates [88]. Graphene-coated Si/SiO_2 substrates supported the proliferation of human MSCs (hMSCs) while accelerating their specific osteogenic differentiation, comparable to that in the presence of growth factors [89]. The 3D graphene foams (Figure 3.2b) have been used as bone TE scaffolds, which can maintain cell viability of hMSCs and promote their osteogenic differentiation as shown by Figure 3.3 [90]. GO nanosheets could be used as 2D coatings. After being modified with peptide nanofibers (PNFs), the GO-PNF nanohybrid could induce biomimetic mineralization, generating an HA layer. The GO-PNF-HA minerals have the capacity to enhance the proliferation of MC3T3-E1 cells [91]. Besides TE scaffolds, GO can also be used to modify other bone TE materials. For example, GO-modified β-TCP bioceramics stimulated osteogenesis both in vitro and in vivo [92]. CNTs and CNFs can also improve the

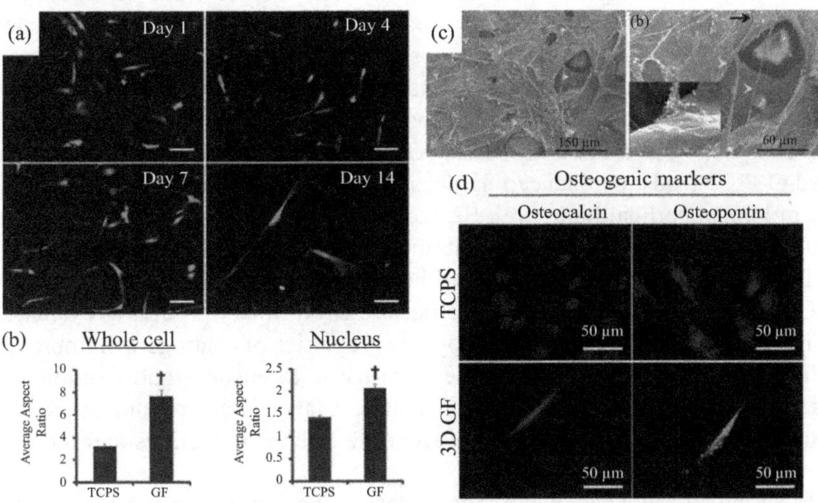

Figure 3.3 *(a) GFs support the attachment and viability of hMSCs over 14 days. The majority of attached hMSCs remained viable (extended) and few dead cells (samll round) were observed. Scale bars = 100 μm. (b) Cellular and nuclear aspect ratio of hMSCs cultured on GFs or tissue culture polystyrene (TCPS) control. †$p < 0.01$ between groups. (c) hMSCs cultured on GFs for 4 days formed protrusions up to 100 microns in length (grey arrowheads) that extended from small cell bodies (black arrows). Interestingly, several of these protrusions spanned large pores in the GF and interacted directly with the material surface. (d) Immunostaining for osteogenic markers for hMSCs cultured on GFs or (TCPS) control for 7 days. Culture of hMSCs on GFs stimulated the expression of osteocalcin and upregulated the expression of osteopontin [90]*

osteogenic activity of bone TE scaffold, especially when using surface functionalized CNTs or CNFs. For example, osteoblast proliferation and differentiation are significantly enhanced on PCL with amine functionalized MWCNTs when compared with pristine PCL [71]. Enhanced in vivo ectopic bone formation has been reported around the human allografts combined with carboxyl-modified SWCNTs. Transcriptional analysis of classical bone markers and increased presence of secreted *osteopontin* protein also clearly reveal the improved osteogenic properties of the carboxyl-modified SWCNTs [93]. BG decorated nanoporous CNF composite also shows distinct promotion in proliferation and osteogenic differentiation of osteoblast cells [83]. Electrical stimuli have been shown to enhance osteogenesis in vitro and in vivo [94,95]. Due to the excellent conductive properties of CNTs, electrical stimuli can be effectively applied on composites with CNT, as shown by the study on conductive CNT/glass/HA ceramics. The bone cell proliferation, DNA content and mRNA expression are all maximized on the conductive CNT/glass/HA substrates when electrical stimuli are applied (as demonstrated in Figure 3.4) [96].

Besides the osteogenic properties, the bone TE scaffold requires proper mechanical properties. CNT- and CNF-based scaffolds have excellent mechanical strengths and can also reinforce composite scaffolds [81,97,98]. PNDs can also improve the mechanical properties and osteogenic activity of scaffolds. For example, octadecylamine-modified nanodiamond (O-PND) has been used to fabricate composite scaffolds with PLLA (O-PND-PLLA). The addition of well-dispersed O-PND leads to enhanced hardness and Young's modulus values similar to those of human cortical bone [99]. Besides, Oxygen-terminated PND modification significantly increases cell attachment and differentiation of MSCs on poly(LLA-co-CL) scaffolds in vitro and enhanced bone formation in vivo [100].

Osteogenic and mechanical properties are two primary criteria to evaluate TE coatings on bone implants. Good osteogenic properties of coatings will improve the bonding strength of implants with the bone tissue to ensure good fixation, while good mechanical properties, including wear resistance, fracture toughness, elastic modulus and bonding strength, can prevent the failure of coatings and reduce the generation of coating debris.

NCD and DLC are widely used for implant coating applications. Specifically, the NCD films are used as coatings to enhance the biocompatibility and also anti-wear properties. Enhanced cell adhesion, proliferation and osteogenic differentiation of osteoblast-like cells and stem cells in vitro [47,101–104], and good osseointegration in vivo [104,105], have been reported on both NCD and DLC coatings. The NCD and DLC coatings, well known for their low-wear characteristics, have also been applied onto the articulating surfaces of artificial joints [53,106].

In addition, the use of graphene, GO and CNTs has been shown to enhance the osteogenic and mechanical properties of implant coatings. For example, GO coatings on bone TE scaffold enhances the proliferation and differentiation of osteoblastic MC3T3-E1 cells in vitro when the concentration of GO is properly controlled to prevent adverse biological effects in vivo [107]. The addition of GO into HA coatings on pure Ti substrates can reduce surface cracks and increase bonding strength, corrosion resistance and biocompatibility of the HA coatings

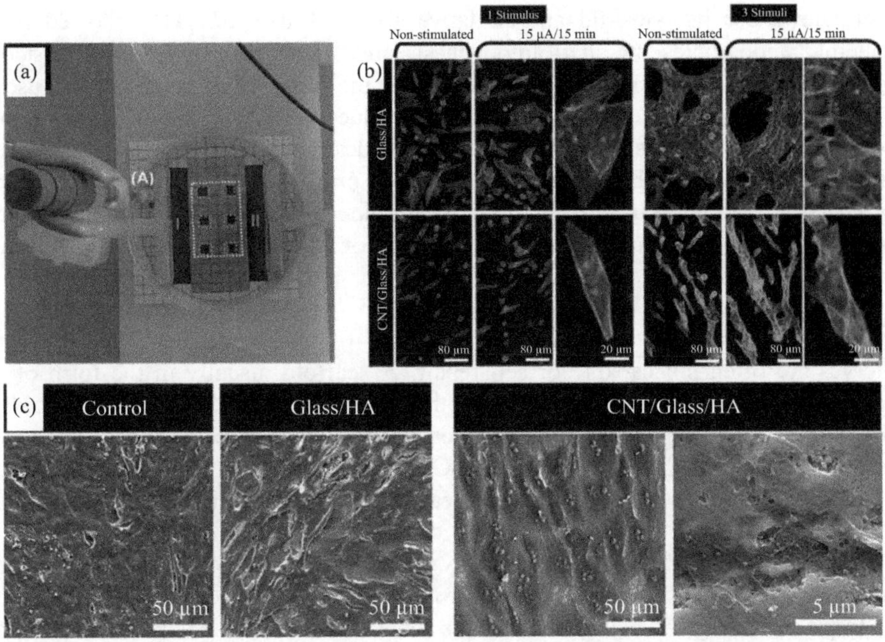

Figure 3.4 (a) A photographic image of a culture dish with parallel disposed salt-bridge electrodes for electrically stimulating cells. (b) CLSM appearance of osteoblastic cell cultures grown over the glass/HA matrix and CNT/glass/HA composite observed 24 h after 1 and 3 daily electrical stimuli under different stimulation conditions. The cells were stained for the F-actin cytoskeleton (the visible, light grey body of the cells) and nucleus (small circles present within the cytoskeleton). (c) SEM appearance of the cultures observed 5 days after the 3 daily electrical stimuli (15 mA/15 min); the panel also displays a high magnification image of the CNT/glass/HA composite showing a close interaction between the osteoblastic cells and the CNT on the composite matrix [96]

[108]. Graphene also enhances the wear resistance of and human marrow stem cell adhesion on calcium silicate (CS) coatings [109]. The CNT coating on TiO_2 nanotubes can also promote the proliferation of MC3T3-E1 osteoblasts [110]. The mechanical properties, such as hardness, fracture toughness, wear resistance and bonding strength of CNT/HA composite coatings on Ti alloys are higher than those of pure HA coatings [111–116]. Furthermore, CNTs can improve the anticorrosive properties of HA coatings [117].

In addition to osteogenic and mechanical properties, some carbon-based nanomaterials have other properties which are attractive for bone tissue repair. Increased oxidative stress is currently considered a crucial cause of corticosteroid-induced osteonecrosis. Fullerol, as a water-soluble fullerene derivative, is well-known for its powerful anti-oxidative ability. Liu *et al.* [118] showed that fullerol nanoparticles inhibited adipogenesis and simultaneously enhanced osteogenesis of a bone marrow MSC, indicating that the fullerol can potentially be used for prevention and treatment of corticosteroid-induced osteonecrosis. The resorption of bone by osteoclasts is another reason for bone destruction. Yudoh *et al.* found that water-soluble C_{60} prohibited the differentiation of precursor cells into osteoclasts and osteoclastic resorption in vitro. In an adjuvant-induced arthritic bone resorption rat model, they showed that intra-articular injection of C_{60} significantly inhibited local inflammation and joint destruction [119].

3.3.1.2 Nerve

Nerves are often not capable of self-repair. A scaffold, usually in the form of a conduit, is needed to facilitate and guide the neural regeneration. For the scaffolds, the capability to induce the neural differentiation of stem cells and to promote neurite outgrowth of neurons is critical. Graphene, CNTs and CNFs have all been actively used for nerve repair in recent years due to their excellent mechanical and electrical properties.

Neural stem cell (NSC)-based therapy provides a promising approach for neural regeneration. A scaffold is required to provide cell growth microenvironments and appropriate synergistic cell guidance cues for NSC. Graphene has been adopted to fabricate the scaffold and positive interaction between NSC and graphene has been reported. For example, Laminin-coated 2D graphene films enhance human NSC differentiation towards neurons rather than glial cells [120]. 3D graphene foams enhance the differentiation of NSCs towards astrocytes and neurons [121]. Rolled 3D GO foam also promotes the differentiation of human NSCs into neurons (rather than glia) under electrical stimulation [122]. A study finds that even MSCs demonstrated neuronal differentiation when cultured on fluorinated graphene [123]. In addition, highly viable and interconnected neural networks can form on 3D free-standing, porous and flexible GOx scaffolds, containing both neurons and glial cells while also being rich in dendrites, axons and synaptic connections [57].

The CNTs and CNFs have also shown their potential for neural regeneration. The CNTs functionalized with 4-hydroxynonenal are found to induce neurite outgrowth and increase neurite branching [124]. CNTs are able to improve neural signal transfer (may be related to the high electrical conductivity of CNTs) while supporting dendrite elongation and cell adhesion [125]. The outgrowth and branching pattern of neuronal processes can be controlled by manipulating the charge carried by functionalized CNTs [126]. Recently, the in vivo functions of CNT-interfaced nerve conduits in the regeneration of transected rat sciatic nerves have been studied (Figure 3.5). Sixteen weeks after implantation of a CNT-phosphate glass microfiber (PGF) nerve conduit into the 10 mm gap of a transected rat sciatic nerve, the number of regenerating axons crossing the scaffold, the cross-sectional area of the re-innervated muscles and the electrophysiological findings were all significantly improved by interfacing with CNTs [127]. CNTs can also be potentially used as coatings on the nerve implants. Neurite outgrowth and FAK expression of dorsal root ganglion (DRG) neurons was improved on the MWCNTs-coated poly (L-lactic

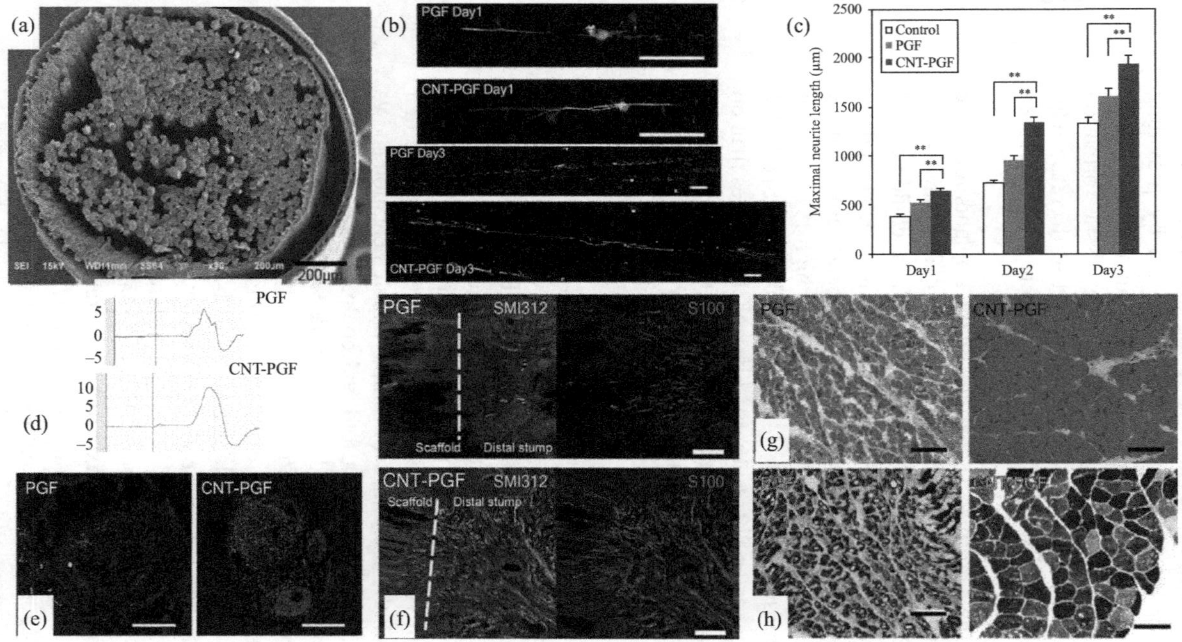

Figure 3.5 (a) *The whole cross-sectional structure of the periphery of the CNT-interfaced PGFs scaffold; (b) representative images of DRG neuronal cell culture on plain dish (Control), PGFs or optimally functionalized MWCNT-interfaced PGFs (CNT–PGF) for 1 and 3 days. Scale bar = 200 μm; and (c) quantitative results of the maximal length of DRG neuritis of each group (plain dish, PGFs and CNT–PGFs). *$p < 0.05$ by the Kruskal–Wallis test with Bonferroni correction; **$p < 0.05$ by one-way ANOVA with the Duncan post hoc test; ***$p < 0.05$ by the Kruskal–Wallis test. (d) Representative images of CMAP following PGF scaffold (PGF) or CNT–PGF scaffold implantation (CNT–PGF). Representative immunohistochemical images of axons (grey) in the transverse section at the distal stump (11 mm from the proximal stump end, e) and axons (SMI312, left), and Schwann cells (S100, right) in the sagittal section at the border between the scaffold and the distal stump (f) of PGF scaffold-implanted sciatic nerve (PGF) or CNT–PGF scaffold implanted sciatic nerve (CNT–PGF) at 16 weeks post-implantation, scale bar of (e) = 500 μm, scale bar of (f) = 200 μm. (g) Representative images of H&E-stained gastrocnemius muscle following PGF scaffold (PGF) or CNT–PGF scaffold implantation (CNT–PGF) at 16 weeks post-implantation (scale bar = 50 μm). (h) Representative images of ATPase stained muscle (PGF and CNT–PGF) and examples of muscle fibre types (right, I = type I, Iia = type IIa, Iib = type IIb, scale bar = 50 μm) [127]*

acid-co-caprolactone) (PLCL) scaffolds when compared to the PLCL group [128]. The CNT coating on conventional tungsten and stainless steel wire electrodes enhance both recording and electrical stimulation of neurons in culture, rats and monkeys by decreasing the electrode impedance and increasing charge transfer [129].

CNF scaffolds are able to enhance and direct nerve growth by providing physical support for cell growth and impulses to stimulate the nerve [130]. One of the main reasons leading to neural implant failures is glial scar tissue formation around the implant. The adhesion of astrocytes has been found to decrease with increasing CNFs in polycarbonate urethane (PCU), which may lead to decreased gliotic scar tissue formation [131].

NCD and DCL have shown their promise as coatings on the implants for nerve repair. For example, spontaneous cell proliferation and differentiation are shown when NSCs are cultured on UNCD films, suggesting the UNCD films could be used for nervous system transplantation and neural TE [49]. Fully differentiated spiking neurons survive well on NCD coating without any protein coating, indicating that the NCD allows direct neuronal interfacing, which is important for high resolution of neuronal stimulation required for visual prostheses [132]. Neuroblastoma SH-SY5Y cells also adhere and proliferate better on differently modified NCD surfaces than on the bare fused silica [133]. Kelly *et al.* show that ultraviolet (UV) treated DLC can be used to support patterned and unpatterned cortical neuron growth [134].

3.3.1.3 Other tissues

Carbon-based nanomaterials have also been studied for cartilage, myocardial tissue repair and wound healing. Articular cartilage has an avascular structure that limits its self-healing capacities once damaged. Exogenous scaffold-based TE coupled with cell therapy is an effective strategy for cartilage repair. In a recent study, GO containing scaffolds for cartilage TE has been investigated in vitro and in vivo. Cartilage cells contacted with the scaffold remain viable and showed growth potential in vitro. In a rabbit cartilage tissue repair model, the scaffold with cellular supplementation shows better repair results when compared with the scaffold and control groups [135]. In addition to GO, C_{60} and derivatives have been used in cartilage degeneration treatment due to their free radical scavenging properties [136].

Cardiac muscle is an electroactive tissue capable of transferring electrical signals and allowing the heart to beat. When damaged, adult heart muscle has a poor capability to repair itself due to the minimal regeneration potential of cardiomyocytes [137]. Selection of a TE scaffold with appropriate mechanical and electrical properties is critical for the generation of functional myocardial tissue, [138]. Due to appropriate mechanical and excellent electrical properties, GO, CNT and CNFs have been studied for myocardial tissue repair. For example, an in vitro study shows that the myogenic protein expression, multinucleate myotube formation and expression of differentiation specific genes (MyoD, myogenin, Troponin T and MHC) of myoblasts are remarkably enhanced on GO, possible due to serum protein adsorption and nanotopographical cues on the GO [139]. Neonatal rat ventricular myocytes cultured on MWCNTs are reported to form tight contacts and showed increased viability and proliferation. The changes in the electrophysiological properties of cardiomyocytes suggest that CNTs can promote cardiomyocyte maturation [140]. Cardiomyocyte density on CNF/PLGA composites increases with increasing amounts of CNF in PLGA [141]. After a 14-day culture of neonatal rat heart cells on CNF/chitosan scaffolds, the pores throughout the construct volume were filled with cells. Significantly higher metabolic cell activity and increased expression of cardiac-specific genes involved in muscle contraction and electrical coupling in chitosan/CNF constructs were found as compared to chitosan scaffolds. The improved properties of cardiac tissue constructs are presumably due to enhanced transmission of electrical signals between the cells [142].

The NCD and DLC, due to their good haemocompatibility, have been used for coating on vascular implants that are in direct contact with blood such as stents, heart valves and blood pumps, [16,53,143]. DLC has also been used as a coating on stents in the urinary tract to drastically reduce susceptibility to the encrustation, resulting in a less painful replacement and prolonged indwelling time up to a year [144]. Several studies have shown good in vitro adhesion of epithelial cells and fibroblasts on nanocrystalline diamonds [48,145–147]. Graphene coated nitinol (a widely used cardiovascular stent material) can lead to better proliferation of smooth muscle and endothelial cells. The serum albumin adsorption on the coated nitinol is greater than that of fibrinogen, which is an important criterion for achieving a lower thrombosis rate [148].

Graphene, due to its reported antibacterial activity, has been used to prepare a composite with chitosan–PVA nanofibers to aid wound healing [149].

In summary, increasing evidence has shown the potential of carbon-based nanomaterials in the field of regenerative engineering of various tissues, which are summarized in Table 3.3. However, in vivo evidence is still lacking. In addition, the long-term toxicity effects of these materials on the tissues should be investigated. Further work should focus on in vivo studies to reveal tissue repair outcome, as well as the long-term toxicity, degradation behavior and excreting mechanisms of these materials.

3.3.2 Drug delivery

Carbon-based nanomaterials are important and attractive biomaterials for drug and gene delivery due to their nanoscale sizes and unique physiochemical properties [11,30,150]. Most studies report the use of carbon-based nanomaterials for the delivery of anti-cancer drugs or agents. However, here we emphasize the applications of carbon-based nanomaterials as drug carriers for tissue repair and regeneration. Table 3.4 gives typical examples of the studies in this field, demonstrating the great potential of carbon nanostructures.

In terms of the orthopaedic applications, the carbon-based nanomaterials have been studied for bone morphogenetic protein-2 (BMP-2) delivery. For example, SWNTs-COOH and SWNTs-CH_3 can bind and induce unfolding of BMP-2 molecules to enhance their bioactivities [151]. It is found that delivery of BMP-2 using GO flakes suspended in fibrin gels (GO/F) resulted in significantly greater osteogenic differentiation of human bone marrow-derived MSCs in vitro and

Table 3.3 The biological properties of carbon-based nanomaterials for the repair and regeneration of different tissues

Target tissue	Materials	Biological properties	Ref.
Bone	Graphene or GO thin films	Enhancing the cellular proliferation of MSCs and accelerating their specific osteogenic differentiation	[88,89]
	Graphene 3D foams	Maintaining cell viability of hMSCs and promoting osteogenic differentiation	[90]
	Peptide-nanofibre modified GO 2D sheets	Inducing biomimetic mineralization and enhancing the proliferation ability of osteoblast cells	[91]
	GO-modified β-TCP bioceramics	Stimulating osteogenesis both in vitro and in vivo	[92]
	PCL with amine functionalized MWCNTs	Enhancing osteoblast proliferation and differentiation	[71]
	Carboxyl-modified SWCNTs	Upregulating the expression of classical bone markers and secretion of osteopontin protein	[93]
	BG-decorated nanoporous composite CNFs	Promoting proliferation and osteogenic differentiation of osteoblast cells	[83]
	CNT/glass/HA ceramics	Maximizing bone cell proliferation, DNA content and mRNA expression upon electrical stimulation	[96]
	Oxygen-terminated PND/poly(LLA-co-CL) composite scaffold	Increasing cell attachment and differentiation of BMSCs on the scaffolds in vitro and enhancing bone formation in vivo	[100]
	Fullerol nanoparticles	Inhibiting adipogenesis and enhancing osteogenesis of a BMSCs	[118]
	Water-soluble C_{60}	Prohibiting differentiation of precursor cells into osteoclasts and osteoclastic resorption in vitro, and inhibiting local inflammation and joint destruction in vivo	[119]
	NCD and DLC coatings	Enhancing cell adhesion, proliferation, and osteogenic differentiation of osteoblast-like cells and stem cells in vitro, and showing good osseointegration in vivo	[101–105]
	Graphene/CS coatings	Enhancing the adhesion of human BMSCs	[109]
Nerve	Laminin-coated 2D graphene films	Enhancing human differentiation of NSCs towards neurons rather than glial cells	[120]
	3D graphene foams	Enhancing differentiation of NSCs towards astrocytes and neurons	[121]

Table 3.3 (Continued)

Target tissue	Materials	Biological properties	Ref.
	Rolled 3D GO foam	Promoting differentiation of hNSCs into neurons under an electrical stimulation	[122]
	Fluorinated graphene	Promoting neuronal differentiation of MSCs	[123]
	3D free-standing, porous and flexible GOx scaffolds	Enhancing the formation of neural networks containing both neurons and glial cells	[57]
	4-Hydroxynonenal functionalized CNTs	Inducing neurite outgrowth and increasing neurite branching	[124]
	CNTs	Improving neural signal transfer while supporting dendrite elongation and cell adhesion	[125]
	MWCNTs-coated poly (L-lactic acid-co-caprolactone) scaffolds	Improving neurite outgrowth of DRG neurons and focal adhesion kinase expression of PC-12 cells	[128]
	CNT-interfaced PGF nerve conduits	Improving the number of regenerating axons crossing the scaffold, the cross-sectional area of the re-innervated muscles and the electrophysiological findings	[127]
	CNF scaffolds	Enhancing and directing nerve growth and providing physical support for cell growth	[130]
	CNFs/PCU	Decreasing adhesion of astrocytes	
	UNCD films	Stimulating spontaneous cell proliferation and differentiation of NSCs	[49]
	NCD coating	Supporting the survival of fully differentiated spiking neurons	[132]
	UV treated DLC	Supporting patterned and unpatterned cortical neuron growth	[134]
Cartilage	CSMA/PECA/graphene porous hybrid scaffold	Supporting survival, growth, and repair of cartilage cells	[135]
	C_{60} and derivatives	Treating cartilage degeneration	[136]
Heart	GO	Enhancing myogenic differentiation of myoblasts	[139]
	MWCNTs	Enhancing viability and proliferation of neonatal rat ventricular myocytes and promoting cardiomyocyte maturation	[140]
	CNF/PLGA	Increasing adhesion of cardiomyocytes	[141]
	CNF/chitosan	Improving metabolic activity of neonatal rat heart cells and increasing their expression of cardiac-specific genes	[142]

Table 3.4 Carbon nanomaterial-based as drug carriers for tissue repair and regeneration

Target tissue	Material	Drug	Functions or benefits	Ref.
Bone	Bisphosphonate and diphosphonate C_{60} derivatives	Themselves	Strong affinity for bone HA; As a drug for osteoporosis and other bone disorders	[35,156]
	SWNTs-COOH and SWNTs-CH_3	rhBMP-2	Increased bioactivity of the drug	[151]
	GO flakes suspended in fibrin gels	BMP-2	Greater osteogenic differentiation of MSCs in vitro and bone regeneration in mouse calvaria	[152]
	GO-coated Ti	BMP-2 and SP	Greater new bone formation in mouse calvaria	[153]
	PNDs	BMP-2 and β-FGF	Induced differentiation and proliferation of osteoblast progenitor cells and promoted bone formation; Aided bone healing and may help in giving surgeons more control during operations	[154]
	NCD coating	BMP-2	Strongly upregulating ALP expression of osteosarcoma cells	[155]
Brain	SWCNTs	Acetylcholine	Treatment of experimentally induced Alzheimer's disease	[157]
Nerve	Amino-functionalized MWCNTs	NGFs	Improved dispersibility, reduced toxicity to PC-12 cells and promoted cell differentiation and growth	[158]

significantly greater amounts of bone regeneration in mouse calvarial defects than using fibrin gel alone [152]. GO-coated Ti is also used for the delivery of BMP-2 and substance P (SP, which is known to recruit MSCs) [153]. The osteogenic differentiation of human bone marrow-derived MSCs and new bone formation in vivo are enhanced by this strategy. The dual delivery of BMP-2 and SP using GO-coated Ti showed the greatest new bone formation in the mouse calvaria compared with other groups [153]. PNDs are capable of simultaneously delivering the BMP-2 and β-FGF, which induces differentiation and proliferation of osteoblast progenitor cells and promotes bone formation [154]. Additionally, delayed or pH-triggered release of proteins have been achieved in PND-aided delivery, which may help surgeons gain more control during operations. Furthermore, the sustained release maintains the proteins in the desired region for a longer duration, aiding bone healing [154]. NCD surfaces are also used as BMP-2 carriers and the bound BMP-2 is found fully active [155]. Functionalized with diphosphonate groups or amide bisphosphonate and multiple hydroxyl groups, the C_{60} derivatives reveal a strong affinity to the bone mineral [35,156]. This C_{60}-based compound conjugated with traditional bone promotion agents can be used for osteoporosis treatment.

In addition, SWCNTs can be used to deliver acetylcholine in mice brains for treatment of experimentally induced Alzheimer's disease with a moderate safety range [157]. MWCNTs treated by the amino-functionalized ethylenediamine have been used to deliver nerve growth factors (NGF) by non-covalent grafting [158]. The amino functionalization improves the dispersibility of the MWCNT-NGF complex and reduces their toxicity to PC12 cells, therefore promoting PC12 cell differentiation and chick embryo DRG nerve growth [158].

3.3.3 Bio-sensing and bio-imaging

As to the biosensors, electrochemical biosensors are currently the most popular biosensors. CNTs, CNFs and fullerene, with their outstanding electrochemical and electrical properties, are ideal materials to fabricate components in electrochemical biosensors [159–161]. Due to their ability to undergo fast electron transfer and the resistance of CNT-modified electrodes to surface fouling, CNTs can overcome most of the disadvantages that exist for conventional glassy carbon or metal electrodes for amperometric or voltammetric analyte detection [162]. CNT-based optical biosensors have also been developed due to the near-infrared (NIR)-fluorescence, Raman scattering and fluorescence quenching properties of the nanomaterials [163]. The use of graphene for bio-sensing applications involves two alternative technologies [36]. One is using a probe molecule on the graphene sheet that interacts with the analyte. For instance, sodium dodecylbenzene sulfonate dispersed graphene (SDBS-graphene) is coated with dye-labelled single stranded oligonucleotide molecules (aptamers) via π–π stacking [164]. The fluorescence resonance energy transfer (FRET) effect between graphene and the dye leads to quenching of its fluorescent signal. Upon interaction with a target molecule (e.g. thrombin) complementary to the aptamer, strong complexation between the two occurs and the FRET effect is cancelled, generating fluorescence to a detectable level (as shown in Figure 3.6) [164]. Another is based on measurement of the changes in electrical properties of the graphene platform during the interaction with an analyte. For example, the anchoring of the cancer cells to the aptamer molecules on GO alters the electrical impedance and cyclic voltammetry readings of the GO-based electrochemical sensor [165]. Similarly, electrical signals from cardiomyocytes have been successfully recorded on single [166] and array type [167] graphene-based field effect transistor (FET) biosensors. An advantage of the graphene-based electrochemical sensors is that the electrode alone is the only factor that affects the readings [168].

The biosensors based on carbon-based nanomaterials have also been used for the detection of cancer biomarkers and glucose (analytes associated with two of the most common diseases in developed world, cancer and diabetes), as well as DNA [163,169,170]. Here we highlight the work on CNT-based biosensors for the

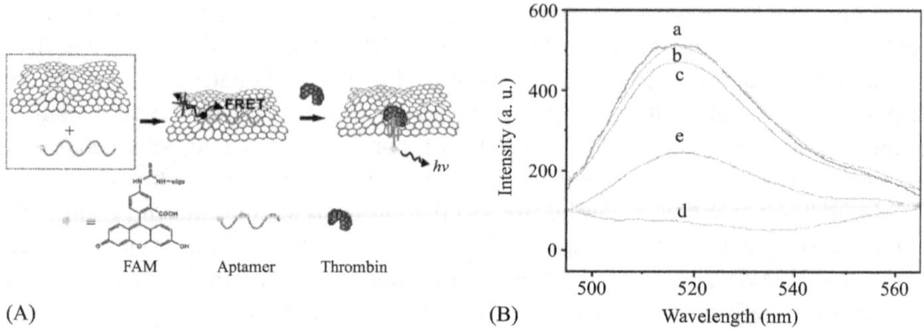

Figure 3.6 *(A) Schematic demonstration of graphene FRET aptasensor and the detection mechanism for thrombin. Fluorescence of a dye labelled aptamer is quenched when the aptamer binds to graphene due to FRET between dyes and graphene. The fluorescence recovers when thrombin combines with aptamers to form quadruplex-thrombin complexes which have much less affinity to graphene, causing the dye to move far away from the graphene surface. (B) Fluorescence spectra of FAM-aptamer (a), and FAM aptamer in the presence of thrombin (b), thrombin and SDBS-graphene in sequence (c), SDBS-graphene (d), SDBS-graphene and thrombin in sequence (e). FAM-aptamer concentration: 20 nM; thrombin concentration: 250 pM; excitation wavelength: 470 nm [164]*

detection of cell behavior and tissue conditions (Table 3.5). Sirivisoot *et al.* [171] designed a bone growth sensor by coating a conductive, biodegradable, polymer layer (polypyrrole and PLGA) onto MWCNT/anodized Ti (a-Ti). When new bone grows onto the implant, the conductive polymer layer degrades and/or is replaced by newly formed bone and, consequently, undergoes a change in conductivity that can be measured by the parallel MWCNTs grown out of the pores of a-Ti [171]. Cyclic voltammograms (CVs) showed an enhanced redox potential for the MWCNT/a-Ti. MWCNT/a-Ti exhibited well-defined and persistent CVs, similar to the ferri/ferrocyanide redox reaction [172]. Their further study confirmed that MWCNTs enhanced the electron transfer or redox reactions of osteoblast extracellular components and ferri-ferrocyanide (control) when compared to glassy carbon and platinum electrodes. Thus, the MWCNT/a-Ti sensor can convert the cellular process of bone mineralization into a readable electrical signal [173]. Furthermore, the MWCNT/a-Ti sensors may determine whether healthy bone, infection or inflammation is occurring on a Ti implant by generating distinct current–voltage profiles or redox peak currents of interested proteins or biomolecules. In addition to the bone tissue sensors, CNT-coated conventional tungsten and stainless steel wire electrodes are demonstrated to enhance both recording and electrical stimulation of neurons in culture, rats and monkeys [129]. These in situ sensors could revolutionize current methods of medical diagnosis and personalized medicine to increase implant success or reduce the cost of medical care.

Table 3.5 CNT-based biosensors for detection of tissue functions

Target organ	Biosensor	Mechanism of detection	Applications	Ref.
Bone	Polypyrrole/ PLGA coated MWCNT/ anodized Ti	Change conductivity of the implant	Detection of new bone growth on the implant	[171]
	MWCNT/ anodized Ti	Convert the cellular process of bone mineralization into a readable electrical signal;Generate distinct current–voltage profiles or redox peak currents of desired proteins or biomolecules	Detection of healthy bone growth, infection, or inflammation	[172,173]
Nervous system	CNT-coated tungsten and stainless steel wire	Record electrical signals from neurons	Studying brain functions and evaluating the function of neuronal networks	[129]

Carbon-based nanomaterials also play an important role in bio-imaging, especially molecular imaging, which introduces molecular imaging agents to measure the expression of indicative markers at different stages of diseases [174]. Metal atoms can be encapsulated inside the fullerene cage, which is called 'endohedral metallofullerenes'. The fullerene cage protects the metal inside both against chemical or enzymatic reaction within the body and the unwarranted release of the metal. Gadolinium-encapsulated fullerenes have been proposed as contrast agents to enhance MRI quality and for applications that require longer residency times [150]. Similarly to fullerene, CNTs can also load gadolinium through sidewall defects or end-of-tube openings [175] to become MRI contrast agents. Besides that, CNTs possess optical transitions in the NIR range that minimize photon interferences and display good photostability. Raman scattering and fluorescence spectroscopy are therefore possible methods for tracking CNTs in cells over long durations of time. For instance, excitation at 660 nm can result in a structured fluorescence spectrum characteristic of SWNT and nucleic acid-encapsulated SWNTs are stable for weeks when ingested by 3T3 fibroblasts and murine myoblast stems cells and observed by Raman spectroscopy [176]. The graphene quantum dots (QDs) also exhibit intrinsic fluorescence but the fluorescence density is too low. Surface functionalization with fluorophores can dramatically enhance the fluorescence density. Directly labelling GO with fluorophores causes quenching of the fluorescence signal by GO, so a linker is usually incorporated to reduce direct interactions between fluorescein and GO. For example, non-targeted rGO is conjugated with graphene QDs via a bridge of bovine serum albumin (BSA) for fluorescence imaging of live HeLa cells [177]. Recently, multifunctional graphene with interesting fluorescence and magnetic properties has been designed and synthesized to take advantages of the synergy of both properties [178]. PND has a very strong fluorescence at 700 nm which is emitted after being excited at 560 nm [179] due to the nitrogen-vacancy (N-V) center within diamonds, which is advantageous for imaging in biological cells. Fu and co-workers found that under the same excitation conditions, the fluorescence of a single 35 nm diamond is significantly brighter than that of a single dye molecule such as Alexa Fluor 546. They also demonstrated that it is possible to conduct a single particle tracking for a 35-nm PND in the cytoplasm of a live HeLa cell (Figure 3.7) [180]. This tracking method can be applied to a drug delivery system of PND, where the interaction between PND and cells can be monitored using fluorescence microscopy.

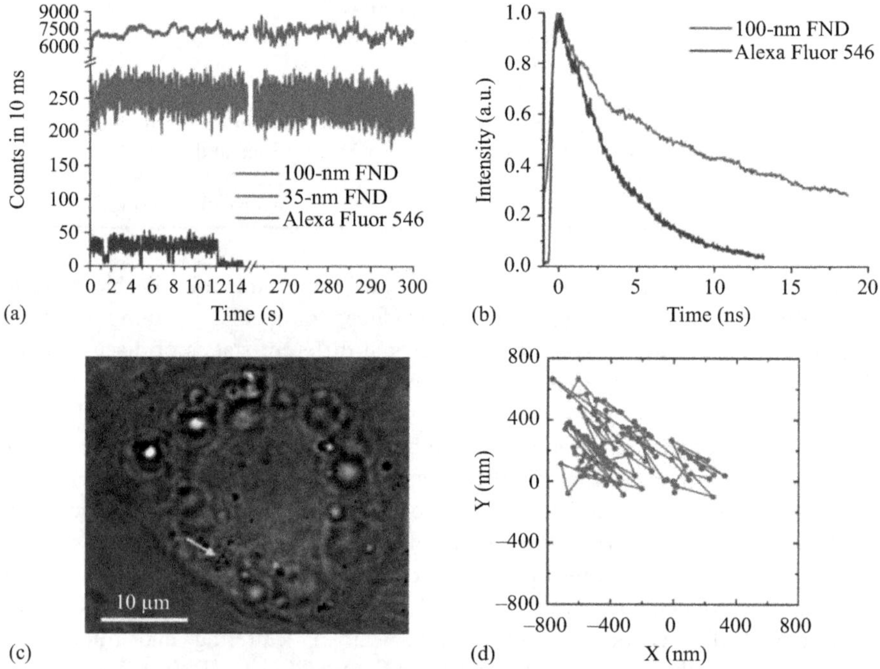

Figure 3.7 (a) Typical time traces of the fluorescence from a single 100-nm PND (top curve), a single 35-nm PND (middle curve), and a single Alexa Fluor 546 dye molecule attached to a single dsDNA molecule (bottom curve). Note that no sign of photobleaching was detected within 300 s of the continuous excitation for the PNDs. (b) Fluorescence lifetime measurements of 100-nm PNDs (top curve) and Alexa Fluor 546 dye molecules (bottom curve). Fitting the time traces with two exponential decays reveals a fast component of 1.7 ns (4%) and a slow component of 17 ns (96%). The latter is four times longer than that (4 ns) of Alexa Fluor 546. (c) Overlay of bright-field and epifluorescence images of a live HeLa cell after uptake of 35-nm PNDs. (d) A 100-step (139 ms per step) trajectory of a single PND, indicated by the white arrow in (c), moving in the cytoplasm of the HeLa cell [180]

3.4 Toxicity concern of carbon-based nano biomaterials

Although carbon-based nanomaterials have demonstrated great potential to dramatically enhanced the performance of biomaterials or biomedical devices, their toxicity to human tissues and organs is a major concern of their ultimate clinical use.

Studies on the toxicity of CNTs have showed different results [181,182]. It is believed that the toxicity of CNTs depends on the size, shape, surface area, dispersion status and intrinsic chemistry of the materials and even the catalyst metal residuals from fabrication process [11]. It should be noted that in vivo toxicity of CNTs also depends on their exposure pathway and location [181,183]. In addition, surface functionalization may either increase or decrease the toxicity of CNTs [184,185], so the toxicity of functionalized CNT should be thoroughly evaluated before discussing its applications in human applications. Similarly, toxicity of graphene and its derivatives is closely associated to their surface functionalization, followed by their size [186]. Although many in vitro studies have reported that the graphene and its derivatives might become a health hazard, there are some in vivo results indicating the acceptable biocompatibility of these materials [36,37]. In addition, chemical manipulation can alleviate their potential risks. How to functionalize graphene and its derivatives to make them less toxic, easier to clear from the body over time, and thereby can further explored for biomedical applications, is currently a major research focus in this field [37].

Similarly to CNTs, studies performed on non-functionalized fullerenes reveal inconsistent toxicity results [187–189]. The toxicity of fullerenes depends on their concentration [190] and will diminish when they become water soluble after modification [191]. An in vitro study shows that nanodiamond-mediated oxidative stress may contribute to DNA damage limiting nanodiamond biocompatibility and the effect was dose-dependent [192]. However, some in vivo studies provide promising information on the biocompatibility of PNDs [193–195]. In addition, depending on different fabrication and surface modification strategies, the toxicity of nanodiamonds can vary significantly. Numerous in vitro and in vivo studies have indicated the biocompatibility of NCL and DLC coatings. However, the debris caused by the wear or peeling off of the coating could be a potential risk to human body. Continued work on the toxicity of carbon-based nanomaterials should examine long-term organ exposure outcomes and the fate of materials in the body.

3.5 Summary and outlook

Carbon-based nanomaterials have shown the potential to change numerous healthcare aspects due to their unparalleled properties. Surface modification of carbon-based nanomaterials is often the first and most critical step before biomedical applications. It alters the biocompatibility and processability of the materials, as well as the final properties of the devices that employ them. Larger scale structures and bulk composites based on carbon nanomaterials have exhibited advantages in tissue repair and regeneration, artificial tissue fabrication, bio-sensing and bio-imaging. In vivo studies are important for developing carbon nanomaterials-based implants and evaluating the long-term degradation and toxicity of the materials. In addition, these nanomaterials have also shown a capacity to achieve targeted, controlled and sustained release of various drugs, while simultaneously maintaining the drugs' activity. However, the toxicity and metabolic pathways of carbon-based nanomaterials are still unclear.

Future research is expected to take further advantage of the unique biological properties of carbon-based nanomaterials to fabricate multi-functional TE scaffolds that induce tissue regeneration, while, for example, simultaneously preventing infection or allowing for the ability to monitor the tissue growth. The development of large-scale carbon-based nanomaterials for TE scaffolds remains an important topic in the TE area. To take advantage of their unique functional properties aside from their structural and biocompatible advantages, making the large structures 'smart' would be an attractive research direction. When using carbon-based nanomaterials for the surface modification of other biomaterials, studies should be focused not only on the biological performance, but also on the mechanical properties of the modification layer (e.g. wear resistance and bonding strengths).

One of the main objectives of drug delivery studies on carbon-based nanomaterials involves improving the efficiency of drug delivery and the bioavailability of the drugs being delivered. Furthermore, future investigations should also focus on improving the modes of administration (e.g. oral versus injection). Carbon-based nanomaterials have shown their tremendous advantage in material properties for sensing cell behavior and tissue conditions. Improving the stability of materials and devices should be a developing direction in this field. High sensitivity and long-term stability of the devices should be achieved. Portable and even wearable devices for bio-detection and disease diagnosis is a current research hotspot. Future development in the field lies not only in material optimization, but also in device design.

Last but not the least, low toxicity and satisfactory bio-safety of carbon-based nanomaterials form the basis of all the aforementioned applications. Analysis of the mechanism governing nanomaterial toxicity and the improvement of the safety evaluation method of carbon-based nanomaterials will both complement and accelerate the development of safer and better carbon-based nanomaterials for a myriad of biomedical applications.

Overall, both opportunities and challenges exist in this field and have drawn a huge interest from both material engineers and clinicians alike. People should always keep in mind that although material development and device optimization can give us hope, the understanding of the toxicity of these materials gives us the confidence to pursue these applications.

Acknowledgements

The authors (X. Lin and L. Yang) would like to thank the Jiangsu Provincial Special Program of Medical Science (BL2012004), National Basic Research Program of China (973 Program, 2014CB748600), the Priority Academic Program Development of Jiangsu High Education Institutions (PAPD), Jiangsu Provincial R&D Innovation Program (BY2014059-07), National Natural Science Foundation of China (51472279 and 81501858), Jiangsu Six Peak of Talents Program (2013-WSW-056), Chinese Ministry of Education Star-up Fund for Oversea Scholars, and Open Fund of State Key Laboratory of Nonlinear Mechanics and National 1000 Young Talents Program of China for supporting this work.

References

[1] Yanhong H, Olga AS, Zushou H, Clifford WP, Donald WB. Carbon nanostructures for advanced composites. *Reports on Progress in Physics* 2006;69:1847.

[2] Kroto HW, Heath JR, O'Brien SC, Curl RF, Smalley RE. C_{60}: Buckminsterfullerene. *Nature* 1985;318:162–3.

[3] Krätschmer W, Lamb LD, Fostiropoulos K, Huffman DR. C_{60}: a new form of carbon. *Nature* 1990;347:354–8.

[4] Hirsch A. *Chemistry of the Fullerenes*. Stuttgart: Thieme; 1994.

[5] Novoselov KS, Geim AK, Morozov SV, *et al.* Electric field effect in atomically thin carbon films. *Science* 2004;306:666–9.

[6] Yan L, Zheng YB, Zhao F, *et al.* Chemistry and physics of a single atomic layer: strategies and challenges for functionalization of graphene and graphene-based materials. *Chemical Society Reviews* 2012;41:97–114.

[7] Li D, Müller MB, Gilje S, Kaner RB, Wallace GG. Processable aqueous dispersions of graphene nanosheets. *Nature Nanotechnology* 2008;3:101–5.

[8] Hummers Jr WS, Offeman RE. Preparation of graphitic oxide. *Journal of the American Chemical Society* 1958;80:1339.

[9] Iijima S. Helical microtubules of graphitic carbon. *Nature* 1991;354:56–8.

[10] Bekyarova E, Ni Y, Malarkey E, *et al.* Applications of carbon nanotubes in biotechnology and biomedicine. *Journal of Biomedical Nanotechnology* 2005;1:3–17.

[11] Tran PA, Zhang L, Webster TJ. Carbon nanofibers and carbon nanotubes in regenerative medicine. *Advanced Drug Delivery Reviews* 2009;61:1097–114.

[12] Danilenko VV. On the history of the discovery of nanodiamond synthesis. *Physics of the Solid State* 2004;46:595–9.

[13] Schrand AM, Hens SAC, Shenderova OA. Nanodiamond particles: properties and perspectives for bioapplications. *Critical Reviews in Solid State and Materials Sciences* 2009;34:18–74.

[14] Schrand AM, Huang H, Carlson C, *et al.* Are diamond nanoparticles cytotoxic? *Journal of Physical Chemistry B* 2007;111:2–7.

[15] Mochalin VN, Shenderova O, Ho D, Gogotsi Y. The properties and applications of nanodiamonds. *Nature Nanotechnology* 2012;7:11–23.

[16] Pennisi CP, Alcaide M. Nanocrystalline diamond films for biomedical applications. *Frontiers in Biomaterials: The Design, Synthetic Strategies and Biocompatibility of Polymer Scaffolds for Biomedical Application* 2014. 70–100.

[17] Williams OA. Nanocrystalline diamond. *Diamond and Related Materials* 2011;20:621–40.

[18] Jakubowski W, Bartosz G, Niedzielski P, Szymanski W, Walkowiak B. Nanocrystalline diamond surface is resistant to bacterial colonization. *Diamond and Related Materials* 2004;13:1761–3.

[19] Medina O, Nocua J, Mendoza F, *et al.* Bactericide and bacterial anti-adhesive properties of the nanocrystalline diamond surface. *Diamond and Related Materials* 2012;22:77–81.

[20] Aisenberg S, Chabot R. Ion-beam deposition of thin films of diamond-like carbon. *Journal of Applied Physics* 1971;42:2953–8.

[21] Cui F, Li D. A review of investigations on biocompatibility of diamond-like carbon and carbon nitride films. *Surface and Coatings Technology* 2000;131:481–7.

[22] Ignatius M, Sawhney N, Gupta A, Thibadeau B, Monteiro O, Brown I. Bioactive surface coatings for nanoscale instruments: effects on CNS neurons. *Journal of Biomedical Materials Research* 1998;40:264–74.

[23] Dai H. Carbon Nanotubes: synthesis, integration, and properties. *Accounts of Chemical Research* 2002;35:1035–44.

[24] Shenderova OA, Zhirnov VV, Brenner DW. Carbon nanostructures. *Critical Reviews in Solid State and Materials Sciences* 2002;27:227–356.

[25] Yang L, Zhang L, Webster TJ. Carbon nanostructures for orthopedic medical applications. *Nanomedicine* 2011;6:1231–44.

[26] Chung C, Kim Y-K, Shin D, Ryoo S-R, Hong BH, Min D-H. Biomedical applications of graphene and graphene oxide. *Accounts of Chemical Research* 2013;46:2211–24.

[27] Hirsch A, Lamparth I, Groesser T, Karfunkel HR. Regiochemistry of multiple additions to the fullerene core: synthesis of a Th-symmetric hexakis adduct of C60 with bis(ethoxycarbonyl)methylene. *Journal of the American Chemical Society* 1994;116:9385–6.

[28] Partha R, Lackey M, Hirsch A, Casscells SW, Conyers JL. Self-assembly of amphiphilic C60 fullerene derivatives into nanoscale supramolecular structures. *Journal of Nanobiotechnology* 2007;5:6.

[29] Partha R, Mitchell LR, Lyon JL, Joshi PP, Conyers JL. Buckysomes: fullerene-based nanocarriers for hydrophobic molecule delivery. *ACS Nano* 2008;2:1950–8.

[30] Bakry R, Vallant RM, Najam-ul-Haq M, *et al.* Medicinal applications of fullerenes. *International Journal of Nanomedicine* 2007;2:639.

[31] Bolskar RD, Benedetto AF, Husebo LO, *et al.* First soluble M@C_{60} derivatives provide enhanced access to metallofullerenes and permit in vivo evaluation of Gd@C60 [C(COOH)2] 10 as a MRI contrast agent. *Journal of the American Chemical Society* 2003;125:5471–8.

[32] Daroczi B, Kari G, McAleer MF, Wolf JC, Rodeck U, Dicker AP. In vivo radioprotection by the fullerene nanoparticle DF-1 as assessed in a zebrafish model. *Clinical Cancer Research* 2006;12:7086–91.

[33] Marchesan S, Da Ros T, Spalluto G, Balzarini J, Prato M. Anti-HIV properties of cationic fullerene derivatives. *Bioorganic & Medicinal Chemistry Letters* 2005;15:3615–8.

[34] TSAI MC, Chen Y, Chiang L. Polyhydroxylated C60, fullerenol, a novel free-radical trapper, prevented hydrogen peroxide- and cumene hydroperoxide-elicited changes in rat hippocampus in-vitro. *Journal of Pharmacy and Pharmacology* 1997;49:438–45.

[35] Gonzalez KA, Wilson LJ, Wu W, Nancollas GH. Synthesis and in vitro characterization of a tissue-selective fullerene: vectoring C60(OH)16AMBP to mineralized bone. *Bioorganic & Medicinal Chemistry* 2002;10:1991–7.

[36] Bitounis D, Ali-Boucetta H, Hong BH, Min DH, Kostarelos K. Prospects and challenges of graphene in biomedical applications. *Advanced Materials* 2013;25:2258–68.

[37] Zhang Y, Nayak TR, Hong H, Cai W. Graphene: a versatile nanoplatform for biomedical applications. *Nanoscale* 2012;4:3833–42.

[38] Goenka S, Sant V, Sant S. Graphene-based nanomaterials for drug delivery and tissue engineering. *Journal of Controlled Release* 2014;173:75–88.

[39] Colvin VL. The potential environmental impact of engineered nanomaterials. *Nature Biotechnology* 2003;21:1166–70.

[40] Liu J, Rinzler AG, Dai H, *et al.* Fullerene pipes. *Science* 1998;280:1253–6.

[41] Balasubramanian K, Burghard M. Chemically functionalized carbon nanotubes. *Small* 2005;1:180–92.

[42] Georgakilas V, Kordatos K, Prato M, Guldi DM, Holzinger M, Hirsch A. Organic functionalization of carbon nanotubes. *Journal of the American Chemical Society* 2002;124:760–1.

[43] Bianco A, Kostarelos K, Partidos CD, Prato M. Biomedical applications of functionalised carbon nanotubes. *Chemical Communications* 2005:571–7.

[44] Petrov P, Stassin F, Pagnoulle C, Jérôme R. Noncovalent functionalization of multi-walled carbon nanotubes by pyrene containing polymers. *Chemical Communications* 2003:2904–5.

[45] Pantarotto D, Singh R, McCarthy D, *et al.* Functionalized carbon nanotubes for plasmid DNA gene delivery. *Angewandte Chemie* 2004;116: 5354–8.

[46] Meinhardt T, Lang D, Dill H, Krueger A. Pushing the functionality of diamond nanoparticles to new horizons: orthogonally functionalized nanodiamond using click chemistry. *Advanced Functional Materials* 2011; 21:494–500.

[47] Liskova J, Babchenko O, Varga M, *et al.* Osteogenic cell differentiation on H-terminated and O-terminated nanocrystalline diamond films. *International Journal of Nanomedicine* 2015;10:869.

[48] Chong KF, Loh KP, Vedula S, *et al.* Cell adhesion properties on photochemically functionalized diamond. *Langmuir* 2007;23:5615–21.

[49] Chen Y-C, Lee D-C, Hsiao C-Y, *et al.* The effect of ultra-nanocrystalline diamond films on the proliferation and differentiation of neural stem cells. *Biomaterials* 2009;30:3428–35.

[50] Huang H, Chen M, Bruno P, *et al.* Ultrananocrystalline diamond thin films functionalized with therapeutically active collagen networks. *Journal of Physical Chemistry B* 2009;113:2966–71.

[51] Wang J, Carlisle JA. Covalent immobilization of glucose oxidase on conducting ultrananocrystalline diamond thin films. *Diamond and Related Materials* 2006;15:279–84.

[52] Christiaens P, Vermeeren V, Wenmackers S, *et al.* EDC-mediated DNA attachment to nanocrystalline CVD diamond films. *Biosensors and Bioelectronics* 2006;22:170–7.

[53] Hauert R, Thorwarth K, Thorwarth G. An overview on diamond-like carbon coatings in medical applications. *Surface and Coatings Technology* 2013;233:119–30.

[54] Roy RK, Choi HW, Yi JW, *et al.* Hemocompatibility of surface-modified, silicon-incorporated, diamond-like carbon films. *Acta Biomaterialia* 2009;5:249–56.

[55] Jiang L, Fan Z. Design of advanced porous graphene materials: from graphene nanomesh to 3D architectures. *Nanoscale* 2014;6:1922–45.

[56] Chen Z, Ren W, Gao L, Liu B, Pei S, Cheng H-M. Three-dimensional flexible and conductive interconnected graphene networks grown by chemical vapour deposition. *Nature Materials* 2011;10:424–8.

[57] Serrano MC, Patiño J, García-Rama C, *et al.* 3D free-standing porous scaffolds made of graphene oxide as substrates for neural cell growth. *Journal of Materials Chemistry B* 2014;2:5698–706.

[58] Yamada T, Ishihara M, Kim J, Hasegawa M, Iijima S. A roll-to-roll microwave plasma chemical vapor deposition process for the production of 294 mm width graphene films at low temperature. *Carbon* 2012;50:2615–9.

[59] Kobayashi T, Bando M, Kimura N, *et al.* Production of a 100-m-long high-quality graphene transparent conductive film by roll-to-roll chemical vapor deposition and transfer process. *Applied Physics Letters* 2013;102:023112.

[60] Srivastava A, Srivastava ON, Talapatra S, Vajtai R, Ajayan PM. Carbon nanotube filters. *Nature Materials* 2004;3:610–4.

[61] Correa-Duarte MA, Wagner N, Rojas-Chapana J, Morsczeck C, Thie M, Giersig M. Fabrication and biocompatibility of carbon nanotube-based 3D networks as scaffolds for cell seeding and growth. *Nano Letters* 2004;4:2233–6.

[62] Li X, Liu H, Niu X, *et al.* The use of carbon nanotubes to induce osteogenic differentiation of human adipose-derived MSCs in vitro and ectopic bone formation in vivo. *Biomaterials* 2012;33:4818–27.

[63] Huang YJ, Wu HC, Tai NH, Wang TW. Carbon nanotube rope with electrical stimulation promotes the differentiation and maturity of neural stem cells. *Small* 2012;8:2869–77.

[64] Serrano MC, Gutiérrez MC, del Monte F. Role of polymers in the design of 3D carbon nanotube-based scaffolds for biomedical applications. *Progress in Polymer Science* 2014;39:1448–71.

[65] Lahiri D, Ghosh S, Agarwal A. Carbon nanotube reinforced hydroxyapatite composite for orthopedic application: a review. *Materials Science and Engineering: C* 2012;32:1727–58.

[66] Puértolas J, Kurtz S. Evaluation of carbon nanotubes and graphene as reinforcements for UHMWPE-based composites in arthroplastic applications: a review. *Journal of the Mechanical Behavior of Biomedical Materials* 2014;39:129–45.

[67] Li CS, Vannabouathong C, Sprague S, Bhandari M. The use of carbon-fiber-reinforced (CFR) PEEK material in orthopedic implants: a systematic review. *Clinical Medicine Insights Arthritis and Musculoskeletal Disorders* 2015;8:33.

[68] Lahiri D, Dua R, Zhang C, *et al.* Graphene nanoplatelet-induced strengthening of ultrahigh molecular weight polyethylene and biocompatibility in vitro. *ACS Applied Materials & Interfaces* 2012;4:2234–41.

[69] Kobayashi S, Kawai W. Development of carbon nanofiber reinforced hydroxyapatite with enhanced mechanical properties. *Composites Part A: Applied Science and Manufacturing* 2007;38:114–23.

[70] Zhang L, Liu W, Yue C, *et al.* A tough graphene nanosheet/hydroxyapatite composite with improved in vitro biocompatibility. *Carbon* 2013;61:105–15.

[71] Kumar S, Bose S, Chatterjee K. Amine-functionalized multiwall carbon nanotubes impart osteoinductive and bactericidal properties in poly (ε-caprolactone) composites. *RSC Advances* 2014;4:19086–98.

[72] MacDonald RA, Laurenzi BF, Viswanathan G, Ajayan PM, Stegemann JP. Collagen–carbon nanotube composite materials as scaffolds in tissue engineering. *Journal of Biomedical Materials Research Part A* 2005;74:489–96.

[73] Wang Z-k, Cai L. Chitosan and multi-walled carbon nanotube composite rods. *Chinese Journal of Polymer Science* 2010;28:801–6.

[74] Xu Y, Wang Y, Cui L. An efficient functionalization method for the multi-walled carbon nanotubes and their applications in PMMA bone cement. *Sensors & Transducers* 2013;21:36.

[75] Boroujeni NM, Zhou H, Luchini TJ, Bhaduri SB. Development of multi-walled carbon nanotubes reinforced monetite bionanocomposite cements for orthopedic applications. *Materials Science and Engineering: C* 2013; 33:4323–30.

[76] Chew K-K, Low K-L, Zein SHS, *et al.* Reinforcement of calcium phosphate cement with multi-walled carbon nanotubes and bovine serum albumin for injectable bone substitute applications. *Journal of the Mechanical Behavior of Biomedical Materials* 2011;4:331–9.

[77] Shi X, Hudson JL, Spicer PP, Tour JM, Krishnamoorti R, Mikos AG. Injectable nanocomposites of single-walled carbon nanotubes and biodegradable polymers for bone tissue engineering. *Biomacromolecules* 2006; 7:2237–42.

[78] Sun Y, Finne-Wistrand A, Waag T, *et al.* Reinforced degradable biocomposite by homogenously distributed functionalized nanodiamond particles. *Macromolecular Materials and Engineering* 2015;300:436–47.

[79] Harrison BS, Atala A. Carbon nanotube applications for tissue engineering. *Biomaterials* 2007;28:344–53.

[80] Abarrategi A, Gutiérrez MC, Moreno-Vicente C, *et al.* Multiwall carbon nanotube scaffolds for tissue engineering purposes. *Biomaterials* 2008; 29:94–102.

[81] Mikael PE, Amini AR, Basu J, *et al.* Functionalized carbon nanotube reinforced scaffolds for bone regenerative engineering: fabrication, in vitro and in vivo evaluation. *Biomedical Materials* 2014;9:035001.

[82] Park S, Park J, Jo I, *et al.* In situ hybridization of carbon nanotubes with bacterial cellulose for three-dimensional hybrid bioscaffolds. *Biomaterials* 2015;58:93–102.

[83] Zhang C, Cheng D, Tang T, Jia X, Cai Q, Yang X. Nanoporous structured carbon nanofiber/bioactive glass composites for skeletal tissue regeneration. *Journal of Materials Chemistry B* 2015;3:5300–9.

[84] Li Q. Carbon nanotube reinforced porous magnesium composite: 3D nondestructive microstructure characterization using x-ray micro-computed tomography. *Materials Letters* 2014;133:83–6.

[85] Firkowska I, Olek M, Pazos-Peréz N, Rojas-Chapana J, Giersig M. Highly Ordered MWNT-based matrixes: topography at the nanoscale conceived for tissue engineering. *Langmuir* 2006;22:5427–34.

[86] Walter C, Barg S, Ni N, *et al.* A novel approach for the fabrication of carbon nanofibre/ceramic porous structures. *Journal of the European Ceramic Society* 2013;33:2365–74.

[87] Boccaccini AR, Chicatun F, Cho J, *et al.* Carbon nanotube coatings on bioglass-based tissue engineering scaffolds. *Advanced Functional Materials* 2007;17:2815–22.

[88] Lee WC, Lim CHYX, Shi H, *et al.* Origin of enhanced stem cell growth and differentiation on graphene and graphene oxide. *ACS Nano* 2011;5: 7334–41.

[89] Nayak TR, Andersen H, Makam VS, *et al.* Graphene for controlled and accelerated osteogenic differentiation of human mesenchymal stem cells. *ACS Nano* 2011;5:4670–8.

[90] Crowder SW, Prasai D, Rath R, *et al.* Three-dimensional graphene foams promote osteogenic differentiation of human mesenchymal stem cells. *Nanoscale* 2013;5:4171–6.

[91] Wang J, Ouyang Z, Ren Z, *et al.* Self-assembled peptide nanofibers on graphene oxide as a novel nanohybrid for biomimetic mineralization of hydroxyapatite. *Carbon* 2015;89:20–30.

[92] Wu C, Xia L, Han P, *et al.* Graphene-oxide-modified β-tricalcium phosphate bioceramics stimulate in vitro and in vivo osteogenesis. *Carbon* 2015; 93:116–29.

[93] Barrientos-Duran A, Carpenter EM, zur Nieden NI, Malinin TI, Rodríguez-Manzaneque JC, Zanello LP. Carboxyl-modified single-wall carbon nanotubes improve bone tissue formation in vitro and repair in an in vivo rat model. *International Journal of Nanomedicine* 2014;9:4277.

[94] Friedenberg ZB, Andrews ET, Sinolenski BI, Pearl BW, Brighton CT. Bone reaction to varying amounts of direct current. *Plastic and Reconstructive Surgery* 1971;48:95–6.

[95] Wiesmann H-P, Hartig M, Stratmann U, Meyer U, Joos U. Electrical stimulation influences mineral formation of osteoblast-like cells in vitro. *Biochimica et Biophysica Acta (BBA) – Molecular Cell Research* 2001;1538:28–37.

[96] Mata D, Oliveira FJ, Neto MA, *et al.* Smart electroconductive bioactive ceramics to promote in situ electrostimulation of bone. *Journal of Materials Chemistry B* 2015;3:1831–45.

[97] Miyazaki M, Toyoda M, Yoshiiwa T, Kawano M, Kaku N, Tsumura H. Enhancement of the effects of exfoliated carbon nanofibers by bone morphogenetic protein in a rat femoral fracture model. *Journal of Orthopaedic Research* 2015;33:185–92.

[98] Gupta A, Main BJ, Taylor BL, *et al.* In vitro evaluation of three-dimensional single-walled carbon nanotube composites for bone tissue engineering. *Journal of Biomedical Materials Research Part A* 2014;102:4118–26.

[99] Zhang Q, Mochalin VN, Neitzel I, *et al.* Fluorescent PLLA-nanodiamond composites for bone tissue engineering. *Biomaterials* 2011;32:87–94.

[100] Xing Z, Pedersen TO, Wu X, *et al.* Biological effects of functionalizing copolymer scaffolds with nanodiamond particles. *Tissue Engineering Part A* 2013;19:1783–91.

[101] Amaral M, Dias A, Gomes P, *et al.* Nanocrystalline diamond: in vitro biocompatibility assessment by MG63 and human bone marrow cells cultures. *Journal of Biomedical Materials Research Part A* 2008;87:91–9.

[102] Kalbacova M, Kalbac M, Dunsch L, *et al.* The effect of SWCNT and nanodiamond films on human osteoblast cells. *Physica Status Solidi (B)* 2007;244:4356–9.

[103] Kalbacova M, Rezek B, Baresova V, Wolf-Brandstetter C, Kromka A. Nanoscale topography of nanocrystalline diamonds promotes differentiation of osteoblasts. *Acta Biomaterialia* 2009;5:3076–85.

[104] Allen M, Myer B, Rushton N. In vitro and in vivo investigations into the biocompatibility of diamond-like carbon (DLC) coatings for orthopedic applications. *Journal of Biomedical Materials Research* 2001;58:319–28.

[105] Metzler P, von Wilmowsky C, Stadlinger B, *et al.* Nano-crystalline diamond-coated titanium dental implants–a histomorphometric study in adult domestic pigs. *Journal of Cranio-Maxillofacial Surgery* 2013;41:532–8.

[106] Amaral M, Maru M, Rodrigues S, *et al.* Extremely low wear rates in hip joint bearings coated with nanocrystalline diamond. *Tribology International* 2014.

[107] Nishida E, Miyaji H, Takita H, *et al.* Graphene oxide coating facilitates the bioactivity of scaffold material for tissue engineering. *Japanese Journal of Applied Physics* 2014;53:06JD4.

[108] Li M, Liu Q, Jia Z, *et al.* Graphene oxide/hydroxyapatite composite coatings fabricated by electrophoretic nanotechnology for biological applications. *Carbon* 2014;67:185–97.

[109] Youtao X, Hongqing L, Chi Z, Xin G, Xuebin Z, Liping H. Graphene-reinforced calcium silicate coatings for load-bearing implants. *Biomedical Materials* 2014;9:025009.

[110] Park JE, Park IS, Bae TS, Lee MH. Electrophoretic deposition of carbon nanotubes over TiO_2 nanotubes: evaluation of surface properties and biocompatibility. *Bioinorganic Chemistry and Applications* 2014;2014:7.

[111] Facca S, Lahiri D, Fioretti F, *et al.* In vivo osseointegration of nano-designed composite coatings on titanium implants. *ACS Nano* 2011; 5:4790–9.

[112] Hahn B-D, Lee J-M, Park D-S, *et al.* Mechanical and in vitro biological performances of hydroxyapatite–carbon nanotube composite coatings deposited on Ti by aerosol deposition. *Acta Biomaterialia* 2009;5:3205–14.

[113] Lahiri D, Benaduce AP, Kos L, Agarwal A. Quantification of carbon nanotube induced adhesion of osteoblast on hydroxyapatite using nano-scratch technique. *Nanotechnology* 2011;22:355703.

[114] Pei X, Zeng Y, He R, *et al.* Single-walled carbon nanotubes/hydroxyapatite coatings on titanium obtained by electrochemical deposition. *Applied Surface Science* 2014;295:71–80.

[115] Abrishamchian A, Hooshmand T, Mohammadi M, Najafi F. Preparation and characterization of multi-walled carbon nanotube/hydroxyapatite nanocomposite film dip coated on Ti–6Al–4V by sol–gel method for biomedical applications: an in vitro study. *Materials Science and Engineering: C* 2013;33:2002–10.

[116] Gopi D, Shinyjoy E, Karthika A, *et al.* Single walled carbon nanotubes reinforced mineralized hydroxyapatite composite coatings on titanium for improved biocompatible implant applications. *RSC Advances* 2015; 5:36766–78.

[117] Gopi D, Nithiya S, Shinyjoy E, Rajeswari D, Kavitha L. Carbon nanotubes/carboxymethyl chitosan/mineralized hydroxyapatite composite coating on Ti–6Al–4V alloy for improved mechanical and biological properties. *Industrial & Engineering Chemistry Research* 2014;53:7660–9.

[118] Liu H, Yang X, Zhang Y, Dighe A, Li X, Cui Q. Fullerol antagonizes dexamethasone-induced oxidative stress and adipogenesis while enhancing osteogenesis in a cloned bone marrow mesenchymal stem cell. *Journal of Orthopaedic Research* 2012;30:1051–7.

[119] Yudoh K, Karasawa R, Masuko K, Kato T. Water-soluble fullerene (C_{60}) inhibits the osteoclast differentiation and bone destruction in arthritis. *International Journal of Nanomedicine* 2009;4:233–9.

[120] Park SY, Park J, Sim SH, *et al.* Enhanced differentiation of human neural stem cells into neurons on graphene. *Advanced Materials* 2011;23: H263–7.

[121] Li N, Zhang Q, Gao S, *et al.* Three-dimensional graphene foam as a biocompatible and conductive scaffold for neural stem cells. *Scientific Reports* 2013;3.

[122] Akhavan O, Ghaderi E, Shirazian SA, Rahighi R. Rolled graphene oxide foams as three-dimensional scaffolds for growth of neural fibers using electrical stimulation of stem cells. *Carbon* 2016; 97:71–7.

[123] Wang Y, Lee WC, Manga KK, *et al.* Fluorinated graphene for promoting neuro-induction of stem cells. *Advanced Materials* 2012;24:4285–90.

[124] Mattson MP, Haddon RC, Rao AM. Molecular functionalization of carbon nanotubes and use as substrates for neuronal growth. *Journal of Molecular Neuroscience* 2000;14:175–82.

[125] Lovat V, Pantarotto D, Lagostena L, *et al.* Carbon nanotube substrates boost neuronal electrical signaling. *Nano Letters* 2005;5:1107–10.

[126] Hu H, Ni Y, Montana V, Haddon RC, Parpura V. Chemically functionalized carbon nanotubes as substrates for neuronal growth. *Nano Letters* 2004;4:507–11.

[127] Ahn H-S, Hwang J-Y, Kim MS, *et al.* Carbon-nanotube-interfaced glass fiber scaffold for regeneration of transected sciatic nerve. *Acta Biomaterialia* 2015;13:324–34.

[128] Jin G-Z, Kim M, Shin US, Kim H-W. Neurite outgrowth of dorsal root ganglia neurons is enhanced on aligned nanofibrous biopolymer scaffold with carbon nanotube coating. *Neuroscience Letters* 2011;501:10–4.

[129] Keefer EW, Botterman BR, Romero MI, Rossi AF, Gross GW. Carbon nanotube coating improves neuronal recordings. *Nature Nanotechnology* 2008;3:434–9.

[130] Zhang H, Yu M, Xie L, Jin L, Yu Z. Carbon-nanofibers-based micro-/nanodevices for neural-electrical and neural-chemical interfaces. *Journal of Nanomaterials* 2012;2012:6.

[131] McKenzie JL, Waid MC, Shi R, Webster TJ. Decreased functions of astrocytes on carbon nanofiber materials. *Biomaterials* 2004;25:1309–17.

[132] Bendali A, Agnès C, Meffert S, *et al.* Distinctive glial and neuronal interfacing on nanocrystalline diamond. *PLOS ONE* 2014;9:e92562.

[133] Vaitkuviene A, McDonald M, Vahidpour F, *et al.* Impact of differently modified nanocrystalline diamond on the growth of neuroblastoma cells. *New Biotechnology* 2015;32:7–12.

[134] Regan EM, Uney JB, Dick AD, *et al.* Differential patterning of neuronal, glial and neural progenitor cells on phosphorus-doped and UV irradiated diamond-like carbon. *Biomaterials* 2010;31:207–15.

[135] Liao J, Qu Y, Chu B, Zhang X, Qian Z. Biodegradable CSMA/PECA/graphene porous hybrid scaffold for cartilage tissue engineering. *Scientific Reports* 2015;5.

[136] Liu Q, Cui Q, Li XJ, Jin L. The applications of buckminsterfullerene C60 and derivatives in orthopaedic research. *Connective Tissue Research* 2014;55:71–9.

[137] Shin M, Ishii O, Sueda T, Vacanti J. Contractile cardiac grafts using a novel nanofibrous mesh. *Biomaterials* 2004;25:3717–23.

[138] Leor J, Landa N, Cohen S. Renovation of the injured heart with myocardial tissue engineering. *Expert Review of Cardiovascular Therapy* 2006;4:239–52.

[139] Ku SH, Park CB. Myoblast differentiation on graphene oxide. *Biomaterials* 2013;34:2017–23.

[140] Martinelli V, Cellot G, Toma FM, *et al.* Carbon nanotubes promote growth and spontaneous electrical activity in cultured cardiac myocytes. *Nano Letters* 2012;12:1831–8.

[141] Stout DA, Basu B, Webster TJ. Poly (lactic–co-glycolic acid): carbon nanofiber composites for myocardial tissue engineering applications. *Acta Biomaterialia* 2011;7:3101–12.

[142] Martins AM, Eng G, Caridade SG, Mano JoF, Reis RL, Vunjak-Novakovic G. Electrically conductive chitosan/carbon scaffolds for cardiac tissue engineering. *Biomacromolecules* 2014;15:635–43.

[143] Narayan R, Wei W, Jin C, *et al.* Microstructural and biological properties of nanocrystalline diamond coatings. *Diamond and Related Materials* 2006;15:1935–40.

[144] Laube N, Kleinen L, Bradenahl J, Meissner A. Diamond-like carbon coatings on ureteral stents—a new strategy for decreasing the formation of crystalline bacterial biofilms? *Journal of Urology* 2007;177:1923–7.

[145] Bajaj P, Akin D, Gupta A, *et al.* Ultrananocrystalline diamond film as an optimal cell interface for biomedical applications. *Biomedical Microdevices* 2007;9:787–94.

[146] Lechleitner T, Klauser F, Seppi T, *et al.* The surface properties of nanocrystalline diamond and nanoparticulate diamond powder and their suitability as cell growth support surfaces. *Biomaterials* 2008;29: 4275–84.

[147] Amaral M, Gomes P, Lopes M, Santos J, Silva R, Fernandes M. Cytotoxicity evaluation of nanocrystalline diamond coatings by fibroblast cell cultures. *Acta Biomaterialia* 2009;5:755–63.

[148] Podila R, Moore T, Alexis F, Rao AM. Graphene coatings for enhanced hemo-compatibility of nitinol stents. *RSC Advances* 2013;3:1660–5.

[149] Lu B, Li T, Zhao H, *et al.* Graphene-based composite materials beneficial to wound healing. *Nanoscale* 2012;4:2978–82.

[150] Partha R, Conyers JL. Biomedical applications of functionalized fullerene-based nanomaterials. *International Journal of Nanomedicine* 2009;4:261.

[151] Li Z, Gan Q, Zhang W, Zhang J, Yuan Y, Liu C. Surface-induced conformational and functional changes of bone morphogenetic protein-2 adsorbed onto single-walled carbon nanotubes. *Biochemical and Biophysical Research Communications* 2013;440:215–21.

[152] La W-G, Jung M-J, Yoon J-K, *et al.* bone morphogenetic protein-2 for bone regeneration–dose reduction through graphene oxide-based delivery. *Carbon* 2014;78:428–38.

[153] La W-G, Jin M, Park S, *et al.* Delivery of bone morphogenetic protein-2 and substance P using graphene oxide for bone regeneration. *International Journal of Nanomedicine* 2014;9:107–16.

[154] Moore L, Gatica M, Kim H, Osawa E, Ho D. Multi-protein delivery by nanodiamonds promotes bone formation. *Journal of Dental Research* 2013;92:976–81.

[155] Steinmüller-Nethl D, Kloss FR, Najam-Ul-Haq M, *et al.* Strong binding of bioactive BMP-2 to nanocrystalline diamond by physisorption. *Biomaterials* 2006;27:4547–56.

[156] Mirakyan AL, Wilson LJ. Functionalization of C60 with diphosphonate groups: a route to bone-vectored fullerenes. *Journal of the Chemical Society, Perkin Transactions* 2002;2:1173–6.

[157] Yang Z, Zhang Y, Yang Y, *et al.* Pharmacological and toxicological target organelles and safe use of single-walled carbon nanotubes as drug carriers in treating Alzheimer disease. *Nanomedicine: Nanotechnology, Biology and Medicine* 2010;6:427–41.

[158] Chen W, Xiong Q, Ren Q, Guo Y, Li G. Can amino-functionalized carbon nanotubes carry functional nerve growth factor? *Neural Regeneration Research* 2014;9:285–92.

[159] Sotiropoulou S, Gavalas V, Vamvakaki V, Chaniotakis N. Novel carbon materials in biosensor systems. *Biosensors and Bioelectronics* 2003;18:211–5.

[160] Vamvakaki V, Chaniotakis NA. Carbon nanostructures as transducers in biosensors. *Sensors and Actuators B: Chemical* 2007;126:193–7.

[161] Gao C, Guo Z, Liu J-H, Huang X-J. The new age of carbon nanotubes: an updated review of functionalized carbon nanotubes in electrochemical sensors. *Nanoscale* 2012;4:1948–63.

[162] Salimi A, Hallaj R, Khayatian GR. Amperometric detection of morphine at preheated glassy carbon electrode modified with multiwall carbon nanotubes. *Electroanalysis* 2005;17:873–9.

[163] Kruss S, Hilmer AJ, Zhang J, Reuel NF, Mu B, Strano MS. Carbon nanotubes as optical biomedical sensors. *Advanced Drug Delivery Reviews* 2013;65:1933–50.

[164] Chang H, Tang L, Wang Y, Jiang J, Li J. Graphene fluorescence resonance energy transfer aptasensor for the thrombin detection. *Analytical Chemistry* 2010;82:2341–6.

[165] Feng L, Chen Y, Ren J, Qu X. A graphene functionalized electrochemical aptasensor for selective label-free detection of cancer cells. *Biomaterials* 2011;32:2930–7.

[166] Cohen-Karni T, Qing Q, Li Q, Fang Y, Lieber CM. Graphene and nanowire transistors for cellular interfaces and electrical recording. *Nano Letters* 2010;10:1098–102.

[167] Hess LH, Jansen M, Maybeck V, *et al.* Graphene transistor arrays for recording action potentials from electrogenic cells. *Advanced Materials* 2011;23:5045–9.

[168] Daly R, Kumar S, Lukacs G, *et al.* Cell proliferation tracking using graphene sensor arrays. *Journal of Sensors* 2012;2012.

[169] Justino CI, Rocha-Santos TA, Duarte AC. Advances in point-of-care technologies with biosensors based on carbon nanotubes. *TrAC Trends in Analytical Chemistry* 2013;45:24–36.

[170] Vamvakaki V, Tsagaraki K, Chaniotakis N. Carbon nanofiber-based glucose biosensor. *Analytical Chemistry* 2006;78:5538–42.

[171] Sirivisoot S, Yao C, Xiao X, Sheldon BW, Webster TJ. Developing biosensors for monitoring orthopedic tissue growth. *MRS Proceedings: Cambridge University Press*; 2006. p. 0950-D15-04.

[172] Sirivisoot S, Webster TJ. Multiwalled carbon nanotubes enhance electrochemical properties of titanium to determine in situ bone formation. *Nanotechnology* 2008;19:295101.

[173] Sirivisoot S, Webster TJ. In situ bone growth detection using carbon nanotubes–titanium sensors. *BioNanoScience* 2013;3:184–91.

[174] Weissleder R, Pittet MJ. Imaging in the era of molecular oncology. *Nature* 2008;452:580–9.

[175] Sitharaman B, Kissell KR, Hartman KB, *et al.* Superparamagnetic gadonanotubes are high-performance MRI contrast agents. *Chemical Communications* 2005:3915–7.

[176] Heller DA, Baik S, Eurell TE, Strano MS. Single-walled carbon nanotube spectroscopy in live cells: towards long-term labels and optical sensors. *Advanced Materials* 2005;17:2793–9.

[177] Chen M-L, Liu J-W, Hu B, Chen M-L, Wang J-H. Conjugation of quantum dots with graphene for fluorescence imaging of live cells. *Analyst* 2011; 136:4277–83.

[178] Gollavelli G, Ling Y-C. Multi-functional graphene as an in vitro and in vivo imaging probe. *Biomaterials* 2012;33:2532–45.

[179] Yu S-J, Kang M-W, Chang H-C, Chen K-M, Yu Y-C. Bright fluorescent nanodiamonds: no photobleaching and low cytotoxicity. *Journal of the American Chemical Society* 2005;127:17604–5.

[180] Fu C-C, Lee H-Y, Chen K, *et al.* Characterization and application of single fluorescent nanodiamonds as cellular biomarkers. *Proceedings of the National Academy of Sciences of the United States of America* 2007; 104:727–32.

[181] Ji S-r, Liu C, Zhang B, *et al.* Carbon nanotubes in cancer diagnosis and therapy. *Biochimica et Biophysica Acta (BBA) – Reviews on Cancer* 2010;1806:29–35.

[182] Yokoyama A, Sato Y, Nodasaka Y, *et al.* Biological behavior of hat-stacked carbon nanofibers in the subcutaneous tissue in rats. *Nano Letters* 2005;5:157–61.

[183] He H, Pham-Huy LA, Dramou P, Xiao D, Zuo P, Pham-Huy C. Carbon nanotubes: applications in pharmacy and medicine. *BioMed Research International* 2013;2013:578290.

[184] Liopo AV, Stewart MP, Hudson J, Tour JM, Pappas TC. Biocompatibility of native and functionalized single-walled carbon nanotubes for neuronal interface. *Journal of Nanoscience and Nanotechnology* 2006;6:1365–74.

[185] Zhou H, Mu Q, Gao N, *et al.* A nano-combinatorial library strategy for the discovery of nanotubes with reduced protein-binding, cytotoxicity, and immune response. *Nano Letters* 2008;8:859–65.

[186] Bianco A. Graphene: safe or toxic? The two faces of the medal. *Angewandte Chemie International Edition* 2013;52:4986–97.

[187] Blickley TM, McClellan-Green P. Toxicity of aqueous fullerene in adult and larval *Fundulus heteroclitus*. *Environmental Toxicology and Chemistry* 2008;27:1964–71.

[188] Shinohara N, Matsumoto K, Endoh S, Maru J, Nakanishi J. In vitro and in vivo genotoxicity tests on fullerene C 60 nanoparticles. *Toxicology Letters* 2009;191:289–96.

[189] Sayes CM, Marchione AA, Reed KL, Warheit DB. Comparative pulmonary toxicity assessments of C_{60} water suspensions in rats: few differences in fullerene toxicity in vivo in contrast to in vitro profiles. *Nano Letters* 2007;7:2399–406.

[190] Nielsen GD, Roursgaard M, Jensen KA, Poulsen SS, Larsen ST. In vivo biology and toxicology of fullerenes and their derivatives. *Basic & Clinical Pharmacology & Toxicology* 2008;103:197–208.

[191] Sayes CM, Fortner JD, Guo W, *et al.* The differential cytotoxicity of water-soluble fullerenes. *Nano Letters* 2004;4:1881–7.

[192] Dworak N, Wnuk M, Zebrowski J, Bartosz G, Lewinska A. Genotoxic and mutagenic activity of diamond nanoparticles in human peripheral lymphocytes in vitro. *Carbon* 2014;68:763–76.

[193] Yuan Y, Wang X, Jia G, *et al.* Pulmonary toxicity and translocation of nanodiamonds in mice. *Diamond and Related Materials* 2010;19:291–9.

[194] Chow EK, Zhang X-Q, Chen M, *et al.* Nanodiamond therapeutic delivery agents mediate enhanced chemoresistant tumor treatment. *Science Translational Medicine* 2011;3:73ra21.

[195] Vaijayanthimala V, Cheng P-Y, Yeh S-H, *et al.* The long-term stability and biocompatibility of fluorescent nanodiamond as an in vivo contrast agent. *Biomaterials* 2012;33:7794–802.

[196] Sureshbabu AR, Kurapati R, Russier J, *et al.* Degradation-by-design: surface modification with functional substrates that enhance the enzymatic degradation of carbon nanotubes. *Biomaterials* 2015;72:20–8.

[197] Thostenson ET, Ren Z, Chou T-W. Advances in the science and technology of carbon nanotubes and their composites: a review. *Composites Science and Technology* 2001;61:1899–912.

[198] Harutyunyan A, Pradhan B, Sumanasekera G, Korobko EY, Kuznetsov A. Carbon nanotubes for medical applications. *European Cells & Materials* 2002;3:84–7.

[199] Gass MH, Bangert U, Bleloch AL, Wang P, Nair RR, Geim A. Free-standing graphene at atomic resolution. *Nature Nanotechnology* 2008; 3:676–81.
[200] Singh SK, Singh MK, Kulkarni PP, Sonkar VK, Grácio JJ, Dash D. Amine-modified graphene: thrombo-protective safer alternative to graphene oxide for biomedical applications. *ACS Nano* 2012;6:2731–40.
[201] Ōsawa E, Ho D. Nanodiamond and its application to drug delivery. *Journal of Medical and Allied Sciences* 2012;2:31–40.

Part II

Nanomaterials for soft tissue engineering

Chapter 4

Usage of stem cells in bone and cartilage tissue engineering

Gorke G. Pekozer[1,2,3] and Gamze T. Kose[1,3]

Abstract

There are a large number of people suffering from organ or tissue loss due to injury, infection, or disease. Currently, major approaches to solve this problem are surgical reconstruction, transplantation, or the use of prosthesis. Although these therapies have saved and improved countless lives, they remain imperfect solutions due to the occurrence of infection, chronic irritation, donor shortages, tissue rejection, and longer term complications such as the development of malignant tumors.

Tissue engineering, the development of cell seeded three-dimensional (3D) biomaterials for introduction to the defect area, is a potential solution for the problems mentioned above.

Mesenchymal stem cells (MSCs) such as bone marrow MSCs (BMMSCs) and adipose derived stem cells are widely being used for clinical applications in regenerative medicine. Stem cells from umbilical cord, synovium, and dental tissues have also been started to be used for the same purposes.

The use of stem cells for promoting the biologic potential of scaffolds in tissue engineering, especially bone and cartilage tissue regeneration, has gained interest within last 10 years. Stem cell delivery has a potential as an effective treatment alternative for cartilage and bone defects with an impaired healing.

4.1 Bone and cartilage defects

Loss of skeletal tissues is one of the most common tissue losses and represents a major concern as life expectancy increases.

More than half of all chronic disorders experienced by people older than 50 years of age in developed countries are bone and joint disorders, and they are the

[1]Genetics and Bioengineering Department, Yeditepe University, Istanbul, Turkey
[2]Molecular Biology – Genetics and Biotechnology Department, Istanbul Technical University, Istanbul, Turkey
[3]Center of Excellence in Biomaterials and Tissue Engineering, BIOMATEN, METU, Ankara, Turkey

most common cause of intense, long-term pain, and disability decreasing life quality. In the Unites States, there are annually 1 million patients, who are hospitalized due to skeletal defects, with a cost of over $26 billion. Also, bone and cartilage defects of the skull and face contribute an additional $1.3 billion. Approximately 10% of these defects were reported for impaired healing [1]. This is becoming an important socioeconomic burden to the healthcare system as the world population ages.

Bone has an inherent ability to heal without leaving a scar. However, critical size defects, which by definition are the smallest size of tissue defects that will not completely heal over the lifetime of a person or an animal [2], fail to heal without intervention. These bone defects can arise from infection, traumatic injury, tumor resection, congenital malformations or diseases and require reconstruction surgeries with tissue grafting or with metal or ceramic prostheses as the conventional methods of the treatment [3].

Bone is the second most transplanted tissue in the world after blood [4]. Half a million bone grafting surgeries are performed each year in the United States and 2.2 million worldwide, and these numbers are expected to increase over the next decade as the population ages [5].

On the other hand, cartilage is an avascular, aneural, and alymphatic tissue and thus has limited ability to self-repair [6]. In the United States, over 30% of the population older than 30 years has cartilage degeneration and 9–10% of the population at age 30 suffers symptoms of osteoarthritis of the hip or knee. Annual costs of osteoarthritis related conditions is estimated at $36 billion dollars in the United States with more than 200,000 knee replacement procedures per year [5]. More than 1 million people each year in the United States suffer from articular cartilage injuries, 600,000 of whom undergo surgical procedures for the treatment of these injuries [7].

4.1.1 Current treatment methods of bone and cartilage defects

As mentioned in section 1, bone grafting and the use of metal and ceramic prostheses are the current methods to treat critical-size bone defects.

Autologous bone grafting, in which bone taken from another site on the patient's own body is transplanted to the defect site, has been accepted as the "gold standard" for the treatment of critical-sized bone defects for many years due to its superior osteogenic capacity, ease of incorporation, and lack of immunological problems [3,8]. It provides essential components of bone regeneration such as osteogenic cells, osteoinductive growth factors, and a bone-supporting matrix [9]. However, there is a complication rate of approximately 30% in autologous bone grafting which includes donor site morbidity, pain, risk of severe infection, and inflammation [10].

Allogenic bone grafting, in which bone taken from another person's body is transplanted to the defect site, is another alternative for the treatment of critical-size bone defects where autografting cannot be applied due to the shortage of healthy bone in the patient's own body. The acceptance of the allograft by the host's body is slower and vascular penetration, which is an important criterion for the success of the graft, is also slower and less dense [8]. These limitations decrease the success of allografts. In addition, allograft bone may lead to infection, immune rejection, and disease transmission from the donor to the host.

The size and extent of the defects limit the applications of bone grafts since grafts, either autologous or allogenic, are not vascular and the supply of nutrients and oxygen depends on diffusion. Furthermore, due to bone resorption the fate of grafted bone can be unpredictable. In large defects, bone grafts can be resorbed by the body before osteogenesis is completed [11].

Metals and ceramics are also used as alternatives to the grafting. Metal prostheses provide the immediate mechanical support but integrate poorly with the tissue at the defect site. Also, they can fail due to infection or fatigue. On the other hand, ceramics mimicking the inorganic composition of bone have low tensile strength and are brittle which makes it impossible to use them at locations of significant torsion, bending, or shear stress [3]. Moreover, the use of metals and ceramics requires revision surgeries in approximately 20% of cases due to their insufficient mechanical properties, slow or absent resorption and infection [12].

Cartilage defects, some partial and full thickness lesions (2–4 mm in diameter), require surgical procedures for treatment. Microfracturing, one of the arthroscopic techniques, is used to stimulate the body's intrinsic repair response by allowing penetration of subchondral bone in full thickness defects. However, microfracturing results in the formation of fibrocartilage which is mostly composed of collagen type I unlike the composition of native cartilage and the clinical value of this approach remains uncertain [6].

Autologous osteochondral grafting, another surgical cartilage repair strategy also called mosaicplasty, involves the removal of cylindrical plugs of osteochondral tissue from minor load bearing regions of the articular cartilage and then transplantation of those plugs to the pre-drilled holes of debrided full depth defect. Although clinical outcomes of this technique in small and medium size full-thickness defects are decreased pain and improved joint function, donor site morbidity and the viability of the chondrocytes from the minor load bearing tissue in stress of a load bearing site remains as a question [13].

Due to the limited availability of donor tissue and donor site morbidity, mosaicplasty cannot be applied when the defect is larger than 2 cm^2. Allogenic osteochondral grafting, in which osteochondral tissue from compatible live donors or cadavers are transplanted, can be used in extensive cartilage defects. Donor shortage and immunological response despite the immunologically privileged position of articular cartilage, due to the dense extracellular matrix and lack of vasculature of the tissue [14], are potential problems associated with this approach [15].

4.2 Tissue engineering

Tissue engineering is an interdisciplinary field that combines the knowledge of materials and cell biology for the development of tissue substitutes that act as "spare-parts" for the body. Tissue engineering research began as a response to the growing need for tissues and organs for transplantation [16]. Tissue engineering offers a life-long solution and a better quality of patients' life reducing the need for repeated hospitalization and health care costs associated with pharmaceutical therapy.

As stated earlier in this chapter, conventional methods for the treatment of bone and cartilage defects are associated with major drawbacks, which limit their use or result in their failure. Tissue engineering can be a potential solution addressing the drawbacks of those conventional methods.

There are three fundamental components of tissue engineering: (i) cells, (ii) scaffolds, and (iii) chemical signals. This chapter will mainly focus on cells for bone tissue engineering (BTE) and cartilage tissue engineering (CTE).

4.2.1 Cells for bone tissue engineering (BTE) and cartilage tissue engineering (CTE)

Choice of cells for tissue engineering applications can be grouped in two categories:

(i) Source of cells: autologous, allogenic, xenogenic
(ii) Type of cells: mature cells, stem cells

4.2.1.1 Cell sources

An ideal cell source for tissue engineering should exhibit the following characteristics: easy access and high availability of the source, minimal donor site morbidity, capacity of the cells for expansion to generate sufficient quantity, the capacity to differentiate readily into cell lineages of interest as a response to differentiation cues, the ability to manipulate the native environment to promote tissue regeneration and integration, and lack of immunogenicity, tumorigenicity, and the risk of disease transmission.

Depending on the choice of source, cells can be grouped as: (i) autologous, when the cells are taken directly from the patient, (ii) allogenic, when they are obtained from a human donor, or (iii) xenogenic, when cells are taken from a species other than human [11].

The simplest source of cells is autologous tissue in which cells are harvested from a patient via biopsy, grown in culture to the required number and then re-implanted into the patient. There are no legal problems with their use and no risk of immune rejection, which are the common problems faced when other cell sources are used. For tissue engineering applications, cells first should be cultivated to reach to a sufficient number to meet the functional needs of the tissue engineered construct within the limited time until implantation. Hence the time required for harvesting tissue from a patient and expanding the cells in culture to the desired number can be a critical limitation for the use of autologous cells. Besides, healthy tissue may be sparse or unavailable to harvest the cells, even when it is not, donor site morbidity is always a problem [3].

When the allogenic source of cells is chosen, it is possible to isolate higher numbers of cells which can be further expanded in culture to a sufficient quantity in a reasonable time. However, immune rejection is an important problem that could be encountered unless immune-suppression is employed and there is also a risk of disease transmission from the donor.

For some cases, neither the autologous nor the allogenic sources may be available. This situation has led to strategies which employ xenogenic tissues. The use of xenogenic cells presents a special safety consideration including the risk of disease transmission and immune rejection. New techniques employing cell encapsulation, immune protection, and genetic manipulation are being developed to make use of xenogenic tissues and organs for human beings [3].

4.2.1.2 Cell types

Choosing the right choice of cell source, either autologous, allogenic, or xenogenic, still remains a challenge.

Osteoblasts are the most obvious choice for BTE applications. They can be isolated from bone obtained by biopsies from the patient via explant culture or enzymatic digestion and expanded in vitro to reach the amount of cells (millions to several hundred millions) required for tissue engineering constructs [12]. Although these cells generally show superior performance regarding tissue-specific characteristics, their use for tissue engineering is often limited by time consuming isolation protocols and limited proliferation capacities in vitro. Also upon transplantation, these cells may be insufficient to heal the defect site in a reasonable time and there is the possibility of senescence which could lead to failure of the tissue engineered construct [17].

Similarly, chondrocytes are the most obvious choice for CTE strategies. However, as stated earlier in this chapter, expansion of the isolated cell type to the desired number is a critical step in the production of tissue engineering constructs. A biopsy from cartilage of patients aged 40 provides on average 4.5×10^5 cells/100 mg of cartilage. Autologous chondrocyte implantation requires $5–30 \times 10^6$ chondrocytes depending on the size (2×10^6 cells/cm^2) of defect to be filled and the procedure used. For restorative treatment of the entire knee joint with a surface area of 120 cm^2, 240×10^6 cells are required [18,19]. Thus, monolayer expansion is required to generate sufficient cells for implantation. This is especially an important concern for chondrocytes which dedifferentiate in monolayer culture upon repeated passaging [20]. Dedifferentiated chondrocytes change their shape, their metabolic state, and their synthesis of extracellular matrix components [21]. Mature chondrocytes express type II collagen and synthesize cartilage-specific proteoglycan. However, after several passages in monolayer culture, this phenotype is replaced by type I collagen expression and a low level of proteoglycan synthesis [20]. After prolonged monolayer culture, dedifferentiated chondrocytes proliferate but lose their chondrogenic potential irreversibly [21]. However, there are approaches to redifferentiate the dedifferentiated chondrocytes.

One approach to redifferentiate the cells into their mature state is suspension culture of the cells in hydrogels such as agarose [22], collagen [23,24], and alginate [24–26]. It was reported complete redifferentiation of the phenotype of dedifferentiated cells to primary chondrocytes was possible when they were cultured in firm gels of 0.5% below the melting point agarose. Approximately 80% of the cells could survive the transition from the flattened morphology of anchorage-dependent monolayer culture to the spherical morphology of anchorage-independent suspension culture and deposited a characteristic collagen type II and proteoglycan consisting matrix [22].

Plating chondrocytes in high density (e.g. 4×10^5 cells/cm^2) is another approach to redifferentiate the chondrocytes in a dedifferentiated state. It was reported that chondrocytes in monolayer culture regained their chondrogenic potential when they were transferred from an early monolayer passage to high density culture since this type of culture mimicked more closely the physiological conditions of cartilage in vivo and supported cell–cell interactions [27] mediated by cell adhesion molecules such as N-cadherin and integrins which played an important role in cell condensation during chondrogenesis [21].

In order for dedifferentiated chondrocytes to regain their mature phenotype, the pellet culture technique is also employed. In the pellet culture technique, expanded chondrocytes are redifferentiated through the aggregation of $1\text{–}5 \times 10^5$ cells into a 3D pellet by centrifugation in polypropylene tubes or v-bottom plates. The resulting pellet is macroscopic, having diameters of 1–2 mm [28]. Cells cultured in this model are shown to upregulate the synthesis of key phenotypic markers of cartilage such as collagen type II and glycosaminoglycans. The ability of the pellet culture system to retain a cartilage phenotype has also been exploited to induce adult stem cells to develop the chondrocyte phenotype in vitro [29].

In most cases, fully differentiated cells present in adult tissues exhibit a very limited proliferation capacity. This restricts their expansion in culture and limits their use for in vitro reconstruction of engineered tissues. On the other hand, undifferentiated cells have higher proliferative capacities and hence they may be more efficient for tissue engineering applications. Differentiation of those cells can be achieved in vitro by applying appropriate culture conditions after expansion or in vivo by taking advantage of the physiological microenvironment in the transplant area [15].

Stem cells are characterized by their ability to self-renew and differentiate into a diverse range of specialized cell types. Stem cells are characterized generally by their potency to differentiate: (i) totipotent, (ii) pluripotent, and (iii) multipotent.

Totipotency refers to a single cell's ability to create a whole organism. Multicellular organisms originate from a single totipotent cell which is called a zygote. The cells remain as totipotent until the blastocyst stage. At the blastocyst stage, embryos develop into a hollow ball of cells in which inner cell mass is present and surrounded by a trophoblast. A trophoblast later develops into the placenta whereas the inner cell mass is the source for the isolation of embryonic stem cells (ESCs) [30].

ESCs are characterized by their self-renewal property, high telomerase activity, and pluripotency. They have the ability to differentiate into a wide variety of specialized cell types in vitro and to all cell lineages (endoderm, mesoderm, ectoderm) in vivo.

ESCs demand strict culture conditions to maintain their normal morphology and genetic integrity. They also require a layer of mitotically inactivated mouse embryonic fibroblasts (MEF) as a feeder layer. This layer provides critical nutrients and trophic factors to keep the cells in an undifferentiated stem cell state or provide the necessary cell–cell contact to do so. Protocols that avoid the use of feeder layers are also suggested for the maintenance of self-renewal of human ESCs such as feeder layer conditioned medium on Matrigel [31], addition of bone morphogenic protein (BMP) antagonists and basic fibroblast growth factor (bFGF) [32], addition of transforming growth factor β1 (TGF-β1), leukemia inhibitory factor (LIF), bFGF, and fibronectin matrix [33], and commercially available medium compositions [34].

ESCs express transcription factors such as Oct4, Sox2, and Nanog, cell surface markers SSEA-3 and SSEA-4, TRA-1-60 and TRA-1-81, and exhibit high telomerase and alkaline phosphatase activity [35].

ESCs are good choice for use in BTE and CTE applications due to their excellent ability for self-renewal, proliferation, and differentiation. β-glycerophosphate, ascorbic acid, dexamethasone, and osteogenic factors have been used for the osteogenic differentiation of human ESCs (hESCs). It was reported that hESCs could be differentiated into osteogenic cells that are positive for osteocalcin and Runx2 by the addition of β-glycerophosphate, ascorbic acid, and also dexamethasone at specific time points. Besides, these cells gave rise to mineralized tissue in vivo when implanted into SCID mice on a poly-D,L-lactide (PDLLA) scaffold without any teratoma formation [36].

Chondrogenic differentiation of ESCs, on the other hand, is generally induced by co-culture of ESCs with chondrocytes [37,38], by pellet culture [39], and/or by the addition of chondrogenic supplements like BMP-2 and BMP-4 [40], TGF-β3 [39], and BMP-7 and TGF-β1 [41].

Despite the promises in terms of plasticity and proliferation capacity, the use of ESCs has also major challenges including appropriate differentiation of the cells to the target cell type in a controlled and homogeneous way and avoiding any undifferentiated ESCs which can form teratomas upon implantation [42].

There are also major ethical problems associated with the derivation and use of ESCs, while the use of stem cells isolated from adult tissues is less risky and are generally well accepted by the society [15].

Stem cells are also found in adult tissues in which their role is to maintain or repair the tissue in case of trauma, disease, or aging. Adult stem cells are multipotent cells that can renew themselves and with certain limitations differentiate to all the specialized cell types of the tissue from which they originate. Adult stem cells are generally multipotent differentiating only into the cells of the same germ layer [43].

MSCs are a type of adult stem cells and are present in adult tissues like bone marrow, adipose tissue, synovium, skeletal muscle, dental pulp, heart, and spleen; from neonatal tissues, such as the placenta, amniotic fluid, and umbilical cord blood; and from fetal tissues, such as liver and blood [44] serving as a cell source for the regeneration of various mesenchymal tissues. MSCs are characterized by their ability to proliferate in culture and by their properties to differentiate into multiple mesenchymal lineages such as bone, cartilage, and adipose under defined culture conditions [45].

Under normal physiological conditions, only a minor amount of MSCs can be detected in circulating blood. However, due to the homing behavior of MSCs to injury sites, they are recruited from the circulatory system after injury and they can migrate to the injury site along the blood vessels by the help of the receptors they express, including P-selectin and vascular cell adhesion molecules (VCAM). The ability of MSCs to migrate to the sites of injury supports their use in tissue engineered constructs, since they can sense and respond to factors and cytokines secreted in an injury environment [46].

MSCs are hypo-immunogenic, thus, attractive for allogenic transplantation, expressing low (MSCs from fetal tissues) to intermediate (MSCs from adult tissues) major histocompatibility complex (MHC) class I molecules on their surface helping to protect them from natural killer cells and lacking MHC class II molecules making them not recognizable to CD4+ T cells [5].

The International Society of Cell Therapy proposed three criteria for defining MSCs: (i) adherence to plastic, (ii) multipotent differentiation potential, and (iii) specific surface antigen expression. MSCs should be positive for CD105, CD73, and CD90 surface antigens and negative for CD45, CD34, CD14 or CD11b, CD79a or CD19, and HLA class II proteins. Cells also should demonstrate the ability to differentiate along osteogenic, adipogenic, and chondrogenic lineages under standard in vitro differentiation conditions [47]. Other studies showed MSCs are also positive for CD44, CD29, and CD166 [5].

MSCs can be induced for osteogenesis in vitro typically by the addition of synthetic glucocorticoid dexamethasone, β-glycerophosphate, and ascorbic acid to the serum containing media. In some protocols, 1,25-dihydroxyvitamin D3, TGF-β, and BMPs are also used as supplements for the osteogenic induction media [48].

In vitro chondrogenesis of MSCs involves a 3D culture of MSCs under the stimulation of a suitable chondrogenic stimulus. Chondrogenic stimulus includes the use of insulin-transferrin-selenium, linoleic acid, and bovine serum albumin (BSA) as serum replacement and TGF-β, dexamethasone, and ascorbic acid addition in the media. Hypoxic conditions can also be used for chondrogenic induction since during in vivo condensation in chondrogenesis processes, a local low oxygen environment is induced within the cell clusters, which is inherent to the adult cartilage due to its avascular nature [49].

Bone marrow is usually collected by aspiration of the iliac crests or femoral heads of patients with a Jamshidi needle. Bone marrow is then plated directly, or after density gradient centrifugation, in culture medium. Since MSCs are identified first by their ability to adhere to plastic, hematopoietic stem cells (HSCs) can be eliminated from the culture after several washes with phosphate buffered saline (PBS) or medium changes. Adherent cells are organized in small clusters of spindle-shaped (fibroblastic) cells within 7–10 days. The frequency and yield of MSCs in bone marrow can vary between donors depending on the age and disease state as well as the harvesting method [50] and estimated as 1 in 50,000–100,000 cells in skeletally mature adults [44]. Large numbers of MSCs can be obtained in culture after a reasonable time of cultivation, making it possible to engineer even large tissue engineered constructs requiring a high numbers of cells.

Since bone marrow is the mostly studied MSC source, the usefulness of BMMSCs in BTE and CTE has been demonstrated both in vitro and in vivo using multiple animal models including mouse, rat, rabbit, and pig. For example, in one study, both in vitro and in vivo osteogenesis of BMMSCs were studied on dexamethasone and BMP-2 loaded electrospun nanofibrous poly(ε-caprolactone)-co-poly(ethylene glycol) scaffolds [51]. In vitro studies demonstrated that the drug-loaded groups exhibited a strong ability to induce differentiation toward osteoblasts shown by ALP activity and mineralization staining. Healing of rat calvarial critical-sized defects was best promoted by dual-drug loaded nanofibrous scaffolds. In another study, porcine bone marrow stem cells (PBMSCs) were seeded on PEG-PCL-PEG (PCL) or PEG-PCL-PEG/hydroxyapatite (PCL/HA) composite scaffolds and induced for osteogenesis in vitro and used for the treatment of full size 2×2 cm^2 temporal defects of pigs in vivo [52]. In vitro osteogenesis of PBMSCs was revealed by the expression of osteogenic mRNAs such as ALP, Runx2, and osteocalcin. PBMSCs showed similar osteogenesis and angiogenesis pattern on both PCL and PCL/HA scaffolds. However, the PCL/HA scaffold, which had a bone-mimetic environment due to the addition of HA like in the native bone, obtained better bone regeneration in vivo than a PCL scaffold.

A recent study by our group demonstrated that rat BMMSCs when co-cultured with endothelial cells could differentiate osteoblast lineage and also contributed to microvessel formation which is important for proper bone healing [53].

In a CTE study, a biphasic 3D scaffold platform was constructed using a chitosan hydrogel and demineralized bone matrix (DBM) and conjugated with a BMMSC affinity peptide, E7 [54]. After 3 days of in vitro culture, significantly more BMMSCs were recruited onto E7-conjugated biphasic scaffolds than non-conjugated biphasic scaffolds. After 4 weeks, BMMSCs on E7-conjugated biphasic scaffolds significantly produced more cartilage specific ECM elements, glycosaminoglycan, aggrecan, and collagen type II. E7-conjugated scaffolds also performed better in vivo in full-thickness osteochondral defects (4 mm diameter, 2.5 mm deep) in rabbit knees exhibiting a more complete, uniform, and consistent regeneration with an enhanced production of a cartilage-like matrix throughout 24 weeks of post-operation.

In another study, BMMSCs isolated from bone marrow of newborn pigs were seeded on acellularized cartilage sheets (ACSs) which were then stacked layer by layer to form a CTE construct [55]. Disks made of PGA/PLA, which have been well characterized for use in CTE, were used as controls. Both experimental and control constructs were cultured with or without chondrogenic induction to test the ability of ACSs to induce chondrogenesis. Gene expression analysis of in vitro cultures for aggrecan, COMP, Sox9, and type II collagen indicated a weak-positive chondrogenic induction in non-induced ACS constructs. After 4 weeks post-implantation to nude mice, no ossification was observed in ACS groups with or without induction indicating that ACSs may possess an anti-angiogenic activity due to factors, such as chondromodulin-1, which remained active after decellularization process.

Adipose tissue is an alternative source for MSCs due to the first isolation of multipotent cells in 2001 from collagenase-digested human adipose tissue [56].

Like MSCs isolated from bone marrow, adipose tissue-derived MSCs (AMSCs) also show extensive self-renewal capacity and potency to differentiate into multiple lineages including adipose [57], bone [58], cartilage [58–60], myogenic [61], cardiomyogenic [62], and endothelial tissue [63].

The yield of AMSCs is approximately 5,000 CFU-F units per gram of adipose tissue compared to 100–1,000 CFU-F units of MSCs per milliliter of bone marrow meaning that adipose tissue is a more efficient source for stem cell isolation [64].

Furthermore, the surface antigen profile of AMSCs and BMMSCs is also similar. AMSCs express mesenchymal surface antigens like CD29, CD44, CD73, CD90, CD105, and CD166 and a lack of the expression of hematopoietic surface antigens like CD14 and CD45. However, differences in surface antigen expression of AMSCs and BMMSCs have also been reported [65–67].

Theoretical advantages of using AMSCs instead of BMMSCs in tissue engineering applications include a less invasive method of tissue extraction and higher number of stem cells obtained from the tissue. Besides, superior capacity of AMSCs for osteogenesis and chondrogenesis has also been investigated by researchers. In one study, in vitro proliferation and osteogenic differentiation capacities of AMSCs and BMMSCs were compared and then confirmed in vivo by implanting differentiated and non-differentiated AMSC or BMMSC loaded cancellous bone chips into the paravertebral musculature of nude mice. In vitro experiments demonstrated a superiority of AMSCs for proliferation, but no significant difference between the time intervals required for complete osteogenesis of both cell types was observed. However, in vivo studies revealed a significantly higher amount of newly formed bone matrix and a significantly higher expression of osteocalcin for allografts supplemented with differentiated AMSCs compared to differentiated BMMSCs [68].

Although AMSCs have been widely employed to repair cartilage defects with satisfactory results, the chondrogenic superiority of AMSCs over BMMSCs is questionable since there are studies that reported both better [69] and worse [70–72] chondrogenic capacity.

Fetal tissues are alternative sources of MSCs and considered as a solution both to the political and ethical controversies surrounding ESCs and low proliferation and differentiation capacities of adult tissue derived MSCs.

The umbilical cord, one of the fetal tissues which serves as a MSC source, is an elastic cord connecting the fetus and placenta during pregnancy. A mature umbilical cord has a mean length of 50–70 cm, in vivo diameter of 12 mm, and weight of 100 g. Anatomically, it consists of two arteries and one vein embedded in a mucoid connective tissue called Wharton's jelly and covered by amniotic epithelium. Having the vasculature, the umbilical cord provides the developing fetus oxygen and nutrients, and allows metabolite exchange and waste removal [73].

Stem cells isolated from umbilical cords show some characteristic pluripotent properties of ESCs and some multipotent properties of MSCs. Like ESCs they have broader multipotent plasticity and higher proliferation rates compared to stem cells derived from post-natal tissues. Besides, due to their resemblance to ESCs, umbilical cord cells are immune-privileged having immunomodularity effects and low immunogenicity [73]. They express MHC class I antigens and low levels of MHC class II (less compared to BMMSCs). These cells do not require HLA matching before transplantation and thus can effectively be used as an allogenic cell source for a patient without the risk of graft-versus-host disease and the need for immunosuppressant drugs [74]. Most important of all, regardless of the tissue in an adult body they are isolated from, the proliferation and differentiation abilities of MSCs decrease with donor age. In this perspective, the cells isolated from umbilical cords can be considered as primitive having better proliferation and differentiation abilities [75].

Umbilical cord blood, which is collected from umbilical veins after the umbilical cord has been detached from the newborn, is generally used for the isolation of HSCs for cord blood banking for future transplantation to treat pediatric genetic, hematologic, immunologic, and oncologic disorders. However, umbilical cord blood was also reported as a source of MSCs [75–77]. Besides, it was stated as an easy, cheap, and non-invasive method to isolate MSCs.

Umbilical cord blood MSCs (UCBMSCs) differentiated to osteogenic and chondrogenic lineages but not to an adipogenic lineage even though they exhibit a similar surface marker expression profile being CD45, CD14, CD34 negative and positive for CD29, CD44, CD73, CD90, CD105, and CD106 [78].

Shetty *et al.* [79] also differentiated UCBMSCs to osteoblasts and chondrocytes but not to adipocytes unlike BMMSCs that were used as controls. They also stated that although MSC yield isolated from cord blood was low, cells expanded faster compared to BMMSCs.

Liu *et al.* [80] isolated UCBMSCs and were seeded on porcine partially DBM (pDBM) to assess the BTE potential of these cells in vitro and then in bilateral critical size cranial defects (5 mm in diameter) of athymic rats in vivo. Osteogenically induced UCBMSCs cultured on pDBM in vitro expressed elevated levels of ALP, osteocalcin, osteopontin, and collagen type I. When pDBM with the osteogenically induced UCBMSCs were implanted onto bone defects, new bone formation was noticed after 6 weeks and greater bone volume and density were found after 12 weeks post-implantation.

Wharton's jelly forms a soft gelatinous skeleton comprised of proteoglycans and collagen fibrils maintaining the tissue architecture, holding the arteries and the vein in place and preventing them from pressure. Wharton's jelly homes primitive MSCs possibly trapped in its fibrillar matrix structure during their migration to and from the placenta during embryogenesis [74]. MSCs can be isolated from Wharton's jelly by enzymatic methods using collagenase alone or in combination with dispase and hyaluronidase and by explant culture method after mechanically scraping it.

Wharton's jelly MSCs (WJMSCs) represent an adherent cell population with a fibroblastic morphology and non-endothelial (CD34 negative) phenotype. They share a similar set of surface markers with adult MSCs, including CD10, CD13, CD29, CD44, CD49e, CD51, CD73, CD90, CD105, CD106, CD117, CD166, and HLA-1/HLA-ABC and are negative for hematopoietic markers CD14, CD31, CD34, CD38, CD45, and HLA-DR. Besides, there are reports stating that WJMSCs express pluripotency markers such as Oct-4 [81–83], Sox-2 [81,82], and Nanog [81–83].

WJMSCs have the ability to differentiate along several cell lineages of mesodermal or ectodermal origin [84].

Schneider *et al.* [85] compared the osteogenic potential of BMMSCs and WJMSCs in collagen I/III gels. They reported that both cell types expressed osteogenic genes, mineralized the extracellular matrix, migrated and colonized the scaffold, and contracted the collagenous matrix as it was needed for fracture healing. WJMSCs developed a higher amount of matrix deposits than BMMSC. However, gene expression analysis by real-time PCR revealed higher synthesis of osteopontin and ALP in BMMSCs indicating the influence of anatomical origin of MSCs on osteogenesis.

Another study presented WJMSCs as an inexhaustible, low-cost, and minimally invasive alternative to the gold-standard BMMSCs since WJMSCs differentiated into the osteogenic lineage, with highly elevated ALP, OC, collagen type I, and osterix expressions and mineralization when they were encapsulated in an injectable form of calcium phosphate cement containing alginate microbeads [86].

WJMSCs were previously differentiated into chondrocytes with or without usage of scaffolds. In one study, WJMSCs were induced for chondrogenesis in 3 weeks of culture in a rotary cell culture system by using TGF-β and dexamethasone [87]. Histochemical analysis of paraffin sections of the pellets showed chondrocyte-like cells within lacunae and synthesis of a cartilaginous matrix.

Chen *et al.* [88] used collagen type I hydrogel as a CTE scaffold and showed the expression of cartilage specific ECM proteins such as collagen type II, aggrecan, and COMP by WJMSCs by real-time PCR, histological, and histochemical staining in 3 weeks of culture in standard chondrogenic medium. Also, Sox-9, a transcription factor with a central role in chondrogenesis, was significantly upregulated in chondrogenically induced WJMSCs on collagen hydrogels.

Human dental pulp has a limited ability to repair itself in case of caries or trauma which do not involve the pulp cavity [89]. This situation led researchers to think that ectomesenchymal progenitor cells remain in the pulp tissue after the development of a tooth is complete.

Dental tissues gained attention as a rich MSC source due to accessibility and multilineage differentiation capacity [90] since the first isolation and characterization of stem cells from dental pulp in 2000 [91].

Up to date, dental stem cells, characterized by their ability to adhere to plastic and high clonogenic properties when grown and expanded under the same conditions as the BMSCs [92], have been isolated from dental pulp of mature teeth (DPSC: dental pulp stem cells), from the pulp of resorbing milk teeth, termed as exfoliated deciduous teeth (SHED: stem cells from exfoliated deciduous teeth), from periodontal ligaments (PDLSCs: periodontal ligament stem cells) which hold the tooth in an alveolar socket [89] and also from immature dental tissues such as the apical papilla of developing teeth (SCAP: stem cells from apical papilla) located at the tips of growing tooth roots [93], dental follicle (DFSCs: dental follicle stem cells) which is a loose connective tissue surrounding the developing tooth (including enamel organ and dental papilla) before eruption and also from dental germs which contain follicle and surrounding tissues (DGSCs: dental germ stem cells) [94].

Researchers generally have cultured stem cells derived from immature tooth tissues by dissecting dental follicle and apical papilla separately and established either DFSC or SCAP cultures. However, stem cells of dental origin, which are responsible for tooth development, are derived from both ectoderm and the underlying mesenchyme. Thus, reciprocal signaling pathways between these cell groups should be considered in designing a culture system. The use of the whole tooth germ is a promising strategy to preserve the stemness of the culture when isolating stem cells (TGSCs: tooth germ stem cells) from immature teeth such as third molars (wisdom teeth) [89].

In humans, tooth germ tissues derived from third molars are quite unique as they undergo organogenesis to give rise to dental structures at around age 6. Until this time, embryonic tissues of dental lamina remain quiescent and undifferentiated [95]. Besides, crown mineralization begins at around age 8 and usually roots are still not complete at age 18 meaning that undifferentiated cells that are required for development of those immature teeth are present in the dental germ [96]. In addition, third molar tooth germs are routinely extracted as an orthodontic treatment and discarded as surgical waste. Therefore, stem cell harvesting from this source causes no ethical controversy [95].

Park *et al.* [97] compared osteogenic potentials of stem cells derived from skin, bone marrow, and dental follicle tissues in vitro on DBM and fibrin glue without using osteogenic induction medium and in vivo by implanting subcutaneously to athymic mice. All cell types differentiated into osteogenic lineage in vitro as shown by von Kossa staining and the expression of bone related genes such as OC, ON, Runx2, and osterix. However, the hDFMSCs-implanted group showed the highest OC expression and calcium content in vivo revealing the potential of stem cells isolated from dental tissues in BTE.

Recently, our group isolated and characterized dental germ stem cells from pig third molars and reported differentiation of porcine dental germs stem cells (pDGSCs) to mature osteoblasts on fibronectin modified polybutylene succinate scaffolds [98].

Our group also reported chondrogenic differentiation of human tooth germ stem cells (hTGSCs) that were positive for MSC surface markers CD44, CD90, and CD105 and negative for HSC surface markers CD34, CD45, and CD117 on PCL, PLLA, and PCL-PLLA macroporous foams as scaffolds [99]. hTGSCs deposited cartilage specific ECM proteins on scaffolds as shown by positive collagen type II and aggrecan staining by confocal microscopy. Collagen type II and aggrecan expressions of hTGSCs that were seeded on PCL-PLLA foams were higher compared to hTGSCs on PCL and PLLA foams revealing the potential of both PCL-PLLA foams and hTGSCs in CTE applications.

The synovium (synovial membrane) is a loose connective tissue that lines the synovial cavity and the surface of joints. The cells of the synovium produce the synovial fluid (SF). SF is a viscous joint lubricant and a source for nourishment and shock absorber for articular cartilage.

Synovium is the closest tissue to articular cartilage and can be harvested in an arthroscopy without any harm to the tissue. Sakaguchi *et al.* [100] demonstrated that an average of 21,000 cells per milligram of synovium were obtained after the nucleated cells were plated at optimal density and cultured for 14 days.

Fibroblast-like cells were first isolated from synovium in 2001 by De Bari *et al.* [101] and defined as multipotent stem cells.

Jo *et al.* [102] characterized synovium derived MSCs (SMSCs) by their surface marker expression profile and found that after first passage SMSCs were positive for CD10, CD13, CD44, CD49a, CD73, CD105, and CD166 and negative for CD14, CD34, CD45, and HLA-DR. They also took advantage of these cells being sourced from the synovium, which is the closest tissue to articular cartilage and recruits cells for cartilage defect repair, to induce them in cell pellets for chondrogenesis. In the presence of TGF-β1, SMSCs pellets produced cartilage specific ECM proteins positively stained for proteoglycans and collagen type II and X.

Koga *et al.* [103] employed SMSCs to repair full-thickness articular cartilage defects in rabbits. They demonstrated that when cells were applied in a collagen gel and covered with periosteum, the cells in the superficial zone differentiated into chondrocyte like cells and produced a cartilage matrix within 24 weeks. The regenerated tissue integrated well to the host tissue, did not become hypertrophic, and showed hyaline cartilage properties.

MSCs have also been found in SF in the knee joint and characterized by their high uridine diphosphoglucose dehydrogenase (UDPGD) activity and CD44 expression, which give SF derived MSCs (SFMSCs) a superior ability to differentiate into chondrocytes [104]. This can be due to shearing forces in the synovium of joints and embryological predetermination of the cells in that zone. These cells are a resident of SF of a normal articular cartilage but increases in number in defective joints [105].

Ando *et al.* [106] compared osteogenic, chondrogenic, and adipogenic performance of MSCs isolated from SF, synovial membrane, bone marrow, and the skin of pigs. They revealed that SFMSCs and SMSCs were superior in terms of chondrogenic potential where BMMSCs showed a better performance for osteogenesis and skin derived MSCs showed more adipogenic potential.

4.3 Conclusion

Skeletal defects represent a major concern being a common cause of severe, long-term pain, and disability and becoming an important socioeconomic burden as the life expectancy increases. Bone and cartilage tissue engineering strategies offer a life-long therapy for patients with skeletal defects. However, choosing the cell source and cell type for tissue regeneration can be challenging and depends on the type and state of injury as well as patient's health and age. Stem cells with different clonogenic properties and varying degrees of differentiation potential toward bone and cartilage can be isolated from embryonic, fetal, and adult tissues. The right source of stem cells can be chosen after a careful consideration of ethical issues, invasiveness of tissue extraction method, time to reach the amount of cells required, and differentiation capacity of cells to the target tissue.

Acknowledgments

G.G. Pekozer acknowledges the support from The Scientific and Technical Research Council of Turkey (TUBITAK – BIDEB) during her PhD education. Authors also acknowledge financial support provided by TUBITAK (Project no: 114S556).

References

[1] Montoro D.T., Derrick C.W., Longaker M.T. 'Skeletal tissue engineering' in Lanza R., Langer R., Vacanti J. (eds.). *Principles of Tissue Engineering.* 4th edn. Waltham, Massachusetts: Academic Press; 2014. pp. 1289–1302.

[2] Spicer P.P., Kretlow J.D., Young S., Jansen J.A., Kasper F.K., Mikos A.G. 'Evaluation of bone regeneration using the rat critical size calvarial defect'. *Nature Protocols.* 2012;**7**:1918–1929.

[3] Salgado A.J., Coutinho O.P., Reis, R.L. 'Bone tissue engineering: state of the art and future trends'. *Macromolecular Bioscience.* 2004;**4**:743–765.

[4] Shegarfi H., Reikeras, O. 'Bone transplantation and immune response'. *Journal of Orthopedic Surgery.* 2009;**17**(2):206–211.

[5] Alexander P.G., Hofer H.R., Clark K.L., Tuan R.S. 'Mesenchymal stem cells in musculoskeletal tissue engineering' in Lanza R., Langer R., Vacanti J (eds.). *Principles of Tissue Engineering.* 4th edn. Waltham, Massachusetts: Academic Press; 2014. pp. 1171–1199.

[6] Chung C., Burdick, J.A. 'Engineering cartilage tissue'. *Advanced Drug Delivery Reviews.* 2008;**60**(2):243–262.

[7] Melero-Martin J.M., Al-Rubeai M. 'In vitro expansion of chondrocytes' in Ashammakhi N., Reis R., Chiellini E. (eds.). *Topics in Tissue Engineering. Expertissues e-book*, 2007;**3**:1–37.

[8] Brown K.L. Cruess, R.L. 'Bone and cartilage transplantation in orthopaedic surgery'. *Journal of Bone and Joint Surgery.* 1982;64-A(2):270–279.

[9] Marolt D., Knezevic M., Novakovic G.V., 'Bone tissue engineering with human stem cells'. *Stem Cell Research & Therapy.* 2010;**1**(2):1–10.

[10] Porter J.R., Ruckh T.T., Popat, K.C., 'Bone tissue engineering: a review in bone biomimetics and drug delivery strategies'. *Biotechnology Progress.* 2009;**25**(6):1539–1560.

[11] Burg K.J., Porter S., Kellam, J.F. 'Biomaterial developments for bone tissue engineering'. *Biomaterials.* 2000;**21**(23):2347–2359.

[12] Marolt D. 'Tissue engineering craniofacial bone products' in Vishwakarma A., Sharpe P., Songtao S., Ramalingam M. (eds.). *Stem Cell Biology and Tissue Engineering in Dental Sciences.* Amsterdam: Elsevier; 2015, pp. 521–539.

[13] Redman S.N., Oldfield S.F., Archer, C.W. 'Current strategies for articular cartilage repair'. *European Cells and Materials.* 2005;**9**:23–32.

[14] Gross A.E., Kim W., Las Heras F., Backstein D., Safir O., Pritzker, K.P. 'Osteochondral allografts for posttraumatic knee defects: long-term followup'. *Clinical Orthopaedics and Related Research.* 2008;**466**(8):1863–1870.

[15] Cancedda R., Dozin B., Giannoni P., Quarto, R. 'Tissue engineering and cell therapy of cartilage and bone'. *Matrix Biology.* 2003;**22**(1):81–91.

[16] Furth M.E., Atala A. 'Tissue engineering: future perspectives' in Lanza R., Langer R., Vacanti J (eds.). *Principles of Tissue Engineering.* 4th edn. Waltham, Massachusetts: Academic Press; 2014. pp. 83–123.

[17] Meyer U., Joos U., Wiesmann, H.P. 'Biological and biophysical principles in extracorporeal bone tissue engineering. Part I'. *International Journal of Oral and Maxillofacial Surgery.* 2004;**33**:325–332.

[18] Andriamanalijaona R., 'Cell therapies for articular cartilage repair: chondrocytes and mesenchymal stem cells'. *Regenerative Medicine and Biomaterials for the Repair of Connective Tissues,* Cambridge: Woodhead Publishing; 2010. pp. 266–300.

[19] Hsieh-Bonassera N.D., Wu I., Lin J.K., *et al.* 'Expansion and redifferentiation of chondrocytes from osteoarthritic cartilage: cells for human cartilage tissue engineering'. *Tissue Eng Part A.* 2009;**15**(11):3513–3523.

[20] Vacanti C.A., Vacanti J.P. 'Engineering cartilage and other structural tissues: principles of bone and cartilage reconstruction' in Lanza R., Langer R., Vacanti J. (eds.). *Principles of Tissue Engineering.* 4th edn. Waltham, Massachusetts: Academic Press; 2014. pp. 1267–1274.

[21] Schulze-Tanzil G., de Souza P., Villegas Castrejon H., *et al.* 'Redifferentiation of dedifferentiated human chondrocytes in high-density cultures'. *Cell and Tissue Research.* 2002;**308**(3):371–379.

[22] Benya P.D., Shaffer, J.D. 'Dedifferentiated chondrocytes reexpress the differentiated collagen phenotype when cultured in agarose gels'. *Cell.* 1982;**30**(1):215–224.

[23] Kawasaki K., Ochi M., Uchio Y., Adachi N., Matsusaki, M. 'Hyaluronic acid enhances proliferation and chondroitin sulfate synthesis in cultured chondrocytes embedded in collagen gels'. *Journal of Cellular Physiology.* 1999;**179**:142–148.

[24] Yamaoka H., Asato H., Ogasawara T., *et al.* 'Cartilage tissue engineering using human auricular chondrocytes embedded in different hydrogel materials'. *Journal of Biomedical Materials Research Part A.* 2006;**78**:1–11.

[25] Xu F., Xu L., Wang Q., Ye Z., Zhou Y., Tan, W.S. '3D dynamic culture of rabbit articular chondrocytes encapsulated in alginate gel beads using spinner flasks for cartilage tissue regeneration'. *BioMed Research International.* 2014;**2014**:1–10.

[26] Homicz M.R., Chia S.H., Schumacher B.L., *et al.* 'Human septal chondrocyte redifferentiation in alginate, polyglycolic acid scaffold, and monolayer culture'. *Laryngoscope.* 2003;**113**:25–32.

[27] Haisch A., Marzahn U., Mobasheri A., Schulze-Tanzil G., Shakibaei, M. 'Development and phenotypic characterization of a high density in vitro

model of auricular chondrocytes with applications in reconstructive plastic surgery'. *Histology and Histopathology.* 2006;**21**(5):467–476.

[28] Babur B.K., Ghanavi P., Levett P., *et al.* 'The Interplay between chondrocyte redifferentiation pellet size and oxygen concentration'. *PLOS ONE.* 2013; **8**(3):e58865.

[29] Tare R.S., Howard D., Pound J.C., Roach H.I., Oreffo, R.O. 'Tissue engineering strategies for cartilage generation – micromass and three dimensional cultures using human chondrocytes and a continuous cell line'. *Biochemical and Biophysical Research Communications.* 2005;**333**(2):609–662.

[30] Bajada S., Mazakova I., Ashton B.A., Richardson J.B., Ashammakhi, N. 'Stem cells in regenerative medicine' in Ashammakhi N., Reis R., Chiellini E. (eds.). *Topics in Tissue Engineering.* Expertissues e-book, 2008;**4**:1–28.

[31] Xu C., Inokuma M.S., Denham J., *et al.* 'Feeder-free growth of undifferentiated human embryonic stem cells'. *Nature Biotechnology.* 2001;**19**(10):971–974.

[32] Xu R.H., Peck R.M., Li D.S., Feng X., Ludwig T., Thomson J.A. 'Basic FGF and suppression of BMP signaling sustain undifferentiated proliferation of human ES cells'. *Nature Methods.* 2005;**2**(3):185–190.

[33] Amit M., Shariki C., Margulets V., Itskovitz-Eldor, J. 'Feeder layer- and serum-free culture of human embryonic stem cells'. *Biology of Reproduction.* 2004;**70**(3):837–845.

[34] Yoon T.M., Chang B., Kim H.T., Jee J.H., Kim D.W., Hwang, D.Y. 'Human embryonic stem cells (hESCs) cultured under distinctive feeder-free culture conditions display global gene expression patterns similar to hESCs from feeder-dependent culture conditions'. *Stem Cell Reviews.* 2010;**6**(3):425–437.

[35] Villa-Diaz L.G., Ross A.M., Lahann J., Krebsbach P.H. 'The evolution of human pluripotent stem cell culture: from feeder cells to synthetic coatings'. *Stem Cells.* 2013;**31**(1):1–7.

[36] Bielby R.C., Boccaccini A.R., Polak J.M., Buttery, L.D. 'In vitro differentiation and in vivo mineralization of osteogenic cells derived from human embryonic stem cells'. *Tissue Engineering.* 2004;**10**(9–10):1518–1525.

[37] Qu C., Puttonen K.A., Lindeberg H., *et al.* 'Chondrogenic differentiation of human pluripotent stem cells in chondrocyte co-culture'. *International Journal of Biochemistry & Cell Biology.* 2013;**45**(8):1802–1812.

[38] Vats A., Bielby R.C., Tolley N., *et al.* 'Chondrogenic differentiation of human embryonic stem cells: the effect of the micro-environment'. *Tissue Engineering.* 2006;**12**(6):1687–1697.

[39] Nakayama, N., Duryea D., Manoukian R., Chow G., Han C.Y. 'Macroscopic cartilage formation with embryonic stem-cell-derived mesodermal progenitor cells'. *Journal of Cell Science.* 2003;**116**:2015–2028.

[40] Kramer J., Hegert C., Guan K., Wobus A.M., Muller P.K., Rohwedel, J. 'Embryonic stem cell derived chondrogenic differentiation in vitro: activation by BMP-2 and BMP-4'. *Mechanisms of Development.* 2000;**92**(2):193–205.

[41] Nakagawa T., Lee S.Y., Reddi, A.H. 'Induction of chondrogenesis from human embryonic stem cells without embryoid body formation by bone

morphogenetic protein 7 and transforming growth factor beta1'. *Arthritis & Rheumatology.* 2009;**60**(12):3686–3692.

[42] Khademhosseini A., Karp J.M., Gerecht-Nir S., *et al.* 'Embryonic stem cells as a cell source for tissue engineering' in Lanza R., Langer R., Vacanti J (eds.). *Principles of Tissue Engineering.* 4th edn. Waltham, Massachusetts: Academic Press; 2014. pp. 609–638.

[43] Bruder S.P., Jaiswal N., Ricalton N.S., Mosca J.D., Kraus K.H., Kadiyala, S. 'Mesenchymal stem cells in osteobiology and applied bone regeneration'. *Clinical Orthopaedics and Related Research.* 1998:**355**:247–256.

[44] Declercq H., Cornelissen, M. 'Stem cell technology for in vitro bone tissue engineering' in Dubruel P., Vlierberghe S.V. (eds.). *Biomaterials for Bone Regeneration.* Cambridge: Woodhead Publishing; 2014. pp. 406–426.

[45] Risbud M.V., Sittinger, M. 'Tissue engineering: advances in in vitro cartilage generation'. *Trends in Biotechnology.* 2002:**20**(8):351–356.

[46] Sundelacruz S., Kaplan, D.L. 'Stem cell- and scaffold-based tissue engineering approaches to osteochondral regenerative medicine'. *Seminars in Cell & Developmental Biology.* 2009;**20**(6):646–655.

[47] Dominici M., Le Blanc K., Mueller I., *et al.* 'Minimal criteria for defining multipotent mesenchymal stromal cells. The International Society for Cellular Therapy position statement.' *Cytotherapy.* 2006;**8**(4):315–317.

[48] Vater C., Kasten P., Stiehler, M. 'Culture media for the differentiation of mesenchymal stromal cells'. *Acta Biomaterialia.* 2001;**7**:463–477.

[49] Bhattacharjee M., Coburn J., Centola M., *et al.* 'Tissue engineering strategies to study cartilage development, degeneration and regeneration.' *Advanced Drug Delivery Reviews.* 2014; Corrected Proof, doi:10.1016/j.addr.2014.08.010

[50] Li J., Wong W.H., Chan S., *et al.* 'Factors affecting mesenchymal stromal cells yield from bone marrow aspiration'. *Chinese Journal of Cancer Research.* 2011;**23**(1):43–48.

[51] Li L., Zhou G., Wang Y., Yang G., Ding S., Zhou, S. 'Controlled dual delivery of BMP-2 and dexamethasone by nanoparticle-embedded electrospun nanofibers for the efficient repair of critical-sized rat calvarial defect'. *Biomaterials.* 2015;**37**:218–229.

[52] Liao H.T., Chen Y.Y., Lai Y.T., Hsieh M.F., Jiang, C.P. 'The osteogenesis of bone marrow stem cells on mPEG-PCL-mPEG/hydroxyapatite composite scaffold via solid freeform fabrication'. *BioMed Research International.* 2014;**2014**:321549.

[53] Pekozer G.G., Kose G.T., Hasirci, V. 'Influence of co-culture on osteogenesis and angiogenesis of bone marrow mesenchymal stem cells and aortic endothelial cells'. *Microvascular Research.* 2016;**108**:1–9.

[54] Huang H., Zhang X., Hu X., *et al.* 'A functional biphasic biomaterial homing mesenchymal stem cells for in vivo cartilage regeneration'. *Biomaterials.* 2014;**35**(36):9608–9619.

[55] Gong Y.Y., Xue J.X., Zhang W.J., Zhou G.D., Liu W., Cao, Y. 'A sandwich model for engineering cartilage with acellular cartilage sheets and chondrocytes'. *Biomaterials.* 2011;**32**(9):2265–2273.

[56] Zuk P.A., Zhu M., Mizuno H., *et al.* 'Multilineage cells from human adipose tissue: implications for cell-based therapies'. *Tissue Engineering.* 2001;**7**(2): 211–228.

[57] Cherubino M., Marra, K.G. 'Adipose-derived stem cells for soft tissue reconstruction'. *Regenerative Medicine.* 2009;**4**(1):109–117.

[58] Dragoo J.L., Samimi B., Zhu M., *et al.* 'Tissue-engineered cartilage and bone using stem cells from human infrapatellar fat pads'. *Journal of Bone and Joint Surgery (British Volume).* 2003;**85**(5):740–747.

[59] Estes B.T., Diekman B.O., Gimble J.M., Guilak, F. 'Isolation of adipose-derived stem cells and their induction to a chondrogenic phenotype'. *Nature Protocols.* 2010;**5**(7):1294–1311.

[60] Heng B.C., Cao T., Lee, E.H. 'Directing stem cell differentiation into the chondrogenic lineage in vitro'. *Stem Cells.* 2004;**22**(7):1152–1167.

[61] Mizuno H., Zuk P.A., Zhu M., Lorenz H.P., Benhaim P., Hedrick M.H. 'Myogenic differentiation by human processed lipoaspirate cells'. *Plastic and Reconstructive Surgery.* 2002;**109**(1):199–209.

[62] Rangappa S., Fen C., Lee E.H., Bongso A., Sim, E.K. 'Transformation of adult mesenchymal stem cells isolated from the fatty tissue into cardiomyocytes'. *Annals of Thoracic Surgery.* 2003;**75**(3):775–779.

[63] Rehman J., Traktuev D., Li J., *et al.* 'Secretion of angiogenic and anti-apoptotic factors by human adipose stromal cells'. *Circulation.* 2004;**109** (10):1292–1298.

[64] Obokata H., Vacanti C.A. 'Stem cells in tissue engineering' in Lanza R., Langer R., Vacanti J. (eds.). *Principles of Tissue Engineering.* 4th edn. Waltham, Massachusetts: Academic Press; 2014. pp. 595–608.

[65] Al-Nbaheen M., Vishnubalaji R., Ali D., *et al.* 'Human stromal (mesenchymal) stem cells from bone marrow, adipose tissue and skin exhibit differences in molecular phenotype and differentiation potential'. *Stem Cell Reviews.* 2013;**9**(1):32–43.

[66] Gimble J.M., Katz A.J., Bunnell, B.A. 'Adipose-derived stem cells for regenerative medicine'. *Circulation Research.* 2007;**100**(9):1249–1260.

[67] Gronthos S., Franklin D.M., Leddy H.A., Robey P.G., Storms R.W., Gimble, J.M. 'Surface protein characterization of human adipose tissue-derived stromal cells'. *Journal of Cellular Physiology.* 2002;**189**(1):54–63.

[68] Schubert T., Xhema D., Vériter S., *et al.* 'The enhanced performance of bone allografts using osteogenic-differentiated adipose-derived mesenchymal stem cells'. *Biomaterials.* 2011;**32**(34):8880–8891.

[69] De Ugarte D.A., Morizono K., Elbarbary A., *et al.* 'Comparison of multi-lineage cells from human adipose tissue and bone marrow'. *Cells Tissues Organs.* 2003;**174**(3):101–109.

[70] Im G.I., Shin Y.W., Lee, K.B. 'Do adipose tissue-derived mesenchymal stem cells have the same osteogenic and chondrogenic potential as bone marrow-derived cells?'. *Osteoarthritis and Cartilage.* 2005;**13**(10):845–853.

[71] Xie X., Wang Y., Zhao C., *et al.* 'Comparative evaluation of MSCs from bone marrow and adipose tissue seeded in PRP-derived scaffold for cartilage regeneration'. *Biomaterials.* 2012;**33**(29):7008–7018.

[72] Lotfy A., Salama M., Zahran F., Jones E., Badawy A., Sobh M. 'Characterization of mesenchymal stem cells derived from rat bone marrow and adipose tissue: a comparative study'. *International Journal of Stem Cells.* 2014;**7**(2):135–142.

[73] Semenov O.V., Breymann, C. 'Mesenchymal stem cells derived from Wharton's jelly and their potential for cardio-vascular tissue engineering'. *Open Tissue Engineering and Regenerative Medicine Journal.* 2011;**4**:64–71.

[74] Taghizadeh R.R., Cetrulo K.J., Cetrulo, C.L. 'Wharton's Jelly stem cells: future clinical applications'. *Placenta.* 2011;**32**(Suppl 4):S311–S315.

[75] Lee O.K., Kuo T.K., Chen W.M., Lee K.D., Hsieh S.L., Chen, T.H. 'Isolation of multipotent mesenchymal stem cells from umbilical cord blood'. *Blood.* 2004;**103**(5):1669–1675.

[76] Zhang X., Hirai M., Cantero S., *et al.* 'Isolation and characterization of mesenchymal stem cells from human umbilical cord blood: reevaluation of critical factors for successful isolation and high ability to proliferate and differentiate to chondrocytes as compared to mesenchymal stem cells from bone marrow and adipose tissue'. *Journal of Cellular Biochemistry.* 2011; **112**(4):1206–1218.

[77] Sibov T.T., Severino P., Marti L.C., *et al.* 'Mesenchymal stem cells from umbilical cord blood: parameters for isolation, characterization and adipogenic differentiation'. *Cytotechnology.* 2012;**64**(5):511–521.

[78] Manca M.F., Zwart I., Beo J., *et al.* 'Characterization of mesenchymal stromal cells derived from full-term umbilical cord blood'. *Cytotherapy.* 2008;**10**(1):54–68.

[79] Shetty P., Cooper K., Viswanathan C. 'Comparison of proliferative and multilineage differentiation potentials of cord matrix, cord blood, and bone marrow mesenchymal stem cells'. *Asian Journal of Transfusion Science.* 2010;**4**(1):14–24.

[80] Liu G., Li Y., Sun J., *et al.* 'In vitro and in vivo evaluation of osteogenesis of human umbilical cord blood-derived mesenchymal stem cells on partially demineralized bone matrix'. *Tissue Engineering Part A.* 2010;**16**(3):971–982.

[81] Yoon J.H., Roh E.Y., Shin S., *et al.* 'Comparison of explant-derived and enzymatic digestion-derived MSCs and the growth factors from Wharton's jelly'. *BioMed Research International.* 2013;**2013**:428726.

[82] Divya M.S., Roshin G.E., Divya T.S., *et al.* 'Umbilical cord blood-derived mesenchymal stem cells consist of a unique population of progenitors co-expressing mesenchymal stem cell and neuronal markers capable of

instantaneous neuronal differentiation'. *Stem Cell Research & Therapy.* 2012;**3**:57.

[83] Nekanti U., Mohanty L., Venugopal P., Balasubramanian S., Totey S., Ta M. 'Optimization and scale-up of Wharton's jelly-derived mesenchymal stem cells for clinical applications'. *Stem Cell Research.* 2010;**5**:244–254.

[84] Wang L., Ott L., Seshareddy K., Weiss M.L., Detamore M.S. 'Musculoskeletal tissue engineering with human umbilical cord mesenchymal stromal cells'. *Regenerative Medicine.* 2011;**6**(1):95–109.

[85] Schneider R.K., Puellen A., Kramann R., *et al.* 'The osteogenic differentiation of adult bone marrow and perinatal umbilical mesenchymal stem cells and matrix remodelling in three-dimensional collagen scaffolds'. *Biomaterials.* 2010;**31**(3):467–480.

[86] Zhao L., Weir M.D., Xu H.H. 'An injectable calcium phosphate-alginate hydrogel-umbilical cord mesenchymal stem cell paste for bone tissue engineering'. *Biomaterials.* 2010;**31**(25):6502–6510.

[87] Liu S., Hou K.D., Yuan M., *et al.* 'Characteristics of mesenchymal stem cells derived from Wharton's jelly of human umbilical cord and for fabrication of non-scaffold tissue-engineered cartilage'. *Journal of Bioscience and Bioengineering*. 2014;**117**(2):229–235.

[88] Chen X., Zhang F., He X., *et al.* 'Chondrogenic differentiation of umbilical cord-derived mesenchymal stem cells in type I collagen-hydrogel for cartilage engineering'. *Injury.* 2013;**44**(4):540–549.

[89] Ramazanoglu M., Schlegel K.A., Kose, G.T. 'Stem cells potential use of dental stem cells for craniofacial tissue regeneration' in Turksen K. (ed.). *Stem Cells: Current Challenges and New Directions. Stem Cell Biology and Regenerative Medicine Series.* New York City: Humana Press; 2013, pp. 105–124.

[90] Lei M., Li K., Li B., Gao L., Chen F., Jin, Y., 'Mesenchymal stem cell characteristics of dental pulp and periodontal ligament stem cells after in vivo transplantation'. *Biomaterials.* 2014;**35**(24):6332–6343.

[91] Gronthos S., Mankani M., Brahim J., Robey P.G., Shi, S. 'Postnatal human dental pulp stem cells (DPSCs) in vitro and in vivo'. *Proceedings of National Academy of Sciences of the United States of America.* 2000;**97**:13625–13630.

[92] Sloan A.J., Waddington, R.J. 'Dental pulp stem cells: what, where, how?'. *International Journal of Paediatric Dentistry.* 2009;**19**(1):61–70.

[93] Arthur A., Shi S., Gronthos, S. 'Dental pulp stem cells' in Vishwakarma A., Sharpe P., Songtao S., Ramalingam M. (eds.). *Stem Cell Biology and Tissue Engineering in Dental Sciences*. Amsterdam: Elsevier; 2015, pp. 279–289.

[94] Yalvac M.E., Ramazanoglu M., Rizvanov A.A., *et al.* 'Isolation and characterization of stem cells derived from human third molar tooth germs of young adults: implications in neo-vascularization, osteo-, adipo- and neurogenesis'. *Pharmacogenomics Journal.* 2010;**10**(2):105–113.

[95] Yalvac M.E., Ramazanoglu M., Tekguc M., *et al.* 'Human tooth germ stem cells preserve neuro-protective effects after long-term cryo-preservation'. *Current Neurovascular Research.* 2010;**7**(1):49–58.

[96] d'Aquino R., De Rosa A., Laino G., *et al.* 'Human dental pulp stem cells: from biology to clinical applications'. *Journal of Experimental Zoology Part B: Molecular and Developmental Evolution.* 2009;**312B**(5):408–415.

[97] Park B.W., Kang E.J., Byun J.H., *et al.* 'In vitro and in vivo osteogenesis of human mesenchymal stem cells derived from skin, bone marrow and dental follicle tissues'. *Differentiation.* 2012;**83**(5):249–259.

[98] Abay N., Pekozer G.G., Ramazanoglu M., Kose G.T. 'Bone formation from porcine dental germ stem cells on surface modified polybutylene cuccinate scaffolds'. *Stem Cells International* 2016;**2016**:879219. doi: http://dx.doi.org/10.1155/2016/8792191.

[99] Koyuncu A.C.C., Pekozer G.G., Ramazanoglu M., Kose G.T., Hasirci, V. 'Cartilage tissue engineering on macroporous scaffolds using human tooth germ stem cells'. *Journal of Tissue Engineering and Regenerative Medicine.* 2015; Corrected Proof, doi:10.1002/term.1975

[100] Sakaguchi Y., Sekiya I., Yagishita K., Muneta, T. 'Comparison of human stem cells derived from various mesenchymal tissues: superiority of synovium as a cell source'. *Arthritis & Rheumatology.* 2005;**52**:2521–2529.

[101] De Bari C., Dell'Accio F., Tylzanowski P., Luyten, F.P. 'Multipotent mesenchymal stem cells from adult human synovial membrane'. *Arthritis & Rheumatology.* 2001;**44**:1928–1942.

[102] Jo, C.H., Ahn, H.J., Kim, H.J., Seong, S.C., Lee, M.C. Surface characterization and chondrogenic differentiation of mesenchymal stromal cells derived from synovium. *Cytotherapy.* 2007;**9**(4):316–327.

[103] Koga H., Muneta T., Ju Y.J., *et al.* 'Synovial stem cells are regionally specified according to local microenvironments after implantation for cartilage regeneration'. *Stem Cells.* 2007;**25**(3):689–696.

[104] Pei M., He F., Vunjak-Novakovic, G. 'Synovium-derived stem cell-based chondrogenesis'. *Differentiation.* 2008;**76**:1044–1056.

[105] Candela M.E., Yasuhara R., Iwamoto M., Enomoto-Iwamoto, M. 'Resident mesenchymal progenitors of articular cartilage'. *Matrix Biology.* 2014;**39**:44–49.

[106] Ando W., Kutcher J.J., Krawetz R., *et al.* 'Clonal analysis of synovial fluid stem cells to characterize and identify stable mesenchymal stromal cell/mesenchymal progenitor cell phenotypes in a porcine model: a cell source with enhanced commitment to the chondrogenic lineage'. *Cytotherapy.* 2014;**16**(6):776–788.

Chapter 5

Nanobiomaterials and 3D bioprinting for osteochondral regeneration

Nathan J. Castro[1] *and Lijie Grace Zhang*[1,2,3]

Abstract

Tissue engineering (TE) has made stark advancements through a greater understanding of the effects of various biomaterials on cell behavior and mediated tissue formation. Human tissue can be readily classified as a nanocomposite owing to the presence of nanoscale organic and/or inorganic constituents found within the extracellular matrix. Research has begun and continues to leverage this understanding in an effort to fabricate more biomimetic nanobiomaterials, which share similar composition and morphology to native tissue components. In addition, novel three-dimension (3D) scaffold manufacturing techniques are being developed to extend the use of these novel materials toward the development of clinically relevant scaffolds exhibiting not only similar composition, but spatial distribution of tissue-specific extracellular components. Interfacial and complex tissue regeneration applications including those related to osteochondral (bone–cartilage interface) regeneration stand to benefit greatly from these advancements. Through the course of this chapter we will explore the most recent developments in nanobiomaterials for osteochondral tissue as well as introduce and discuss 3D bioprinting platforms and techniques for the fabrication of biomimetic scaffolds.

5.1 Introduction

Clinically, osteoarthritis (OA) is the gradual degeneration of cartilage within articulating joints which manifests symptomatically as joint pain, impinged movement, and overall dysfunction [1]. In 2005, 47.8 million Americans were

[1]Department of Mechanical and Aerospace Engineering, The George Washington University, Washington DC, USA
[2]Department of Biomedical Engineering, The George Washington University, Washington DC, USA
[3]Department of Medicine, The George Washington University, Washington DC, USA

diagnosed with OA with updated projections growing to 96 million by 2050 [2]. Of this increasing population, 33% will be workforce contributors aged between 45 and 64 years [3] where adverse effects, including decreased patient quality of life, direct and indirect global economic impact through decreased productivity, reduced contribution to the work force, and increased disability compensation. Another contributing factor currently included in the 2050 projections is the increasing national obesity epidemic. Increased weight leads to increased joint loading thus further exacerbating existing joint injury and degeneration.

Joint degeneration is induced by chronic or acute damage to cartilaginous tissue at the joint surface. Available surgical methods are predicated on the severity and size of the defect [4] as shown in Figure 5.1 ranging from focal sized (<2 cm^2 area)

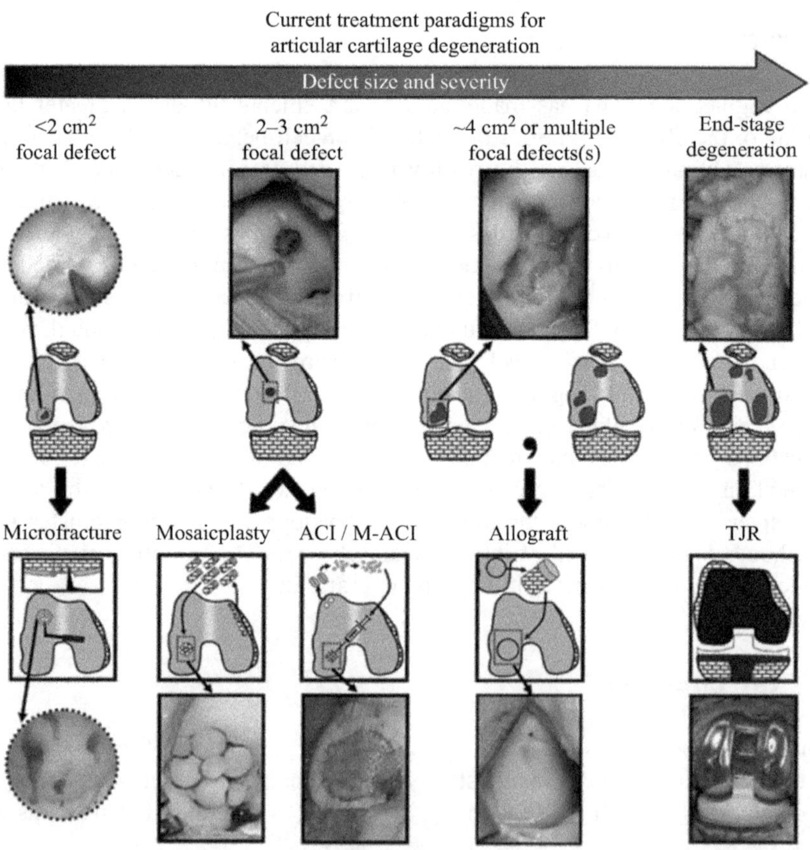

Figure 5.1 Current clinical treatment options for articular joint damage. Adopted from Williams, et al. [4]. "Reprinted from Journal of Biomechanics, 43(1), Williams, et al., Shape, loading, and motion in the bioengineering design, fabrication, and testing of personalized synovial joints, 156–165, (2010), with permission from Elsevier"

defects where no efficient treatment options exist to complete joint degeneration requiring highly invasive complete joint arthroplasty. Less invasive methods to modulate disease progression exist, but are ineffective in recapitulating the native extracellular matrix niche. This mismatch leads to increased defect severity and, in most severe cases, to total joint replacement. Currently, large defect tissue repair is conducted through the use of autografts or allografts when available and feasible, but have inherent limitations with regards to graft thickness inconsistencies [5], donor tissue harvest site [6], and graft incongruency [7]. If the damage is too severe and the supply of healthy tissue is not available for a sufficient auto or allograft, total joint replacement is necessary to mobilize the patient.

Tissue engineering (TE) holds great promise for developing methods of repairing tissue damage thus preventing the need for more invasive orthopedic procedures. TE applies the principles of engineering and life sciences for the development of biological substitutes to restore, maintain, or improve tissue function [8]. For orthopedic tissue regeneration, autologous stem cells can be harvested from the patient, expanded in vitro and subsequently cultivated upon a 3D scaffold. The scaffold can effectively be transplanted into the patient after cell differentiation for enhanced tissue integration [9,10]. Although traditional scaffold fabrication techniques such as particle leaching [11–13] and gas foaming [14–17] have been used to manufacture porous scaffolding for tissue regeneration, limitations exist with regards to insufficient pore interconnectivity leading to poor tissue integration. 3D bioprinting technologies have garnered greater attention for the fabrication of highly ordered tissue engineered scaffolds. The scope of this book chapter will serve to present novel nanobiomaterials used in conjunction with different 3D bioprinting modalities for osteochondral tissue regeneration. In addition, discussions on the current challenges inhibiting efficient and long-term efficacy of scaffolds for articulating joint lesion treatment are included.

5.2 Osteochondral tissue regeneration

Interfacial tissue engineering (ITE) is an approach that addresses the complex bi- or multiphasic nature of tissue defects where two or more disparate tissues are proximally located. One of the main caveats of ITE is the presence of shared biochemical and physical characteristics of the tissues being connected (bone–cartilage, tendon/ligament-bone, muscle-tendon) as well as retention of regions of distinct composition and biological function [18]. Current osteochondral ITE strategies commonly employ a combination of cells, a biocompatible/biodegradable 3D support structure, and chemical or biological factor(s) to promote cell adhesion and proliferation during de novo tissue formation. The basic premise relies on the introduction or elicitation of cells to the defect site by means of an appropriate scaffold, resulting in directed spatial and temporal tissue remodeling. Several key scaffold parameters shown to affect the success of osteochondral regeneration are: (1) biocompatibility and bioactivity to maximize tissue growth, (2) appropriate mechanical properties to enable the patient's rapid return to mobility while protecting developing tissue, (3) controllable biodegradability where degradation rates closely match the rate of new tissue formation, and (4) interconnected porous structure to improve nutrient diffusion and waste transport [19,20].

In an effort to address the complex nature of osteochondral tissue, advances in scaffold design have also included biochemical cues that mimic those found within articular cartilage and subchondral bone exhibiting improved cell adhesion, proliferation, directed differentiation, and phenotypic expression [21,22]. In addition to scaffold design, biomaterial limitations of traditional TE approaches with respect to spatial and temporal control of tissue formation, the mechanical properties of single tissue-specific constructs have also proven challenging. Therefore, applying nanotechnology (i.e. nanobiomaterials and advanced 3D fabrication) to manufacture novel biomimetic osteochondral constructs linking two distinct tissues within certain biological and mechanical constraints merits considerable focused attention. In the following sections, we will explore the role of cutting-edge nanotechnology and 3D bioprinting for osteochondral regeneration.

5.2.1 Nanobiomaterials for osteochondral regeneration

Recent advancement in the area of nanobiomaterials has greatly impacted and influenced the design of biomimetic and bioactive osteochondral tissue scaffolds with improved biocompatibility and functional properties [23,24]. One of the underlying advantages of nanomaterials lies in the materials' ability to mimic the native tissue ECM environment, as well as favorably modulate cell function. Several proposed mechanisms for the improved nanomaterial–cell interaction have been correlated to structural (surface topography and surface area) and physicochemical (surface chemistry, energy, and wettability) cues have been shown to regulate cellular behavior, but the basic principles have yet to be identified [25]. In the following, we will discuss several promising nanobiomaterials for osteochondral regeneration.

5.2.1.1 Self-assembling nanobiomaterials

Since natural tissues are constructed via a bottom-up self-assembly method, self-assembling supramolecular nanobiomaterials hold great potential to facilitate the formation of complex tissue environments [26]. Advances in nanotechnology have greatly increased the design of these sophisticated nanobiomaterials. For instance, Hartgerink *et al.* reported the synthesis and regenerative potential of a peptide-amphiphile (PA) conjugated with the cell-adherent RGD (Arg-Gly-Asp) peptide which can self-assemble into supramolecular nanofibers and align nanocrystalline hydroxyapatite nHA; the primary inorganic of bone) along their long axis similar to the pattern of mineralized/calcified ECM [27]. Hosseinkhani *et al.* [28] showed significantly enhanced osteogenic differentiation of stem cells in a 3D PA scaffold when compared to 2D static tissue culture while Shah *et al.* [29,30] recently designed PA nanofibers which display a high density of transforming growth factor β1 (TGF-β1) binding sites for improved cartilage regeneration. Although the work previously described focused on the use on single PA systems in combination with traditional scaffold fabrication techniques, through the use of

3D bioprinting the fabrication of complex structures exhibiting various peptides can be realized.

In addition to the promising results obtained with PA nanofibers, another promising direction is the development of new self-assembling nanotubes which mimic cellular (such as DNA) and ECM components while displaying trophic signals in a spatiotemporally controlled manner. We have developed these types of highly innovative self-assembling rosette nanotubes (RNTs, Figure 5.2) with controllable surface chemistry. Specifically, RNTs are a new class of biologically inspired supramolecular nanobiomaterials obtained through the self-assembly of low-molecular-weight DNA base pair motifs (Guanine^Cytosine, G^C) in an aqueous solution. In our previous pioneering work, we designed multiple DNA-based RNTs with varying peptide and amino acid side chains via a bottom-up self-assembly method, which has shown great potential for cartilage and bone regeneration [31–37]. Figure 5.2 illustrates the morphology of two types of twin DNA-based RNTs. For the twin DNA-based RNTs, two covalently linked G^C bases can self-assemble into a six member twin rosette maintained by 36 hydrogen bonds. The RNTs have a very stable nanotubular

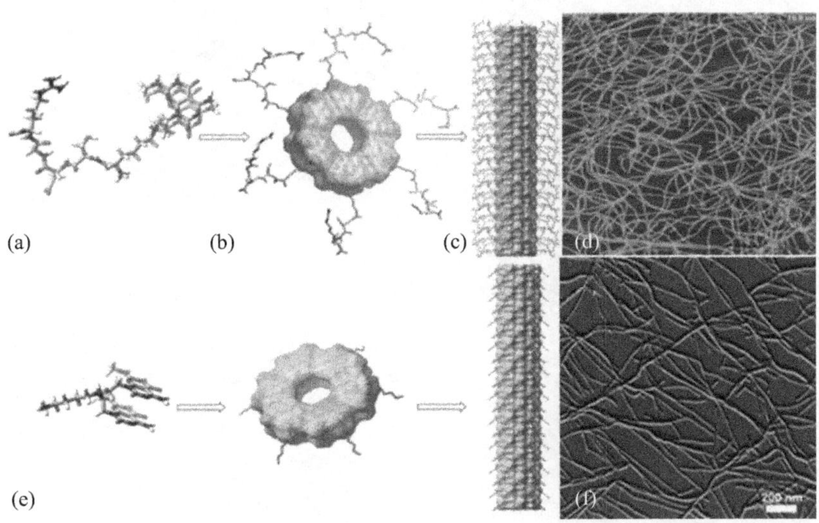

Figure 5.2 Schematic illustration of the self-assembly process of RNTs. (a) Twin G^C motifs with a RGDSK peptide; (b) rosette-like supermacrocycle assembled from six motifs; and (c) rosettes stack up into stable helical nanotubes with a 11 Å hollow core, 3–4 nm in diameter and up to several μm long. Atomic force microscopy image of (d) twin DNA-based RNTs with a RGDSK peptide. (e, f) Twin DNA-based RNTs with an aminobutane linker (TBL). "Reprinted from Journal of Biomedical Materials Research, 95A(2), Zhang, et al., *Tuning cell adhesion on titanium with osteogenic rosette nanotubes, 550–563, (2010), with permission from John Wiley and Sons"*

structure with a hydrophobic core and hydrophilic outer surface via electrostatic forces, base stacking interactions, and hydrophobic effects. The outer diameter and the length of all nanotubes are ~3–4 nm and several hundred nanometers, respectively. Another intriguing feature of RNTs is their flexibility in design, which makes their length, diameter, and surface chemistry tunable. In our lab's recent work, we explored human bone marrow mesenchymal stem cell (MSC) adhesion, proliferation, and 4-week chondrogenic differentiation in twin-based RNTs conjugated with a cell adherent RGDSK (TB-RNT-RGDSK) peptide adsorbed upon poly-l-lactic acid (PLLA) scaffolds [37]. Our results demonstrated that these biomimetic twin-based nanoutbes can significantly enhance MSC adhesion, proliferation, and chondrogenic differentiation [such as glycosaminoglycan (GAG), collagen, and protein synthesis] when compared to controls. Histological examination (Figure 5.3) confirmed that

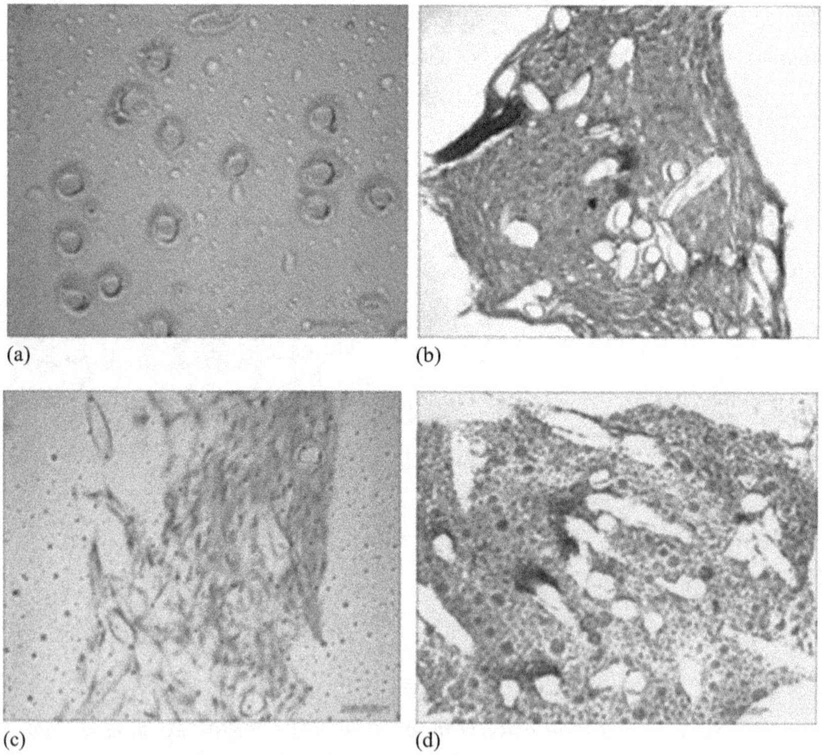

Figure 5.3 Hemotoxylin and eosin staining of (a, c) PLLA controls; and (b, d) TB-RNT-RGDSK PLLA scaffolds for MSC chondrogenic differentiation at weeks 1 and 2. "Reprinted from Biomedical Materials, 8(6), Childs, et al., *Novel biologically-inspired rosette nanotube PLLA scaffolds for improving human mesenchymal stem cell chondrogenic differentiation, 1–12, (2013), with permission from IOP Publishing"*

TB-RNT-RGDSK can greatly improve tissue formation when compared to controls without nanotubes.

Self-assembling RNTs have several unique features which renders them ideal for osteochondral tissue regeneration, including: (1) nanostructure and nanotubular architecture creating an environment similar to the dimensions of natural osteochondral ECM; (2) high density of functionalizable surface groups allowing for the incorporation of bioactive peptides with tunable spatial distribution and density via coassembly; and (3) collagen-like soft nature for improved protein adsorption and cell function. Theoretically, any cell-favorable short peptide can be conjugated onto the G^C motifs to modulate surface chemistry, rendering RNTs as a biomimetic nanotemplate for tissue regeneration. Based on the current available studies, supramolecular biomimetic self-assembling nanobiomaterials have great potential in regenerating osteochondral tissue.

5.2.1.2 Carbon-based nanobiomaterials

Carbon-based nanobiomaterials such as carbon nanotubes (CNTs) (Figure 5.4) have also garnered great attention for osteochondral tissue regeneration owing to their excellent mechanical properties, outstanding electrical, and surface properties. Although the use of CNTs in tissue regeneration is still in its infancy, they are considered an exciting alternative for osteochondral tissue growth. Morphologically, CNTs exhibit dimensions similar to those of native ECM components (i.e., collagen nanofibers), which cells are accustomed to interacting with. This property renders them an excellent candidate to induce positive cellular responses when combined with traditional biomaterials [38]. In addition, their superior mechanical properties are efficient for use as a nanoscale reinforcing material for high-load-bearing applications [39–43]. The unique chemical properties they possess permit them to be functionalized with varying chemical groups for the promotion of cell growth [44].

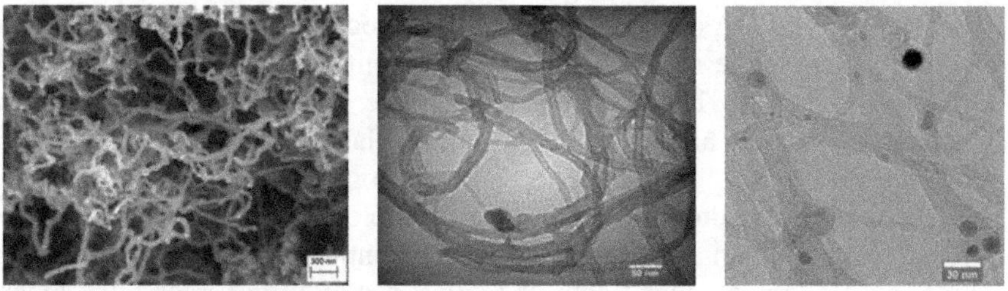

(a) (b) (c)

Figure 5.4 Scanning electron microscopy (SEM) image of (a) hydrogen purified MWCNTs; transmission electron microscopy (TEM) images of (b) untreated MWCNTs and (c) magnetically treated SWCNTs. Reprinted from Biomedical Materials, 8(6), Childs, et al., *Novel biologically-inspired rosette nanotube PLLA scaffolds for improving human mesenchymal stem cell chondrogenic differentiation, 1–12, (2013), with permission from IOP Publishing"*

To date, CNTs have primarily been employed in bone tissue engineering where rigid and robust scaffolds must be able to withstand considerable mechanical load. Notwithstanding, CNTs have been shown to be beneficial in cartilage regeneration applications as well [45]. For instance, recently Holmes *et al.* used H_2-purified multiwalled carbon nanotubes (MWCNTs, Figure 5.4(a)) coated with poly-L-lysine within a biocompatible PLLA polymer to electrospin CNT-doped microfibers for cartilage regeneration [45]. It was found that scaffolds made from these fibers had mechanical properties quite similar to native articular cartilage. When compared to a pure PLLA control, scaffolds with MWCNTs showed no adverse effects on MSC proliferation and, more importantly, displayed enhanced MSC chondrogenesis. The purified H_2-treated nanotubes also showed enhanced differentiation over scaffolds containing untreated tubes (Figure 5.4(b)).

Although cytotoxicity concerns have been raised, carbon-based nanobiomaterials have exhibited great potential for a variety of tissue regenerative applications. With the rapid development of highly reproducible synthesis methods, greater controlled production, and experimental analysis, highly purified and novel functionalized CNTs/fibers with various bioactive molecules have been prepared and have shown better cell or tissue responses. Moving forward, the underlying mechanism of interaction between CNTs/fibers or composites and cells particularly at the molecular level will be thoroughly clarified in order to design ever more cytocompatible nanotubes/fibers in an effort to advance the clinical application of carbon-based nanobiomaterial.

5.2.1.3 Therapeutic encapsulated nanobiomaterials

In addition to the increased attention and implementation of the aforementioned nanobiomaterials, extended delivery of tissue-specific growth factors through the manufacture of growth factor encapsulated nanospheres has also shown great promise for improved osteochondral tissue growth. Nanoparticle fabrication for therapeutic encapsulation has long been an area of research with widespread applications [46–50]. As in the case with osteochondral tissue regeneration, availability of candidate nanomaterials for efficient growth factor encapsulation is still sparse. Our lab has begun exploring the use of electrosprayed core–shell nanospheres for highly efficient encapsulation of growth factors and therapeutics.

Electrospraying, as depicted in Figure 5.5, employs the same basic principle as traditional electrospinning wherein a voltage potential is generated between one electrode (needle) and another electrode (collector plate) to overcome the inherent surface tension of the polymer solution. Unlike electrospinning where the polymer solution is a fairly viscous liquid, polymer solutions employed in electrospraying are quite dilute allowing for the formation of a droplet stream composed of nanospheres. This technique allows for better regulation of particle size with similar or greater encapsulation when compared to traditional emulsion techniques. Although traditional emulsion-based micro/nanosphere fabrication techniques have exhibited positive results, limitations regarding initial burst and uncontrolled release have inhibited their full clinical potential partially due to the

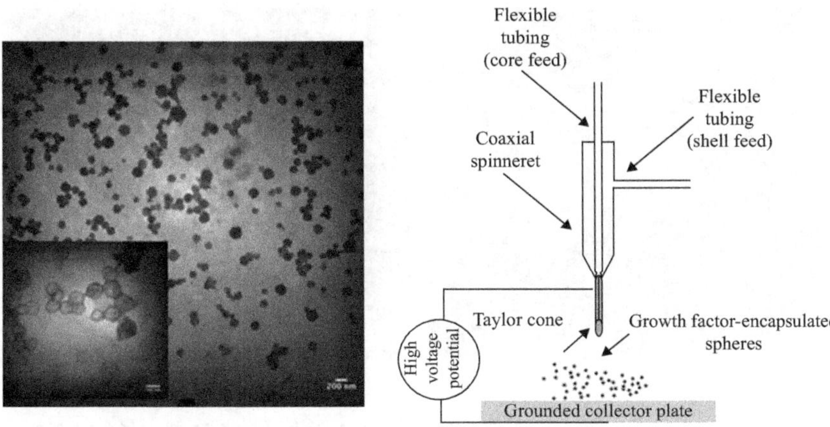

(a) (b)

Figure 5.5 (a) TEM image of bone morphogenetic protein-2 encapsulated polydioxanone nanospheres. (b) Graphical representation of a coaxial electrospray technique for the manufacture of growth factor-encapsulated nanospheres. "Reprinted from AIChE Journal, 60(2), Castro, et al., Biomimetic biphasic 3-D nanocomposite scaffold for osteochondral regeneration, 432–442, (2014), with permission from John Wiley and Sons"

disparity in particle size. Coaxial electrospraying can produce nanospheres with good size distribution (Figure 5.5(a)). Our recent study has revealed that electrosprayed core–shell polydioxanone nanospheres can steadily release encapsulated therapeutics over 18 days without initial burst release [51]. In addition, the flexibility of this technique allows for a wide range of polymer materials to be used as controlled delivery vesicles. Another commonly used nanoparticle biomaterial for osteochondral regeneration is nHA. It is the key bioactive and osteoconductive chemical component in subchondral bone and calcified cartilage. Through a hydrothermal treatment method [31,35,52–55], nHA can be readily synthesized with excellent control of the crystallinity and surface morphology at the nanoscale (Figure 5.6(a)). It has been shown that nHA particles can directly nucleate and align along the long axis of RNTs (Figure 5.6(b)) [35] similar to the self-assembled pattern of nHA and collagen in subchondral bone. Owing to the disparate mechanical properties of native osteochondral tissue, osteogenic nHA particles have been used to reinforce various natural and synthetic "soft" materials such as collagen and hydrogels to create a stiffness gradient emanating from the dense, rigid subchondral region to the ceramic-free superficial articular cartilage zone [56].

In summary, several examples have been provided illustrating the effectiveness of nanobiomaterials for osteochondral tissue regeneration. Although nanobiomaterials have been widely investigated in many other regenerative applications

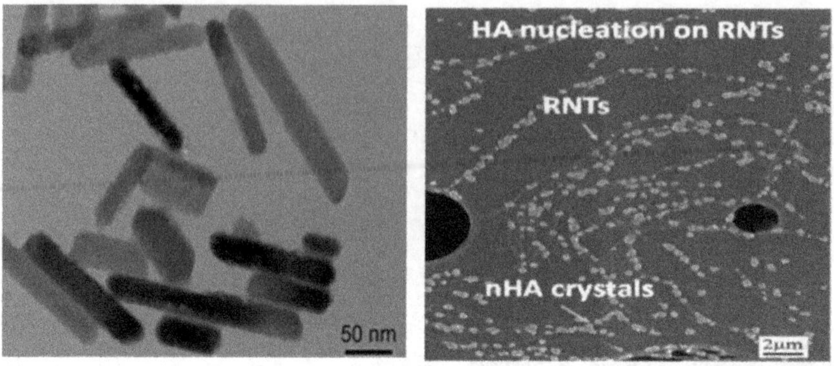

(a) (b)

Figure 5.6 (a) TEM image of rod-like nHAs. (b) SEM image of nHA nucleation on RNTs. Reprinted from the International Journal of Nanomedicine, 3(3), Zhang, et al., *Biomimetic helical rosette nanotubes and nanocrystalline hydroxyapatite coatings on titanium for improving orthopedic implants, 323–333, (2008), with permission from DOVE Medical Press*

to include bone [24], there are still limited studies for complex osteochondral tissue regeneration. Since cells directly interact with (and create) nanostructured osteochondral ECM, the biomimetic features and excellent physiochemical properties of nanomaterials play a key role in stimulating cell growth as well as guided tissue regeneration, thus holding great potential for clinical applications. The following section will further explore 3D spatiotemporal nanocomposite scaffold manufacturing techniques for osteochondral defect repair.

5.2.2 3D printing techniques

Rapid prototyping or 3D printing is an enabling technology whose main attribute is to reduce product development and manufacturing times, as well as cost, resulting in an increase in market competitiveness [57]. Though this definition focuses on manufacturing, the fundamental ideas presented hold true for TE applications of 3D printing. 3D printing can readily enable scientists and doctors to efficiently engineer personalized scaffolds for patient-specific treatment. In addition, patients will not have to wait for a viable donor, or surrender to total joint replacement at an early age due to disease progression, but will instead be treated with cell-laden or bioactive 3D scaffolds. Advances in biomaterials research is a critical component to the realization of functional and efficient scaffolding where processability and biocompatibility must be taken into consideration. Several technologies for 3D bioactive scaffold fabrication exist including: stereolithography (SL), bioplotting/inkjet bioprinting, and selective laser sintering as shown in Figure 5.7. In the following section, we will discuss each technology more in depth as they relate to TE scaffold fabrication.

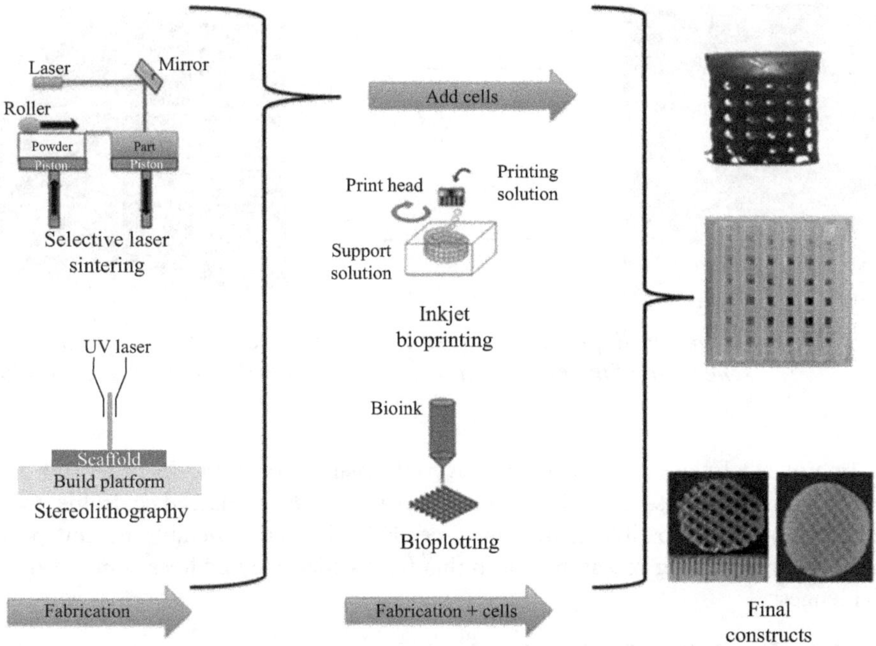

Figure 5.7 Schematics of 3D printing techniques discussed with resulting scaffolds

5.2.2.1 Stereolithography

As one of the oldest RP technologies, SL can be readily employed to manufacture stratified, multilayered scaffolds through UV photopolymerization of photocurable resins; examples of SL setup and printed scaffolds are shown in Figure 5.8. The geometry and architecture of the cured resin can be readily controlled by a focused laser beam [58–63] or digital masks projected from computer aided design interfaces [64–67]. This technology remains one of the most powerful and versatile, exhibiting the highest fabrication accuracy. It is largely a bottom-up process although research does exist on top-down methods for implementation fabrication.

Some challenges associated with stereolithography include available biomaterials, material toxicity complications due to incorporated photoinitiators, and inadequate cell adhesion. Though the breadth of materials is steadily increasing, the resulting mechanical properties of fabricated scaffolds are equally broad. Additionally, for implantation a compounding complication is the presence of photoinitiators which are added to the liquid resin in order to expedite crosslinking [68]. These materials must be carefully controlled and monitored to ensure that in vivo degradation is not marred with excessive build-up leading to cytotoxicity, inadequate tissue integration and resultant failure. Scaffolds fabricated by SL are largely based on poly(ethylene glycol) hydrogels which are inherently non-cell adherent. Often, micro/nanoparticles and growth factors must be incorporated

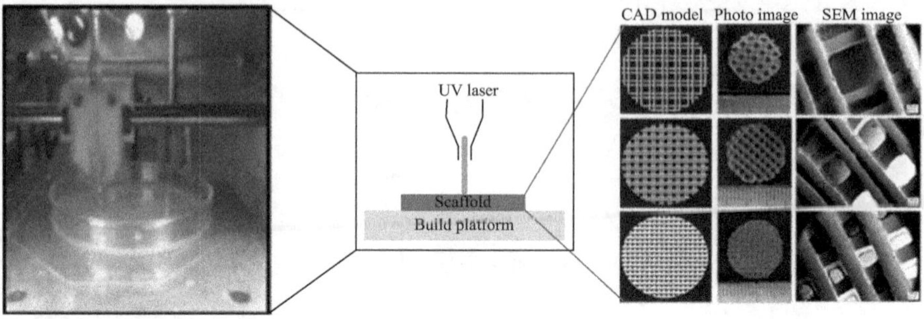

Figure 5.8 Images of SL printer and resulting scaffolds as well as a system schematic of the process in the center. Adopted from Castro et al. *[69]*

to improve cellular attachment and overall tissue formation. In order for this technology to progress toward clinically relevant tissue engineered scaffolds, more research on biocompatible materials is needed. Photocrosslinkable natural polymers are an emerging research area in this field which may address some of these challenges.

5.2.2.2 Bioplotting/inkjet bioprinting

Bioplotting is an additive manufacturing technique for the manufacture of 3D TE scaffolds where cell-laden and/or non-cell-laden materials are extruded drop-wise as a viscous fluid through a syringe into tubular or spherical shapes and cured using ultraviolet radiation, chemical crosslinking agents, or time dependent solidification [70]. Bioplotters allow for the use of one or multiple syringes leading to the formation of more compositionally complex constructs. This capability allows for the integration of cells and biomaterials to replicate physiologically relevant spatial orientation and, geometries, and complexities [71]. The main caveat of this technology is in the necessity for high viscosity printing materials providing adequate surface tension to self-support or be submerged in a liquid media of matching density to provide support while crosslinking [72].

The ability to directly impregnate biomaterials with cells while printing is certainly an advantage of these systems thus reducing material processing. However, bioplotted scaffolds commonly exhibit a very smooth surface which is attributed to the extrusion and curing process. Surface morphology has extensively been linked to cell adhesion and proliferation [73,74]. As such, additional scaffold surface modification is often necessary to create surface roughness to improve cell adhesion [75]. Furthermore, the materials currently available for syringe-based extrusion often do not possess mechanical properties suitable for implantation [70]. Although bioplotters show promise for the rapid fabrication of cell impregnated biomaterials, more focused research is necessary for the development of extrudable, biocompatible materials possessing favorable mechanical and physical properties.

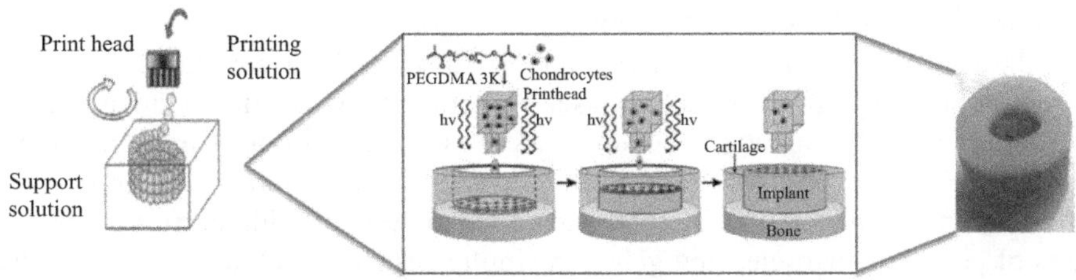

Figure 5.9 Schematic of inkjet bioprinting system, scaffold production schematic, and image of final scaffold created using inkjet bioprinting. Adopted from Wust, S., et al. *[77]. Reprinted from the Journal of Functional Biomaterials, 2(3), Wüst,* et al., *Controlled positioning of cells in biomaterials—approaches towards 3D tissue printing, 119–154, (2008), with permission from MDPI Open Access Publishing*

Inkjet bioprinting is a method of depositing bioinks containing cells and/or biological materials via commercial inkjet printers. Much like the preceding technology, inkjet bioprinting is a bottom-up scaffold fabrication technique in which cells and hydrogels can be directly and precisely printed into 3D tissue engineered scaffolds as, illustrated in Figure 5.9. Additionally, inkjet printers containing multiple print cartridges have the capability of allowing simultaneous printing of multiple types of cells [76]. Owing to the relative accessibility and availability of commercial printers, very precise non-contact deposition of cells can be directly applied upon pre-designed scaffolds.

The biggest challenge with inkjet bioprinting devices is cell settling and aggregation at the printer nozzle within the printer cartridge which can lead to obstruction and non-uniform distribution during printing. Some solutions have been evaluated in an attempt to resolve this problem, such as agitation of medium contained in the cartridge which directly impacts cell viability [78]. More research is needed to find a stand-alone material capable of acting as the cell-carrying medium within the cartridges, and solidifying to a supportable, biocompatible scaffold upon deposition.

Most of the current research has focused on hybrid printing techniques using multiple printing materials wherein one contains a cell suspension and the other serves to deposit the printing substrate or "biopaper." One source acts as the crosslinking agent after contact in an effort to drive material solidification. This technique has shown positive results in vitro and in vivo illustrating normal proliferation rates and good cell viability. Deposited cells survived and matured into functional tissue showing adequate de novo tissue formation [76]. This type of success is driving more advanced research and in vivo studies using hybrid devices, moving focus away from directly printing cell-laden materials. Further research in bioink formulations, especially for direct printing, could lead to continued advancement in cell printing for TE [78].

5.2.2.3 Selective laser sintering

For the manufacture of hard tissue and rigid supporting structures for tissue regeneration applications, selective laser sintering (SLS) is often a suitable technique for scaffold fabrication. This process uses a focused laser similar to stereolithography to fuse nano/micro particles/powders together into a solid component as shown in Figure 5.10. This technique can also be used with slurries, combinations of powdered particles and gels in a similar manner. Both techniques involve the spreading of a thin layer of powder/slurry, exposure of deposited layer with the laser for fusion/solidification and repeating as necessary to obtain the pre-designed 3D structure [79–84]. Solidification of the slurry is achieved through evaporation of the liquid phase which allows for fusion of the solid phase material. The process of bonding/fusing creates surfaces with favorable roughness at scales desired for biocompatibility.

Benefits of this technique include the ability to integrate materials already approved for orthopedic implantation. Research exists on selective laser sintering of pure and alloyed titanium (Ti) powder with silica sol, cobalt chrome (CoCr), and Ti6Al4V alloys in a biphasic lay-up, and HA with silica sol [86,87]. These studies illustrated successful fabrication and evaluation of SLS-based components constructed with materials previously approved as safe for implantation. This holds promise for constructing smaller; individualized components for implantation as a preventative measure before larger, more invasive procedures or full joint replacement are warranted. Further evolution of this technique can lead to efficient and viable scaffolds for bone and other hard tissues by facilitating clinical approval and increasing integration based on the extensive use of FDA approved materials. This is a huge benefit not shared by most other bioprinting techniques at the present time.

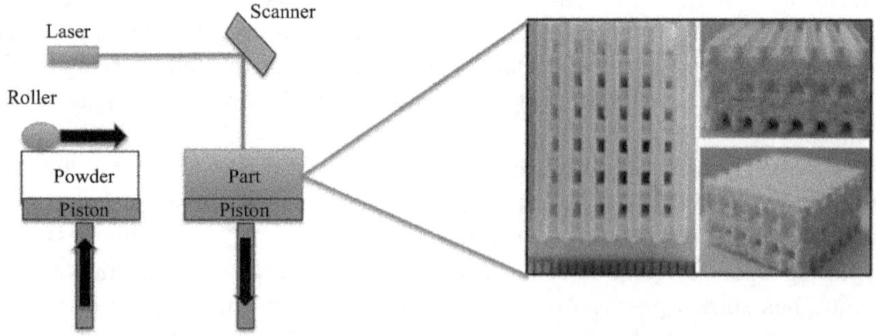

Figure 5.10 Schematic of selective laser sintering system and images of printed scaffolds. Scaffold images are adopted from Gao, C.D., et al. *[85]. Reprinted from the International Journal of Molecular Sciences, 15(3), Gao,* et al., *Current progress in bioactive ceramic scaffolds for bone repair and regeneration, 4714–4732, (2014), with permission from MDPI Open Access Publishing*

Some drawbacks of SLS printing are largely predicated on equipment cost; SLS-based printers are considerably more expensive when compared to other 3D printing technologies. Proper manufacturing of components is a careful balance between laser strength, speed, and spacing to achieve desired scaffold bulk properties. Similar to other 3D bioprinting techniques, SLS processing parameters and inherent material properties of powders have significant implications on the overall performance of the sintered construct [88].

5.3 Conclusions and future directions

3D printing of tissue engineered scaffolds is rapidly advancing with diverse research occurring on numerous printing platforms. The largest challenge in this field remains the ability to balance all performance objectives for successful scaffold integration within areas of biological and physical complexity such as bone, cartilage, and the osteochondral interface. A large extent of current research has shown marked advancement. It should be evident through the course of this discussion that nanomaterials, material properties of printing medium, and microscale geometry of printed scaffolds are critical components driving the overall success or failure of 3D printed scaffolds. Materials that can closely mimic the properties of the natural tissue are essential. More research is needed to better match biocompatible materials with the various printing platforms. In addition, the benefits of directly printing cells with novel bioinks certainly warrant further investigation.

From traditional homogenous constructs to contemporary functionally graded scaffolds, osteochondral ITE has evolved as novel nanobiomaterials, 3D nano/microfabrication techniques, and a greater understanding of cell–ECM interactions have progressed in a concerted effort to direct multitissue regeneration. Despite the challenges that lie ahead, significant evidence now exists elucidating that nanotechnology represents an important growing and transformative area of research that will most certainly aid in improving osteochondral implant efficacy. By extending the capabilities of current TE strategies and nanotechnology in a concerted effort toward an often overlooked and complex research area, advancements in the development of clinically relevant implantable osteochondral constructs may soon be realized in the not too distant future.

Acknowledgments

The authors would like to thank NSF BME program grant # 1510561 for financial support.

References

[1] Buckwalter, J.A. and J.A. Martin, Osteoarthritis. *Advanced Drug Delivery Reviews*, 2006. **58**(2): 150–167.

[2] Fontaine, K.R., S. Haaz, and M. Heo, Projected prevalence of US adults with self-reported doctor-diagnosed arthritis, 2005 to 2050. *Clinical Rheumatology*, 2007. **26**(5): 772–774.

[3] Hootman, J.M. and C.G. Helmick, Projections of US prevalence of arthritis and associated activity limitations. *Arthritis Rheumatism*, 2006. **54**(1): 226–229.

[4] Williams, G.M., E.F. Chan, M.M. Temple-Wong, *et al.*, Shape, loading, and motion in the bioengineering design, fabrication, and testing of personalized synovial joints. *Journal of Biomechanics*, 2010. **43**(1): 156–165.

[5] Thaunat, M., S. Couchon, J. Lunn, *et al.*, Cartilage thickness matching of selected donor and recipient sites for osteochondral autografting of the medial femoral condyle. *Knee Surgery Sports Traumatology Arthroscopy*, 2007. **15**(4): 381–386.

[6] Birman, M.V., T. Le Dan, S.K. Ismaily, *et al.*, The humeral head as a potential donor source for osteochondral allograft transfer to the knee. *Journal of Knee Surgery*, 2009. **22**(2): 99–105.

[7] Johnson, M.R. and R.F. LaPrade, Tibial plateu "Kissing Lesion" from a proud osteochondral autograft. *American Journal of Orthopedics (Belle Mead NJ)*, 2011. **40**(7): 359–361.

[8] Langer, R. and J.P. Vacanti, Tissue engineering. *Science*, 1993. **260**(5110): 920–926.

[9] Ge, Z., X. Tian, B.C. Heng, *et al.*, Histological evaluation of osteogenesis of 3D-printed poly-lactic-co-glycolic acid (PLGA) scaffolds in a rabbit model. *Biomedical Materials*, 2009. **4**(2): 021001.

[10] Yu, H., P.J. Vandevord, W. Gong, *et al.*, Promotion of osteogenesis in tissue-engineered bone by pre-seeding endothelial progenitor cells-derived endothelial cells. *Journal of Orthopaedic Research*, 2008. **26**(8): 1147–1152.

[11] Guan, L. and J.E. Davies, Preparation and characterization of a highly macroporous biodegradable composite tissue engineering scaffold. *Journal of Biomedical Materials Research Part A*, 2004. **71**(3): 480–487.

[12] Lebourg, M., R. Sabater Serra, J. Mas Estelles, *et al.*, Biodegradable polycaprolactone scaffold with controlled porosity obtained by modified particle-leaching technique. *Journal of Materials Science: Materials in Medicine*, 2008. **19**(5): 2047–2053.

[13] Wang, W., B. Li, Y.L. Li, *et al.*, In vivo restoration of full-thickness cartilage defects by poly(lactide-co-glycolide) sponges filled with fibrin gel, bone marrow mesenchymal stem cells and DNA complexes. *Biomaterials*, 2010. **31**(23): 5953–5965.

[14] Yoo, H.S., E.A. Lee, J.J. Yoon, *et al.*, Hyaluronic acid modified biodegradable scaffolds for cartilage tissue engineering. *Biomaterials*, 2005. **26**(14): 1925–1933.

[15] Mathieu, L.M., T.L. Mueller, P.E. Bourban, *et al.*, Architecture and properties of anisotropic polymer composite scaffolds for bone tissue engineering. *Biomaterials*, 2006. **27**(6): 905–916.

[16] Salerno, A., S. Iannace, and P.A. Netti, Graded biomimetic osteochondral scaffold prepared via CO_2 foaming and micronized NaCl leaching. *Materials Letters*, 2012. **82**: 137–140.

[17] Prieto, E.M., J.M. Page, A.J. Harmata, *et al.*, Injectable foams for regenerative medicine. *Wiley Interdisciplinary Reviews: Nanomedicine and Nanobiotechnology*, 2014. **6**(2): 136–154.

[18] Temenoff, J.S. and P.J. Yang, Engineering orthopedic tissue interfaces. *Tissue Engineering Part B-Reviews*, 2009. **15**(2): 127–141.

[19] Hutmacher, D.W., Scaffolds in tissue engineering bone and cartilage. *Biomaterials*, 2000. **21**(24): 2529–2543.

[20] Zhang, L., J. Hu, and K.A. Athanasiou, The role of tissue engineering in articular cartilage repair and regeneration. *Critical Reviews in Biomedical Engineering*, 2009. **37**(1–2): 1–57.

[21] Chen, J., H. Chen, P. Li, *et al.*, Simultaneous regeneration of articular cartilage and subchondral bone in vivo using MSCs induced by a spatially controlled gene delivery system in bilayered integrated scaffolds. *Biomaterials*, 2011. **32**(21): 4793–4805.

[22] Guo, X., H. Park, S. Young, *et al.*, Repair of osteochondral defects with biodegradable hydrogel composites encapsulating marrow mesenchymal stem cells in a rabbit model. *Acta Biomaterialia*, 2010. **6**(1): 39–47.

[23] Zhang, L., S. Sirivisoot, G. Balasundaram, *et al.*, Nanoengineering for bone tissue engineering, in *Micro and Nanoengineering of the Cell Microenvironment: Technologies and Applications*, A. Khademhosseini, *et al.*, Editors. 2008, Artech House. pp. 431–460.

[24] Zhang, L. and T.J. Webster, Nanotechnology and nanomaterials: promises for improved tissue regeneration. *Nanotoday*, 2009. **4**(1): 66–80.

[25] Streicher, R.M., M. Schmidt, and S. Fiorito, Nanosurfaces and nanostructures for artificial orthopedic implants. *Nanomedicine* (London), 2007. **2**(6): 861–874.

[26] Huebsch, N. and D.J. Mooney, Inspiration and application in the evolution of biomaterials. *Nature*, 2009. **462**(7272): 426–432.

[27] Hartgerink, J.D., E. Beniash, and S.I. Stupp, Self-assembly and mineralization of peptide-amphiphile nanofibers. *Science*, 2001. **294**(5547): 1684–1688.

[28] Hosseinkhani, H., M. Hosseinkhani, F. Tian, *et al.*, Osteogenic differentiation of mesenchymal stem cells in self-assembled peptide-amphiphile nanofibers. *Biomaterials*, 2006. **27**(22): 4079–4086.

[29] Shah, R.N., N.A. Shah, M.M. Del Rosario Lim, *et al.*, Supramolecular design of self-assembling nanofibers for cartilage regeneration. *Proceedings of the National Academy of Sciences of the United States of America*, 2010. **107**(8): 3293–3298.

[30] Aida, T., E.W. Meijer, and S.I. Stupp, Functional supramolecular polymers. *Science*, 2012. **335**(6070): 813–817.

[31] Sun, L., L. Zhang, U.D. Hemraz, *et al.*, Bioactive rosette nanotube-hydroxyapatite nanocomposites improve osteoblast functions. *Tissue Engineering. Part A*, 2012. **18**(17–18): 1741–1750.

[32] Zhang, L., U.D. Hemraz, H. Fenniri, *et al.*, Tuning cell adhesion on titanium with osteogenic rosette nanotubes. *Journal of Biomedical Materials Research Part A*, 2010. **95**(2): 550–563.

[33] Zhang, L., F. Rakotondradany, A.J. Myles, *et al.*, Arginine-glycine-aspartic acid modified rosette nanotube-hydrogel composites for bone tissue engineering. *Biomaterials*, 2009. **30**(7): 1309–1320.

[34] Zhang, L., S. Ramsaywack, H. Fenniri, *et al.*, Enhanced osteoblast adhesion on self-assembled nanostructured hydrogel scaffolds. *Tissue Engineering Part A*, 2008. **14**(8): 1353–1364.

[35] Zhang, L., J. Rodriguez, J. Raez, *et al.*, Biologically inspired rosette nanotubes and nanocrystalline hydroxyapatite hydrogel nanocomposites as improved bone substitutes. *Nanotechnology*, 2009. **20**(17): 175101.

[36] Fine, E., L. Zhang, H. Fenniri, *et al.*, Enhanced endothelial cell functions on rosette nanotube-coated titanium vascular stents. *International Journal of Nanomedicine*, 2009. **4**: 91–97.

[37] Childs, A., U.D. Hemraz, N.J. Castro, *et al.*, Novel biologically-inspired rosette nanotube PLLA scaffolds for improving human mesenchymal stem cell chondrogenic differentiation. *Biomedicine Materials*, 2013. **8**(6): 065003.

[38] Zhang, L., B. Ercan, and T.J. Webster, Carbon nanotubes and nanofibers for tissue engineering applications, in *Carbon*, C. Liu (ed) 2009, Research Signpost.

[39] Abarrategi, A., M.C. Gutierrez, C. Moreno-Vicente, *et al.*, Multiwall carbon nanotube scaffolds for tissue engineering purposes. *Biomaterials*, 2008. **29**(1): 94–102.

[40] Edwards, S.L., J.A. Werkmeister, and J.A. Ramshaw, Carbon nanotubes in scaffolds for tissue engineering. *Expert Review of Medical Devices*, 2009. **6**(5): 499–505.

[41] Kawaguchi, M., T. Fukushima, T. Hayakawa, *et al.*, Preparation of carbon nanotube-alginate nanocomposite gel for tissue engineering. *Dental Materials Journal*, 2006. **25**(4): 719–725.

[42] MacDonald, R.A., B.F. Laurenzi, G. Viswanathan, *et al.*, Collagen-carbon nanotube composite materials as scaffolds in tissue engineering. *Journal of Biomedical Materials Research Part A*, 2005. **74**(3): 489–496.

[43] Pan, L., X. Pei, R. He, *et al.*, Multiwall carbon nanotubes/polycaprolactone composites for bone tissue engineering application. *Colloids Surfaces B Biointerfaces*, 2012. **93**: 226–234.

[44] Tran, P.A., L. Zhang, and T.J. Webster, Carbon nanofibers and carbon nanotubes in regenerative medicine. *Advances Drug Delivery Reviews*, 2009. **61**(12): 1097–1114.

[45] Holmes, B., N.J. Castro, J. Li, *et al.*, Enhanced human bone marrow mesenchymal stem cell functions in novel 3D cartilage scaffolds with hydrogen treated multi-walled carbon nanotubes. *Nanotechnology*, 2013. **24**(36): 365102.

[46] Parveen, S., R. Misra, and S.K. Sahoo, Nanoparticles: a boon to drug delivery, therapeutics, diagnostics and imaging. *Nanomedicine-Nanotechnology Biology and Medicine*, 2012. **8**(2): 147–166.

[47] Anchordoquy, T. and L. Xu, Drug delivery trends in clinical trials and translational medicine: challenges and opportunities in the delivery of nucleic acid-based therapeutics. *Journal of Pharmaceutical Sciences*, 2011. **100**(1): 38–52.

[48] Foldvari, M., M. Elsabahy, and A. Nazarali, Non-viral nucleic acid delivery: key challenges and future directions. *Current Drug Delivery*, 2011. **8**(3): 235–244.

[49] Grund, S., M. Bauer, and D. Fischer, Polymers in drug delivery-state of the art and future trends. *Advanced Engineering Materials*, 2011. **13**(3): B61–B87.

[50] Park, K., J.Y. Kim, S. Kim, *et al.*, Hydrotropic polymer micelles as versatile vehicles for delivery of poorly water-soluble drugs. *Journal of Controlled Release*, 2011. **152**(1): 13–20.

[51] Castro, N.J., C. O'Brien, and L.G. Zhang, Biomimetic biphasic 3D nano-composite scaffold for osteochondral regeneration. *AICHE Journal*, 2014. **60**(2): 432–442.

[52] Zhang, L., Y. Chen, J. Rodriguez, *et al.*, Biomimetic helical rosette nano-tubes and nanocrystalline hydroxyapatite coatings on titanium for improving orthopedic implants. *International Journal of Nanomedicine*, 2008. **3**(3): 323–333.

[53] Im, O., J. Li, M. Wang, *et al.*, Biomimetic three-dimensional nanocrystalline hydroxyapatite and magnetically synthesized single-walled carbon nanotube chitosan nanocomposite for bone regeneration. *International Journal of Nanomedicine*, 2012. **7**: 2087–2099.

[54] Wang, M., X. Cheng, W. Zhu, *et al.*, Design of biomimetic and bioactive cold plasma-modified nanostructured scaffolds for enhanced osteogenic differentiation of bone marrow-derived mesenchymal stem cells. *Tissue Engineering Part A*, 2013.

[55] Wang, M., N.J. Castro, J. Li, *et al.*, Greater osteoblast and mesenchymal stem cell adhesion and proliferation on titanium with hydrothermally treated nanocrystalline hydroxyapatite/magnetically treated carbon nanotubes. *Journal for Nanoscience and Nanotechnology*, 2012. **12**(10): 7692–7702.

[56] Khanarian, N.T., J. Jiang, L.Q. Wan, *et al.*, A hydrogel-mineral composite scaffold for osteochondral interface tissue engineering. *Tissue Engineering Part A*, 2011.

[57] Pham, D.T. and R.S. Gault, A comparison of rapid prototyping technologies. *International Journal of Machine Tools & Manufacture*, 1998. **38**(10–11): 1257–1287.

[58] Fisher, J.P., J.W. Vehof, D. Dean, *et al.*, Soft and hard tissue response to photocrosslinked poly(propylene fumarate) scaffolds in a rabbit model. *Journal of Biomedical Materials Research*, 2002. **59**(3): 547–556.

[59] Cooke, M.N., J.P. Fisher, D. Dean, *et al.*, Use of stereolithography to manufacture critical-sized 3D biodegradable scaffolds for bone ingrowth. *Journal of Biomedical Materials Research Part B-Applied Biomaterials*, 2003. **64B**(2): 65–69.

[60] Arcaute, K., B.K. Mann, and R.B. Wicker, Stereolithography of three-dimensional bioactive poly(ethylene glycol) constructs with encapsulated cells. *Annals of Biomedical Engineering*, 2006. **34**(9): 1429–1441.

[61] Lee, K.W., S. Wang, B.C. Fox, *et al.*, Poly(propylene fumarate) bone tissue engineering scaffold fabrication using stereolithography: effects of resin formulations and laser parameters. *Biomacromolecules*, 2007. **8**(4): 1077–1084.

[62] Castro, N., P. Goldstein, and M.N. Cooke, Synthesis and manufacture of photocrosslinkable poly(caprolactone)-based three-dimensional scaffolds for tissue engineering applications. *Advances in Bioscience and Biotechnology*, 2011. **2**(3): 167–173.

[63] Beke, S., F. Anjum, H. Tsushima, *et al.*, Towards excimer-laser-based stereolithography: a rapid process to fabricate rigid biodegradable photopolymer scaffolds. *Journal of Royal Society Interface*, 2012.

[64] Peele, B.N., T.J. Wallin, H. Zhao, *et al.*, 3D printing antagonistic systems of artificial muscle using projection stereolithography. *Bioinspiration & Biomimetics*, 2015. **10**(5): 055003.

[65] Hribar, K.C., D. Finlay, X. Ma, *et al.*, Nonlinear 3D projection printing of concave hydrogel microstructures for long-term multicellular spheroid and embryoid body culture. *Lab Chip*, 2015. **15**(11): 2412–2418.

[66] Grogan, S.P., P.H. Chung, P. Soman, *et al.*, Digital micromirror device projection printing system for meniscus tissue engineering. *Acta Biomaterialia*, 2013. **9**(7): 7218–7226.

[67] Lu, Y. and S. Chen, Projection printing of 3-dimensional tissue scaffolds. *Methods in Molecular Biology*, 2012. **868**: 289–302.

[68] Melchels, F.P., J. Feijen, and D.W. Grijpma, A review on stereolithography and its applications in biomedical engineering. *Biomaterials*, 2010. **31**(24): 6121–6130.

[69] Castro, N.J., J. O'Brien, and L.G. Zhang, Integrating biologically inspired nanomaterials and table-top stereolithography for 3D printed biomimetic osteochondral scaffolds. *Nanoscale*, 2015. **7**(33): 14010–14022.

[70] O'Brien, C.M., B. Holmes, S. Faucett, *et al.*, Three-dimensional printing of nanomaterial scaffolds for complex tissue regeneration. *Tissue Engineering Part B: Reviews*, 2015. **21**(1): 103–114.

[71] Skardal, A., J. Zhang, and G.D. Prestwich, Bioprinting vessel-like constructs using hyaluronan hydrogels crosslinked with tetrahedral polyethylene glycol tetracrylates. *Biomaterials*, 2010. **31**(24): 6173–6181.

[72] Pfister, A., R. Landers, A. Laib, *et al.*, Biofunctional rapid prototyping for tissue-engineering applications: 3D bioplotting versus 3D printing. *Journal of Polymer Science Part A-Polymer Chemistry*, 2004. **42**(3): 624–638.

[73] Liu, H., E.B. Slamovich, and T.J. Webster, Increased osteoblast functions among nanophase titania/poly(lactide-co-glycolide) composites of the highest nanometer surface roughness. *Journal of Biomedical Materials Research Part A*, 2006. **78**(4): 798–807.

[74] Marszalek, J.E., C.G. Simon, Jr., C. Thodeti, *et al.*, 2.5D constructs for characterizing phase separated polymer blend surface morphology in tissue engineering scaffolds. *Journal of Biomedical Materials Research Part A*, 2013. **101**(5): 1502–1510.

[75] Peltola, S.M., F.P.W. Melchels, D.W. Grijpma, *et al.*, A review of rapid prototyping techniques for tissue engineering purposes. *Annals of Medicine*, 2008. **40**(4): 268–280.

[76] Xu, T., W.X. Zhao, J.M. Zhu, *et al.*, Complex heterogeneous tissue constructs containing multiple cell types prepared by inkjet printing technology. *Biomaterials*, 2013. **34**(1): 130–139.

[77] Wust, S., R. Muller, and S. Hofmann, Controlled positioning of cells in biomaterials-approaches towards 3D tissue printing. *Journal of Functional Biomaterials*, 2011. **2**(3): 119–154.

[78] Ferris, C.J., K.J. Gilmore, S. Beirne, *et al.*, Bio-ink for on-demand printing of living cells. *Biomaterials Science*, 2013. **1**(2): 224–230.

[79] Liu, F.H., Y.K. Shen, and J.L. Lee, Selective laser sintering of a hydroxyapatite-silica scaffold on cultured MG63 osteoblasts in vitro. *International Journal of Precision Engineering and Manufacturing*, 2012. **13**(3): 439–444.

[80] Long, J.P., S.J. Hollister, and S.A. Goldstein, A paradigm for the development and evaluation of novel implant topologies for bone fixation: in vivo evaluation. *Journal of Biomechanics*, 2012. **45**(15): 2651–2657.

[81] Kang, H., J.P. Long, G.D. Urbiel Goldner, *et al.*, A paradigm for the development and evaluation of novel implant topologies for bone fixation: implant design and fabrication. *Journal of Biomechanics*, 2012. **45**(13): 2241–2247.

[82] Rasperini, G., S.P. Pilipchuk, C.L. Flanagan, *et al.*, 3D-printed bioresorbable scaffold for periodontal repair. *Journal of Dental Research*, 2015. **94**(9 Suppl): 153S–157S.

[83] Hollister, S.J., C.L. Flanagan, D.A. Zopf, *et al.*, Design control for clinical translation of 3D printed modular scaffolds. *Annals of Biomedical Engineering*, 2015. **43**(3): 774–786.

[84] Coelho, P.G., S.J. Hollister, C.L. Flanagan, *et al.*, Bioresorbable scaffolds for bone tissue engineering: optimal design, fabrication, mechanical testing and scale-size effects analysis. *Medical Engineering & Physics*, 2015. **37**(3): 287–296.

[85] Gao, C., Y. Deng, P. Feng, *et al.*, Current progress in bioactive ceramic scaffolds for bone repair and regeneration. *International Journal of Molecular Sciences*, 2014. **15**(3): 4714.

[86] Liu, F.H., R.T. Lee, W.H. Lin, *et al.*, Selective laser sintering of bio-metal scaffold. *First Cirp Conference on Biomanufacturing*, 2013. **5**: 83–87.

[87] Lodererova, M., J. Rybnicek, J. Steidl, *et al.*, Biocompatibility of metal sintered materials in dependence on multi-material graded structure. in *13th International Conference on Biomedical Engineering*, 2009. **23**(1–3): 1204–1207.

[88] Hao, L., M.M. Savalani, R.A. Harris, *et al.*, Characterisation of HA/polymer bio-composite structure fabricated by selective laser sintering. *Virtual and Rapid Manufacturing*, 2008: 121–127.

Chapter 6

Self-assembling peptide hydrogels for biomedical applications

Seren Hamsici[1] and Mustafa O. Guler[1]

Abstract

The extracellular matrix (ECM) consists of a complex three-dimensional (3D) fibrous network with a wide variety of fibers and pores, which presents complex biochemical and physical signals. Synthetic biomaterials can be produced from a wide range of materials to form a nanofibrous network similar to collagen in the ECM. Self-assembly is a powerful technique to design and synthesize dynamic complex structures at the nanoscale with atomic control due to noncovalent interactions between the molecules. Self-assembled peptide nanostructures consisting of chemical and physical properties related to the ECM can be used for regenerative medicine and drug delivery applications. Therefore, the self-assembled peptide nanostructure hydrogels have capacity to form a bioactive environment such as biocompatibility, biodegradability, and low toxicity for biomedical applications. The focus of this chapter is on the self-assembled peptide hydrogel synthesis and their applications for tissue engineering and regenerative medicine.

6.1 Introduction

Nanomaterials provide many important opportunities to biomedical applications in terms of diagnosis and treatment of diseases. The most important common feature of nanomaterials is that they enable a deeper understanding of materials at the nanoscale including natural materials [1]. Proteins are the most abundant organic biomacromolecules in the cell and they can fulfill many tasks such as transportation, defense, regulation, and catalytic action in living systems. They also have molecular recognition capability by selective pairing of receptor–ligand, antigen–antibody, and enzyme–substrate interactions. These characteristics utilize the self-assembly process, which is the organization of molecules by noncovalent interactions such as van der Waals and electrostatic interactions in a spontaneous way. The self-assembly

[1]Institute of Materials Science and Nanotechnology, National Nanotechnology Research Center (UNAM), Bilkent University, Ankara, Turkey

process is present in the formation of many biological structures such as DNA double helix formation as well as secondary, tertiary, and quaternary forms of proteins [2]. Amino acids, which are the building blocks of the proteins, have different characteristics from each other due to their functional groups in the side chains and these groups are important for the function and structure of the proteins [3].

Self-assembled peptide nanofibers are developed for 3D cell culture studies. They mimic the microenvironment of the cells due to their resemblance to the ECM. They support cell migration, differentiation, and survival [4]. These scaffolds are biologically compatible, which means that they are adaptable to the body without inducing any secondary damage or inflammatory response. The peptide nanofibers can mimic the ECM and enables cell attachment, survival, and proliferation. The ECM is important not only as a structural component to support cells but also as a suitable environment for cell function. The ECM is composed of a highly hydrated network consisting of various protein fibrils and glycosaminoglycan (GAG) chains. Collagen, which is the main structural protein found in the ECM, is widely used as a scaffold for regenerative medicine due to its ability to provide an environment for cells. However, animal-derived collagen can cause immunogenic responses to humans and also it is difficult to modify, which limits its clinical applications [5]. Therefore, self-assembled peptide hydrogels are promising materials to overcome these issues. The functional groups on the surface of the peptide nanostructures provide chemical and biological diversity. In addition to structural properties, it can also provide biological functions similar to other proteins in the ECM. In this chapter, the design and synthesis of self-assembling of peptides and factors to control self-assembly and gel formation are discussed along with their potential applications for biomedical materials and regenerative medicine.

6.2 Molecular design principles

The self-assembly process utilizes noncovalent interactions such as hydrogen bonding, hydrophobic, electrostatic, van der Waals, and π–ð interactions [6]. The self-assembly process is a bottom-up materials fabrication technique, where small building blocks form higher-ordered systems [7]. Therefore, engineering supramolecular nanosystems requires a deeper understanding of molecular assembly mechanisms.

Hydrogels containing self-assembling peptides, which consist of natural amino acids have advantages including biocompatibility, biodegradability, and non-immunogenicity [8]. Synthesizing peptides with the solid phase technique offers not only control at the molecular level but also obtaining peptides with high yield and purity without the need of harsh chemical reaction conditions [8,9]. The process starts with the attachment of the first amino acid to the resin (solid support) from its C-terminal and then further amino acids are attached to the N-terminal end of the α-amino group [10]. Functional side chains of the amino acids are protected in order to prevent side reactions [10]. The resulted peptide is organized into ordered structures by noncovalent interactions such as hydrogen bonding facilitated by amide containing peptide group, electrostatic interactions formed by charged side chains,

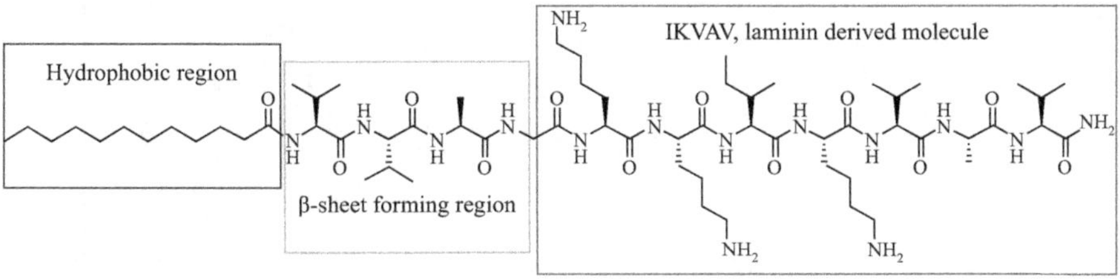

Figure 6.1 Chemical structure of a IKVAV containing PA molecule

and hydrophobic interactions [11]. These interactions drive the self-assembly mechanism, which is the key for the gelation process. The design considerations and self-assembly mechanism of peptide nanostructures forming hydrogels will be discussed below through relevant examples.

Peptide amphiphile (PA) molecules contain hydrophobic fatty acids attached to the peptide such as palmitic, lauric acid, which enables hydrophobic collapse and hydrophilic peptide regions to interact with water to form micellar structures [12]. This micelle is usually ordered to be a high-aspect-ratio nanosystem due to hydrogen bonding interactions among the peptide molecules. PA nanofiber gels can mimic the ECM since they have common features such as high water content and a porous network [13]. The synthetic PA nanofiber scaffolds provide cells to adhere, proliferate, migrate, and differentiate for a desired purpose [14,15]. Laminin is one of the abundant proteins in the ECM responsible for cell adhesion [16]. In order to mimic the adhesive behavior of laminin, the laminin derived Ile-Lys-Val-Ala-Val (IKVAV) peptide epitope (Figure 6.1) has been used for neurite extension, which can result in the recovery of spinal cord injuries in cell-based approaches [17]. Stupp and colleagues showed that the use of IKVAV-PA nanofibers allows reduced scar formation and increased axon regeneration in spinal cord injury models [18].

In addition, they have demonstrated the effect of alignment for guiding neurites along the nanofibers by bathing PAs in a $CaCl_2$-containing solution [19]. Aligned nanofibers enhanced neurite growth and encapsulated neural progenitor cells differentiated into the neurons which shows electrical activity and synaptic connections [19].

Our group previously showed that self-assembled heparin mimetic PA nanofibers (Figure 6.2) can induce angiogenesis (new blood vessel formation) and tissue regeneration without the need of any external heparin or growth factors [20]. The functional groups at the surface of the PA nanofibers enabled binding of growth factors secreted from cells with high affinity and these nanofibers provided a suitable environment for angiogenesis. In vivo experiments were conducted in the rat cornea because it is a special tissue that normally lacks blood vessels, therefore vascularization would be a positive sign for treatment. Vascular endothelial growth factor (VEGF) with or without the heparin mimetic PA gel was injected to the center of the cornea. De novo formed capillaries were observed in blue. It was concluded that growth factors were not sufficient to form blood vessels alone,

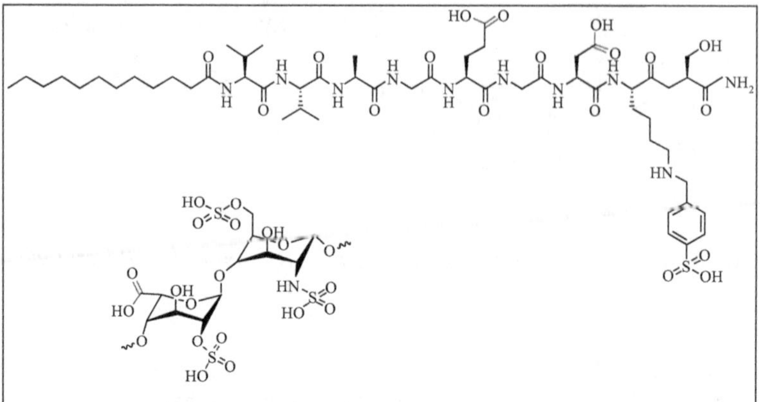

Figure 6.2 Heparin mimetic PA: in order to mimic heparin, PA is functionalized with chemical groups such as carboxylic acid, hydroxyl, and sulfonate groups. Reprinted with permission from Reference 20. Copyright 2011, American Chemical Society

newly formed capillaries were integrated with other blood vessels and also the need of heparin was avoided by mimicking the structure with peptide nanofibers specialized with chemical groups attached to amino acid side chains [20]. Further studies showed that the efficiency of the interaction between heparin mimetic PA nanofibers and a variety of growth factors such as VEGF, hepatocyte growth factor (HGF), and fibroblast growth factor-2 (FGF-2) was necessary in order to understand the ability of mimicking heparin present in the ECM [21]. The heparin mimetic PA nanofibers provided an optimal condition for growth factor binding without any change in bioactivity of growth factors proved by in vitro studies [21].

Another peptide motif derived from fibronectin, Arg-Gly-Asp (RGD), can be used for cell adhesion purposes by interacting with cell surface integrin receptors and it can be incorporated into the PAs for bioactive functionality [22]. After the adhesion process, cells need to migrate to the matrix and for that purpose enzymes emerge and create a pathway for them [23]. The most extensive and abundant one is the matrix metalloproteinase (MMP) family that is able to degrade several types of proteins located in the ECM [24]. Koda *et al.* designed palmitoyl-GGGHGPLGLARK-Am where PLG is the cleavage site recognized by the cancer-associated enzyme matrix metalloproteinase-7 (MMP-7) [25]. After enzyme treatment, the PA molecule changed its morphological characteristic and turned into long fibers from spherical micelles. Maruyama *et al.* used the same cleavage site in order to demonstrate the selectivity of cellular uptake is able to kill cancer cells while not harming normal cells [26]. Hamley *et al.* designed six alanine residues attached to the RGD tripeptide and investigated the self-assembly mechanism changes and affect of cell viability based on the concentration difference [27]. Oligomer formation was observed at 0.1 wt% while short and long fibril morphologies were observed at 1.0 and 2.5 wt% of A_6RGD peptide. Cell studies were conducted with human cornea stromal fibroblasts (hCSFs), both adhesion and proliferation were seen at 0.1 and 1.0 wt%, whereas higher concentrations of A_6RGD peptide caused toxicity for cells.

Mata *et al.* showed the self-assembled peptide nanostructured gels for bone regeneration [28]. The combination of both fibronectin derived epitope palmitoyl-AAALLLKKRGDS (RGDS-PA) and a phosphorylated serine residue palmitoyl-AAALLLEE(P)SG(S(P)-PA) for hydroxyapatite nucleation allows one to mimic the mineralized ECM of bone tissue. Application of gels to defect sites showed the highest amount of bone formation [28]. Peptide nanostructures have attracted attention in terms of the diversity of features mentioned above and the number of studies added to the literature has been increasing.

It is possible to produce peptides for various applications due to arranging amino acids end to end by controlling chemical structure and sequence. The â-structures specialized with sheets, hairpins, and turns are the most popular secondary structures in peptide hydrogels after the α-helical forms [8]. Schneider and colleagues designed β-hairpin hydrogels called MAX peptides which consist of two blocks of valine and lysine residues [29] combined by β-turn forming spacer consists of four residue sequence (Val^D-Pro-Pro-Tyr) [30–32]. Acidic conditions protonate the lysine residues and folding is prevented because of the charge repulsion [29]. When the pH of the solution is increased, peptide strands are folded into β-hairpins and self-assemble in lateral and facial form [33].

Hartgerink's group designed multidomain peptides, which was illustrated as the ABA motif where A is composed of positively charged residues while B consists of alternating hydrophilic and hydrophobic amino acids [15] to stabilize hydrophobic packing and make antiparallel α-sheet formation [34]. ABA structure has high surface area and MMPs are able to attack more easily to degrade the gel and cells are favored by that mechanism due to the cleavage of the hydrogel and then migrating of cells to that area [35].

The $(RADA)_4$ peptide, an ionically complementary peptide, has 16 amino acid residues that consist of hydrophobic (alanine) and hydrophilic (arginine and aspartic acid) groups, which are characterized by forming stable α-sheet structures and forms hydrogel scaffolds by undergoing self-assembly [36,37]. The Ac-RADARADARADARADA-Am (RADA) peptides are used for cell culture that are commercialized as Puramatrix in order to promote differentiation of progenitor cells [38]. Frech and Morgan modified Puramatrix with short peptide sequences; one includes laminin-derived peptide SDPGYIGSR, another one consisted of the GGPFSSTKT peptide sequence, which enhances neural stem cell adhesion and as a result, the formulation of Puramatrix increases the differentiation of human neural progenitor cells (hNPCs) to neuronal cells [39]. Dissanayaka *et al.* used Puramatrix for encapsulation of both dental pulp stem cells (DPSCs) and human umbilical vein endothelial cells (HUVECs) to obtain both angiogenesis and mineralization, which enhance dental pulp regeneration [40].

In bone regeneration studies, signaling peptide sequences derived from osteogenic growth peptides (OGPs) play a crucial role for differentiation of mesenchymal stem cells into osteoblasts [41]. Zhang and coworkers demonstrated that functionalization of RADA with OGP by synthesizing, as solid phase at the C-termini is significantly effective for osteoblastic differentiation [42]. In order to evaluate the level of differentiation, ALP activity, a marker for osteoblastic differentiation, can be measured by a fluorometric assay and a significant effect of cellular proliferation and differentiation was observed [42].

6.3 Factors affecting the self-assembly process

The synthesis of self-assembled peptide hydrogels occurs at room temperature and without the need of specialized equipment [43]. The variations such as charge, enzymatic effects, molecular geometry, pH, and ion addition between PAs can lead to different mechanical and physical properties as well as changing gelation kinetics. Among them, stability, durability, and biocompatibility are crucial considerations for designing PAs to be used in cell-delivery applications. Mechanical properties also have an important effect on cell migration and differentiation [44]. Therefore, external and internal factors for the gelation process should be well considered for the design of self-assembling peptides.

6.3.1 Co-assembly of peptide molecules

Typically a PA molecule is composed of two regions, which are a hydrophobic tail, which packs in the core of the fiber and a hydrophilic peptide sequence presented on the fiber's surface [45]. If two oppositely charged PAs are combined in an aqueous solution at physiological pH, fiber formation is favored by hydrophobic attractions and hydrogen bonding between peptide units, in this regard PAs form assemblies by secondary interactions [46,47]. Electrostatic interactions promote self-assembly and a 3D network, which triggers hydrogel formation [48]. Our group demonstrated that using injectable, synthetic scaffold systems could enhance corneal stroma regeneration. In that study, lauryl-VVAGKYIGSR-Am (YIGSR-PA) and lauryl-VVAGKKRGD-Am (KKRGD-PA), which are laminin and fibronectin mimetic peptides respectively, were used. In order to neutralize positively charged YIGSR-PA and KKRGD-PA, E-PA (lauryl-VVAGE-OH) was added. For in vitro studies, experiments were carried out with human corneal keratocyte (HCK) cells and culturing them with a laminin mimetic peptide scaffold showed similar adhesion and viability with cells cultured on collagen [49]. Toksoz *et al.* [50] showed the effect of electrostatic interactions on nanofiber formations due to the charged groups on peptides. In this study, three **PA** groups which are Lauryl-VVAGK-Am (Lys-PA), Lauryl-VVAGD (Asp-PA), and Lauryl-VVAGH-Am (His-PA) were synthesized in order to understand the self-assembly mechanism. At acidic conditions, a lysine residue is positively charged and when a negatively charged molecule such as Asp-PA was added to the system, gel formation, which is a sign that a 3D network was seen at neutral pH. The aspartic acid residue contributes the negative charge of the Asp-PA system at neutral and basic conditions, so lower pH that can be provided by positively charged molecules like calcium ions enable gel formation and it provides potential applications for mineral rich environments. Furthermore, two negatively charged biomacromolecules, GAG (heparin) and oligonucleotides (DNA), were chosen for gelation with Lys-PA at a neutral pH and their injectable gelation behavior enables using this system for cell-material studies.

6.3.2 *Effect of temperature*

The temperature effect on the β-sheet formation can be considered as two approaches. First, higher temperatures lead to a higher driving force for peptide self-assembly since hydrophobic interactions are increased. Therefore the equilibrium between electrostatic repulsion and self-assembly tend the shift through β-sheet formation. Second, higher temperatures enhance peptide interactions with intermolecular collisions due to the increment of the diffusion of the peptides [51]. The $(RADA)_4$ peptide solution was heated up to 80ºC and α-sheet formation was observed between 25ºC and 75ºC. However, after 75ºC, the β-sheet content decreased while the unordered structure was increased [52]. When the temperature was decreased to 25ºC, the $(RADA)_4$ peptide was able to regain its α-sheet content at a significant level. Therefore, the temperature increase can disturb the ionic and hydrogen interactions [52].

6.3.3 *Effect of pH*

The functionality of PAs is controlled by its amphiphilic nature, which is derived from the sequence of natural amino acids and lipid chains. Hamley *et al.* [53] studied the influence of pH on the self-assembly mechanism of collagen mimetic PA used commercially in anti-wrinkle creams and composed of palmitoic acid attached with a KTTKS pentapeptide group. Since the pH of the skin is around 5 and the self-assembly mechanism of PA is pH dependent, skin care applications can be crucial due to the risk of skin cancer, which occurs in reduced pH values. Two lysine residues and carboxy terminus in the pentapeptide group make a change in net charge when the environment is acidic or neutral and morphological changes of PAs were observed at pH 3, 4, and 7 such as flat tape structures and twisted nanofibrils. Due to the importance of the designed PA in regenerative medicine, it is crucial to understand the pH triggered self-assembly mechanism. The effect of charged groups on the PA molecule was analyzed in another research study conducted by our group. Toksoz *et al.* [50] designed three PA molecules that have in common lauric acid chain conjugated with VVAG which is α-sheet forming region. On the contrary, the difference among the three is their charged amino acids, which are lysine, aspartic acid, and histidine that give a name of Lys-PA, Asp-PA, and His-PA. Since Lys-PA is positively charged at acidic conditions, it can form a gel by increasing pH that can be done with the addition Asp-PA. Due to the presence of a histidine residue in His-PA, gelation occurs above pH 6.5. Neutral and basic conditions makes Asp-PA negatively charged while low pH conditions trigger the gel formation. Physiological pH was intended to be used in this study in order to make the system possible for future cell experiments. Neutralization of positively charged Lys-PA was conducted with Asp-PA and negatively charged biomacromolecules such as GAG and oligonucleotides, which are found in living systems. His-PA has an ability to form a gel at neutral pH without introducing any external factors. However, Asp-PA needs the increase in pH such as adding Lys-PA or calcium ions that makes it a good candidate for bone regeneration studies.

6.3.4 *Effect of salt concentration*

While temperature and pH are the most effective external conditions for making hydrogels, the effect of ions cannot be discarded. Hofmeister series are the series of ions to understand their ability about salting out and salting in proteins that have an effect on their secondary structure [54] and the effect of anions is greater than cations. Salting out decreases the protein solubility so hydrophobic interactions are stronger while salting in increases the solubility in water and hydrophobic interactions become weaker [55]. The anions can be ordered as $F^- > HPO_4^{2-} > Cl^- > NO_3^- > Br^- > I^-$ and cations can be ordered as $NH_4^+ > K^+ > Na^+ > Li^+ > Mg^{2+} > Ca^{2+}$. Shi *et al.* [56] showed that adding calcium ions increases the mechanical strength of nanofibers in hydrogels that consists of peptides of multiple carboxylic acid and aromatic groups and they combine the interactions between calcium-ion bridges and π–π interactions. Three different molecules with structures of N`FGLDD, N`FFCGLD, and N`FFCGLDD where N` is naphthyl acetic acid were studied. The presence of carboxylic acid groups on aspartic acid side chains, which have the affinity to bind Ca^{2+}, determines the pH for the hydrogelation process. Therefore, molecules, which have two aspartic acids, need slightly higher pH than a molecule which has one aspartic acid.

Micklitsch *et al.* [57] reported the design of peptide hydrogels specialized with the response of zinc ions, which is important for protein synthesis as well as cell proliferation by being a cofactor for enzymes responsible for these processes. In their study, zinc-binding heparin peptide (IKVKIKVKV^{D}PPTKIKVKIK-1-NH_2) was synthesized where 1 represents 3-amidoethoxyaminodiacetoxy-2-aminopropionic acid for metal response. If the environment is free form Zn^{2+} ions, the peptide is unfolded. However, if Zn^{2+} ions attach to the negative side chain of 1, it makes it neutral instead of negative charge (2−) in the absence of Zn^{2+} ions. The addition of ions allows for forming β-sheet networks composed of folded β-hairpins. The same study was repeated with other metals which are Ca^{+2}, Cu^{+2}, and Hg^{+2} as their salt form but among them Zn^{2+} was the only metal ion to form a secondary structure and self-assembly mechanism. For further studies, this mechanism can be used for wound healing applications due to important features of zinc ions such as conferring resistance to cell apoptosis induced by bacterial toxins.

6.3.5 *Influence of chirality*

The gelation process is highly dependent on external factors and molecular chirality is one of the stimuli, which specifies supramolecular chirality of the nanostructures due to impact on the self-assembly process [58,59]. Hartley and colleagues [59] demonstrated that tri-peptides, namely, ^{D}Val-Phe-Phe, ^{D}VFF, and ^{D}Phe-Phe-Val, ^{D}FFV form hydrogel at physiological pH whereas their L form was not capable of showing the same behavior under the same conditions. For their further research, they tested eight tripeptides with the combinations of L, D forms in terms of mechanical strength, morphological differences, and cell viability [60]. It was concluded that the gelation process of Phe-^{D}Phe-^{D}Val is the fastest and has the highest elastic modulus according to rheology measurement results. For nanostructure morphology, cryo-TEM and AFM revealed that ^{D}Phe-Phe-Val and Phe-^{D}Phe-^{D}Val have a network of fibers whereas ^{D}Phe-Phe-Val, Phe-^{D}Phe-^{D}Val, and ^{D}Phe-^{D}Phe-^{D}Val did not show any ordered nanostructure because of the instability. Furthermore, in vitro studies showed that two enantiomers, which are ^{D}Phe-Phe-Val and Phe-^{D}Phe-DVal, have the highest cell viability and proliferation that did not cause any toxic effects on mammalian cells [60].

6.3.6 Variations in amino acid chain

The self-assembly mechanism of PAs is derived from hydrophobic interactions that resulted from alkyl tails, hydrogen-bonding interactions caused by a peptide backbone, and electrostatic interactions induced by hydrophilicity [45]. Zhau and colleagues made comparisons about the self-assembly mechanism between leucine and isoleucine residues and they proposed that while isoleucine forms nanotubes, leucine forms spherical aggregates. These morphological changes are related with having different secondary structures of them, which are β-sheets and random coil formations, respectively [61]. Their further research was related to how the number of amino acid residues affect the self-assembly mechanism. From that point of view, they synthesized I_mK and L_mK where *m* is the number of amino acids that are present in the peptide sequence ranging from 3 to 5. It was found that increasing the number leucine residues in L_mK residues lead the conformational change from micelles to long nanofibers, which show the crucial impact on hydrogen bonding in the axial growth. For the isoleucine case in the I_mK residue, the increased number of isoleucine causes additional increases in hydrophobicity, which leads to a decrease in the size of diameter of nanofibers [62]. The Stupp group [63] designed four PAs that consist of hydrophobic palmitoyl attached to a tetrapeptide composed of two glutamic acid (E) and two valine (V) residues in order to show the importance of side chain interactions for self-assembled architectures of small peptides. The supramolecular structures were different from each other by changing the amino acid order of the tetrapeptide sequence and morphologies had different characteristics such as nanobelts, single and bundled nanofibers, twisted ribbons, helical ribbons, and nanotubes.

6.3.7 Hydrophobic interactions

The hydrophobic tail region of PA has the ability to drive self-assembly and enable the functional groups located at the surface. Since the alkyl tail is responsible for hydrophobicity, the chain length of the alkyl tail is an important aspect for the driving force resulting from the self-assembly process. Zhang and colleagues [64] investigated the effect of alkyl tail lengths by selecting four different PAs, which have the same hydrophilic region consisting of the VRGDV peptide sequence with a variety of alkyl chains namely C9, C11, C13, and C15. This study showed that with the increasing length of alkyl tails, hydrophobic interactions were enhanced and it makes the nanofibers tightly packed. Changing environmental conditions from acidic to basic medium proved it. All PAs in this work showed β-sheet conformation resulting in tightly packed nanofibers in an acidic environment. However, when the pH of the medium is increased, self-assembled nanofibers of PA, which have the shortest length of alkyl tail (C9), changed their conformation and turned into spherical micelles. Since the longer chains (C13 and C15) have stronger hydrophobicity, the morphology of them did not change even if the medium is adjusted to basic pH. Among others, PA has an C11 alky tail defined as moderate length and both micellar and nanofibers were formed at a basic medium [64].

6.4 Biomedical applications of the hydrogels

Self-assembled peptide hydrogels have been exploited due to the features of peptide chemistry such as being biocompatible, biodegradable. and non immunogenic. The peptide hydrogel field for biomedical applications has many challenges and intense research has been going on for their application in drug delivery and regenerative medicine.

6.4.1 Drug delivery

One of the most important uses of peptide hydrogels in medicine is controllable drug delivery. It is highlighted that the desired dose of drug should be given under the certain times and hydrogels play a role to release drugs at a constant speed. These gels, which are sensitive to the different conditions in the body, are capable of storing drug levels at the appropriate level in the blood by changing the drug release rate [8]. Recently, Gazit and coworkers [8] demonstrated that short peptide sequences can form hydrogels and due to their low molecular weight, low bioaccumulation occurs for in vivo applications. However in some cases, cytotoxicity can be observed in these systems. Xu and coworkers [65] designed dipeptides that form hydrogels by exposing N-terminal amine, which can inhibit the growth of cancer cells that can lead to applications in cancer therapies. Pharmaceutical drugs can be encapsulated in peptide hydrogels before gelation or they can be mixed during the self-assembly process. Hartley and colleagues [66] designed a hydrophobic tripeptide (^{D}Leu-Phe-Phe) where ciprofloxacin, an antibacterial drug used against several types of gram-positive and negative bacteria, was encapsulated during the self-assembly of peptide hydrogels at physiological pH. Drug loading capacity with 30% weight of the peptide was achieved and also release kinetics showed that sustainability reached an equilibrium state in 2–3 days after encapsulation. In order to see the success of the released drug against bacteria, antimicrobial activity was tested and it was shown that the drug was sufficient to inhibit the growth of bacteria. Furthermore, using red blood cells for live/dead assays eliminated the potential doubts about cytotoxicity for human and MTS assays suggested the ability of cells for proliferation even after 3 days of treatment with ciprofloxacin. Xu's group designed naproxen (Npx), a nonsteroidal anti-inflammatory drug, contained hydrogel that consists of small peptides with D-amino acids. The self-assembly mechanism of the diphenylalanine (Phe-Phe) residue makes it a good candidate for conjugation with Npx that improves the stability of supramolecular structures. On the other hand, there are some limitations based on these methods such as low encapsulation efficiency, leakage of drugs, and disruption of gel formation by drug molecules while the self-assembly process takes place and also control over drug delivery may be difficult [67]. More recently, pro-drugs based hydrogels are a new approach for drug delivery. The advantage of this technique is an increase in drug loading since every molecule in the gel has its own drug molecule [68]. Due to the presence of proteolytic enzymes in the intestine, orally taken drugs are not preferred. Therefore, the preferred way for drug–peptide conjugates is administrating via the parenteral route [69]. In order to use peptides for conjugation, it is necessary to have a special sequence for binding to cell surfaces selectively and circulation time within body should be sufficient to reach the target cells, which have more receptors than other non target ones for distinguishing from each other. One of the epithelial cell kinases, Ephrin type-A receptor 2 (EphA2), is overexpressed in human cancer types such as metastatic melanoma and breast, lung, colon carcinomas [70,71] by affecting the cell cytoskeleton. Wang and colleagues [72] demonstrated that the conjugation of EphA2 targeting peptide YSA (YSAYPDSVPMMS) with an anticancer drug, paclitaxel (PTX), showed promising results by preventing tumor growth in prostate cancer models. Treating mice with YSA-PTX (at equimolar dose with PTX alone) remarkably affected tumor growth inhibition when compared with PTX alone. At the end of three weeks, nearly a 50% decrease was achieved in peptide conjugated drug models. They also suggested that the YSA-PTX conjugate has sufficient plasma and microsomal half-life to release the drug after intravenous injection which is one of the critical factors for effective targeting. Wang and colleagues made another study with Paclitaxel. They synthesized a folic acid conjugated peptide, which consists of phosphorylated and succinated GYK (Gly-Tyr-Lys) with Paclitaxel [73]. In this design, folic acid not only offers to target cancer cells but also allows forming nanospheres by its self-assembly mechanism. The phosphorylated and succinated form of tripeptide is a substrate for phosphatase enzyme, which converts it non-phosphorylated form with less solubility. The gelation kinetics was studied by rheological analysis and it was observed that after adding a phosphatase enzyme to the system, immediate gel formation occurred, which was proved by the increase of elasticity, as occurs during this process.

6.4.2 Regenerative medicine

A biomimetic environment aims to replace defective parts in the human body by combining scaffolds, cells, and biologically active molecules, thus pretends to be an ECM [74]. For this purpose, scaffolds should show key properties such as supporting cell attachment, proliferation, differentiation, and enhancing new tissue formation [75].

6.4.2.1 Cartilage tissue engineering

Self-assembled peptide hydrogels can be used for cartilage tissue repair. In order to prevent cartilage defects, chondrocytes are encapsulated or attached to a scaffold, which provides a suitable microenvironment to produce their own ECM. The scaffold system must be completely biocompatible, stable and has a porous structure to enable chondrocyte migration. Zhang and coworkers used the peptide KLD-12 with sequence Ac-KLDLKLDLKLDL-Am for encapsulation of chondrocytes by providing a 3D environment. During culturing for 4 weeks, any disruption of the hydrogel was not observed and it created cartilage-like ECM niches in type II collagen in parallel with an enhancement in material stiffness [76]. Our group previously designed GAG mimetic self-assembled peptide nanofibers decorated with hydroxyl, carboxylate, and sulfanete in order to construct a chondrogenesis triggering environment with chondroprogenitor cells isolated from mouse embryonic carcinoma cells [77]. In this study, Lauryl-VVAGEGD-K(p-sulfobenzoyl)-S-Am (SO3-PA) assembled with a Lauryl-VVAGK-Am (K-PA) system showed the highest cartilaginous matrix deposition which was determined by safranin-O staining indicating GAG deposition in chondroprogenitor cells. In addition, gene expression analysis was carried out with specific genes responsible for expression of cartilage such as aggrecan, collagen I&II, and Sox-9 and the highest expression was seen in the nanofiber network which consisted of SO3-PA/E-PA/K-PA.

6.4.2.2 Bone repair and regeneration

The Stupp group recently showed research for osteogenesis by PA nanofibers binding to bone morphogenetic protein-2 (BMP-2), which plays a crucial role in the development of bone and cartilage [78]. The designed molecules consist of a BMP-2 binding epitope attached to carboxylate rich peptide (E_3) and β-sheet-forming region (A_3V_3) conjugated to lauric acid and the complementary sequence includes only an E_3 domain linked to A_3V_3 and a palmitic acid alkyl tail. These two molecules co-assembled together in order to promote the presence of the BMP-2 binding domain at the surface of nanofibers. Their results showed that this system is able to mimic the natural environment of osteoblasts without the need of exogenous BMP-2.

Our group showed that GAG mimetic peptide nanofibers are important because of the presence of GAG in high amounts in the ECM and their effect on cellular behavior [79]. BMP-2 interactions were promoted by the peptide nanofibers, which are functionalized with sulfonate and carboxylate groups at the surface. The bioactivity of the designed PAs was proven by an ELISA assay with two different concentrations and a 3.3-fold increment was observed when compared to the non-bioactive control group. In order to test the suitability of the environment, cell viability and proliferation studies were conducted with osteoblast-like Saos-2 cells and not only live cells on the peptides remained the same at day 1, but also a 4-fold increment in cell number was observed with respect to day 1. Since mineral deposition is important for bone-tissue engineering applications, Alizarin red-S staining was performed and large and dense mineralized nodules were seen. For further experiments, surface analysis techniques were used in order to prove the presence of Ca^{+2}.

In another study in our group, we showed the design of PA nanofibers functionalized with osteoinductive collagen I mimetic peptide (Lauryl-VVA-GEGDGEA), mussel adhesive protein epitope (Lauryl-VVAGKDopa-Am), and a bioactive sequence of the non collagenous matrix protein (Lauryl-VVAGEEE-Am), DGEA-PA, DOPA-PA, and E_3PA [80]. Differentiation of human mesenchymal stem cells into functional osteoblasts was observed by elevated alkaline phosphatase activity, which is the initial indication of differentiation to osteoprogenitor cells, and the highest amount of deposition was observed in the combination of DGEA-PA/DOPA-PA nanofibers. However, according to quantification of calcium deposition on the mature osteoblasts, the amount of calcium was significantly higher for the combination of E_3PA/DOPA-PA rather than DGEA-PA/Dopa-PA. Therefore, it was highlighted in this study that the initial commitment of osteoprogenitor cells do not prove the maturation of functional osteoblasts.

6.4.2.3 Neural regeneration studies

The PA system presents a versatile platform, where molecules can be designed and developed in order to allow for neural regeneration, and optimized for cell adhesion. The Stupp group cultured Schwann cells, which are known for having roles in supporting nerve regeneration with bioactive PAs (IKVAV-PA and RGDS-PA) in order to determine the proliferation of cells, and it was observed that the increased proliferation of Schwann cells was observed within these systems when compared with non bioactive PA [81]. Liu *et al.* [81] tested self-assembled $K_2(QL)_6K_2$ (QL6) peptides that led to reduce inflammation, which is a major obstacle for neural regeneration and enhances tissue preservation. Furthermore, they investigated the electrophysiological effects of the self-assembled peptides by the Basso, Beattie, and Bresnahan (BBB) locomotion score in order to determine long-term functional recovery after spinal cord injury including joint movements and the QL6-treated rats reached the highest score after 6 weeks of injury.

6.4.2.4 Regeneration in other systems

Another application of self-assembled peptide nanofibers is an alternative approach to root canal therapy that is placing of synthetic materials after removal of dental pulp. In this research, MDP hydrogels were selected due to its viscoelastic properties which is needed for injection to root canal. In order to promote angiogenesis of dental pump stem cells, VEGF is incorporated into the hydrogel. The SLRG sequence and RGD sequence are also present in the PA for enzyme cleavage and cell adhesion, respectively [82]. Another application of the peptide hydrogel, the RADA system, which is an injectable hydrogel, is to supply cells for cardiac tissue regeneration. Hsieh *et al.* [83] demonstrated that plated derived growth factor (PDGF) affects cardiac endothelial cells (myocardium) and by using self-assembled RADA hydrogels, they can encapsulate the growth factor with sustained delivery for 14 days to reduce the infract size of the myocardium.

6.5 Conclusions and future perspectives

The self-assembly technique is useful in the synthesis of dynamic and complex structures through noncovalent interactions and self-assembled peptide hydrogels are potentially appropriate materials for biomedical applications especially for tissue engineering and drug delivery. However, the mechanism behind these activities should be well understood to investigate the mechanical and chemical properties. In this chapter, parameters affecting the gelation process and different self-assembly processes were discussed. The recent advances in the field show that these nanostructures provide a suitable platform for tissue regeneration and regenerative medicine due to self-supporting and ECM mimicking properties.

Acknowledgments

We would like to thank the TUBITAK 213M406 for supporting this work.

References

[1] Lakshmanan, A., Zhang, S. & Hauser, C. A. E. Short self-assembling peptides as building blocks for modern nanodevices. *Trends Biotechnol.* **30**, 155–165 (2012).

[2] Mandal, D., Shirazi, A. & Parang, K. Self-assembly of peptides to nanostructures. *Biomed. and Pharm. Science* 1–57 (2014).

[3] Gazit, E. Self-assembled peptide nanostructures: the design of molecular building blocks and their technological utilization. *Chem. Soc. Rev.* **36**, 1263–1269 (2007).

[4] Matson, J. B. & Stupp, S. I. Self-assembling peptide scaffolds for regenerative medicine. *Chem. Commun. (Camb).* **48**, 26–33 (2012).

[5] Luo, J. & Tong, Y. W. Self-assembly of collagen-mimetic peptide amphiphiles into biofunctional nanofiber. *ACS Nano* **5**, 7739–7747 (2011).

[6] Cavalli, S., Albericio, F. & Kros, A. Amphiphilic peptides and their cross-disciplinary role as building blocks for nanoscience. *Chem. Soc. Rev.* **39**, 241–263 (2010).

[7] Petkau-Milroy, K. & Brunsveld, L. Supramolecular chemical biology; bioactive synthetic self-assemblies. *Org. Biomol. Chem.* **11**, 219–232 (2013).

[8] Seow, W. Y. & Hauser, C. A. E. Short to ultrashort peptide hydrogels for biomedical uses. *Biochem. Pharmacol.* **17**, 381–388 (2014).

[9] Nagarkar, R. P. & Schneider, J. P. Synthesis and primary characterization of self-assembled peptide-based hydrogels. *Methods Mol Biol.* **474**, 61–77 (2008).

[10] Palomo, J. M. Solid-phase peptide synthesis: an overview focused on the preparation of biologically relevant peptides. *RSC Adv.* **4**, 32658 (2014).

[11] Rughani, R. V. & Scheneider, J. P. Molecular design of β-hairpin peptides for material construction. *MRS Bull.* **33**, (2008).

[12] Zhang, S. Emerging biological materials through molecular self-assembly. *Biotechnol. Adv.* **20**, 321–339 (2002).

[13] Kisiday, J., Jin, M., Kurz, B. *et al.* Self-assembling peptide hydrogel fosters chondrocyte extracellular matrix production and cell division: implications for cartilage tissue repair. *Proc. Natl. Acad. Sci. U. S. A.* **99**, 9996–10001 (2002).

[14] Nosoudi, N., Holman, D., Karamched, S., Mai, L. & Devora, J. Engineered extracellular matrix: current accomplishments and future trends. *Int. J. Biomed. Eng. Sci.* **1**, 1–15 (2014).

[15] Zhou, M., Ulijn, R. V. & Gough, J. E. Extracellular matrix formation in self-assembled minimalistic bioactive hydrogels based on aromatic peptide amphiphiles. *J. Tissue Eng.* **5**, 2041 (2014).

[16] Fernandes, H., Moroni, L., van Blitterswijk, C. & de Boer, J. Extracellular matrix and tissue engineering applications. *J. Mater. Chem.* **19**, 5474 (2009).

[17] Ghasemi-Mobarakeh, L., Prabhakaran, M. P., Morshed, M., Nasr-Esfahani, M.-H. & Ramakrishna, S. Electrospun poly(epsilon-caprolactone)/gelatin nanofibrous scaffolds for nerve tissue engineering. *Biomaterials* **29**, 4532–4539 (2008).

[18] Tysseling-Mattiace, V. M., Sahni, V., Niece, K., *et al.* Self-assembling nanofibers inhibit glial scar formation and promote axon elongation after spinal cord injury. *J. Neurosci.* **28**, 3814–3823 (2008).

[19] Berns, E. J. Sur, S., Pan, L., *et al.* Aligned neurite outgrowth and directed cell migration in self-assembled monodomain gels. *Biomaterials* **35**, 185–195 (2014).

[20] Mammadov, R., Mammadov, B., Toksoz, S. *et al.* Heparin mimetic peptide nanofibers promote angiogenesis. *Biomacromolecules* **12**, 3508–3519 (2011).

[21] Mammadov, R., Mammadov, B., Guler, M. O. & Tekinay, A. B. Growth factor binding on heparin mimetic peptide nano fibers. *Biomacromolecules* **13**, 3311–3319 (2012).

[22] Hartgerink, J. D., Beniash, E. & Stupp, S. I. Peptide-amphiphile nanofibers: a versatile scaffold for the preparation of self-assembling materials. *Proc. Natl. Acad. Sci. U. S. A.* **99**, 5133–5138 (2002).

[23] Hubbell, J. A. Materials as morphogenetic guides in tissue engineering. *Curr. Opin. Biotechnol.* **14**, 551–558 (2003).

[24] Jun, H.-W., Yuwono, V., Paramonov, S. E. & Hartgerink, J. D. Enzyme-mediated degradation of peptide-amphiphile nanofiber networks. *Adv. Mater.* **17**, 2612–2617 (2005).

[25] Koda, D., Maruyama, T., Minakuchi, N., Nakashima, K. & Goto, M. Proteinase-mediated drastic morphological change of peptide-amphiphile to induce supramolecular hydrogelation. *Chem. Commun. (Camb).* **46**, 979–981 (2010).

[26] Tanaka, A., Fukuoka, Y., Morimoto, Y., *et al.* Cancer cell death induced by the intracellular self-assembly of an enzyme-responsive supramolecular gelator. *J. Am. Chem. Soc.* **137**, 770–775 (2015).

[27] Castelletto, V., Gouveia, R. M., Connon, C. J., *et al.* Biomaterials science alanine-rich amphiphilic peptide containing the RGD cell adhesion motif: a coating material for human fibroblast attachment and culture. *Biomater. Sci.* **2**, 362–369 (2014).

[28] Mata, A., Geng, Y., Henrikson, K. J., *et al.* Bone regeneration mediated by biomimetic mineralization of a nanofiber matrix. *Biomaterials* **31**, 6004–6012 (2010).

[29] Guvendiren, M., Lu, H. D. & Burdick, J. A. Shear-thinning hydrogels for biomedical applications. *Soft Matter* **8**, 260 (2012).

[30] Yan, C., Mackay, M. E., Czymmek, K., Nagarkar, R. P., Scheider, J. P. & Pochan, D. J. *et al.* Injectable solid peptide hydrogel as a cell carrier: effects of shear flow on hydrogels and cell payload. *Langmuir* **28**, 6076–6087 (2012).

[31] Haines, L., Rajagopal, K., Branco, M., *et al.* Controlling hydrogelation kinetics by peptide design for three-dimensional encapsulation and injectable delivery of cells. *Proc. Natl. Acad. Sci. U. S. A.* **104**, 7791–7796 (2007).

[32] Altunbas, A., Lee, S. J., Rajasekaran, S. A, Schneider, J. P. & Pochan, D. J. Encapsulation of curcumin in self-assembling peptide hydrogels as injectable drug delivery vehicles. *Biomaterials* **32**, 5906–5914 (2011).

[33] Rajagopal, K., Ozbas, B., Pochan, D. J. & Schneider, J. P. Probing the importance of lateral hydrophobic association in self-assembling peptide hydrogelators. *Eur. Biophys. J.* **35**, 162–169 (2006).

[34] Bakota, E. L., Wang, Y., Danesh, F. R. & Hartgerink, J. D. Injectable multidomain peptide nanofiber hydrogel as a delivery agent for stem cell secretome. *Biomacromolecules* **12**, 1651–1657 (2011).

[35] Galler, K. M., Aulisa, L., Regan, K. R., D'Souza, R. N. & Hartgerink, J. D. Self-assembling multidomain peptide hydrogels: designed susceptibility to enzymatic cleavage allows enhanced cell migration and spreading. *J. Am. Chem. Soc.* **132**, 3217–3223 (2010).

[36] Yokoi, H., Kinoshita, T. & Zhang, S. Dynamic reassembly of peptide RADA16 nanofiber scaffold. *Proc. Natl. Acad. Sci. U. S. A.* **102**, 8414–8419 (2005).

[37] Arosio, P., Owczarz, M., Wu, H., Butté, A. & Morbidelli, M. End-to-end self-assembly of RADA 16-I nanofibrils in aqueous solutions. *Biophys. J.* **102**, 1617–1626 (2012).

[38] Akiyama, N., Fukuda, T., Takahashi, H. & Koji, T., In situ tissue engineering with synthetic self-assembling peptide nanofiber scaffolds, PuraMatrix, for mucosal regeneration in the rat middle-ear. *Int. J. Nanomed.* **8**, 2629–2640 (2013).

[39] Liedmann, A., Frech, S., Morgan, P. J., Rolfs, A. & Frech, M. J. Differentiation of human neural progenitor cells in functionalized hydrogel matrices. *Biores. Open Access* **1**, 16–24 (2012).

[40] Dissanayaka, W. L., Hargreaves, K. M., Jin, L., Samaranayake, L. P. & Zhang, C. The interplay of dental pulp stem cells and endothelial cells in an

injectable peptide hydrogel on angiogenesis and pulp regeneration in vivo. *Tissue Eng. Part A* 1–14 (2014).

[41] Rubert Pérez, C. M., Stephanopoulos, N. Sur, S., Lee, S., Newcomb, C. & Stupp, S. I. The powerful functions of peptide-based bioactive matrices for regenerative medicine. *Ann. Biomed. Eng.* **43**, 501–514 (2014).

[42] Horii, A., Wang, X., Gelain, F. & Zhang, S. Biological designer self-assembling peptide nanofiber scaffolds significantly enhance osteoblast proliferation, differentiation and 3-D migration. *PLoS One* **2**, e190 (2007).

[43] Castillo-león, J., Andersen, K. B. & Svendsen, W. E. Self-assembled peptide nanostructures for biomedical applications: advantages and challenges. *Biomaterials Science and Engineering* (2011).

[44] Anderson, J. Modulating the gelation properties of self-assembling peptide amphiphiles. *ACS Nano* **3**, 3447–3454 (2010).

[45] Paramonov, S. E., Jun, H. & Hartgerink, J. D. Self-assembly of peptide amphiphile nanofibers: the roles of hydrogen bonding and amphiphilic packing. *J. Am. Chem. Soc.* **128**, 7291–7298 (2006).

[46] Behanna, H. A., Donners, J. J. J. M., Gordon, A. C. & Stupp, S. I. Coassembly of amphiphiles with opposite peptide polarities into nanofibers. *J. Am. Chem. Soc.* **127**, 1193–1200 (2005).

[47] Niece, K. L., Czeisler, C., Sahni, V., *et al.* Modification of gelation kinetics in bioactive peptide amphiphiles. *Biomaterials* **29**, 4501–4509 (2008).

[48] Niece, K. L., Hartgerink, J. D., Donners, J. J. J. M. & Stupp, S. I. Self-assembly combining two bioactive peptide-amphiphile molecules into nanofibers by electrostatic attraction. *J. Am. Chem. Soc.* **125**, 7146–7147 (2003).

[49] Uzunalli, G., Soran, Z., Erkal, T. S., *et al.* Bioactive self-assembled peptide nanofibers for corneal stroma regeneration. *Acta Biomater.* **10**, 1156–1166 (2014).

[50] Toksoz, S., Mammadov, R., Tekinay, A. B. & Guler, M. O. Electrostatic effects on nanofiber formation of self-assembling peptide amphiphiles. *J. Colloid Interface Sci.* **356**, 131–137 (2011).

[51] Lamm, S. M. Molecular self-assembly: design, synthesis, and characterization of peptidic materials for bio-and nano-technologies. *Nanotechnology* 63–66 (2007).

[52] Ye, Z., Zhang, H., Luo, H., *et al.* Temperature and pH effects on biophysical and morphological properties of self-assembling peptide RADA16-I. *J. Pept. Sci.* **14**, 152–162 (2008).

[53] Deshorkhi, A., Castelletto, V., Hamley, I. W., Adamcik, J. & Mezzenga, R. The effects of pH on the self-assembly of a collagen derived peptide amphiphile. *Soft Matter* **2**, 1–6 (2013).

[54] Baldwin, R. L. How Hofmeister interactions affect protein stability. *Biophys. J.* **71**, 2056–2063 (1996).

[55] Li, R., Horgan, C. C. & Long, B. Tuning the mechanical and morphological properties of self-assembled peptide hydrogels via control over the gelation mechanism through regulation of ionic strength and the rate of pH change. *RSC Adv.* **5**, 301–307 (2014).

[56] Shi, J., Gao, Y., Zhang, Y., Pan, Y. & Xu, B. Calcium ions to cross-link supramolecular nanofibers to tune the elasticity of hydrogels over orders of magnitude. *Langmuir* **27**, 14425–14431 (2011).

[57] Micklitsch, C. M., Knerr, P. J., Branco, M. C., Nagarkar, R., Pochan, D. J. & Schneider, J. P. Zinc-triggered hydrogelation of a self-assembling β-hairpin peptide. *Angew. Chemie* **123**, 1615–1617 (2011).

[58] Marchesan, S., Easton, C. D., Kushkaki, F., Waddington, L. & Hartley, P. G. Tripeptide self-assembled hydrogels: unexpected twists of chirality. *Chem. Commun. (Camb).* **48**, 2195–2197 (2012).

[59] Marchesan, S., Waddington, L., Easton, C. D., *et al.* Unzipping the role of chirality in nanoscale self-assembly of tripeptide hydrogels. *Nanoscale* **4**, 6752–6760 (2012).

[60] Marchesan, S., Easton, C. D., Styan, K. E., *et al.* Chirality effects at each amino acid position on tripeptide self-assembly into hydrogel biomaterials. *Nanoscale* **6**, 5172–5180 (2014).

[61] Xu, H., Wang, Y., Ge, X., *et al.* Twisted nanotubes formed from ultrashort amphiphilic peptide I3K and their templating for the fabrication of silica nanotubes. *Chem. Mater.* **22**, 5165–5173 (2010).

[62] Han, S., Cao, S., Wang, Y., *et al.* Self-assembly of short peptide amphiphiles: the cooperative effect of hydrophobic interaction and hydrogen bonding. *Chem. Eur. J.* **17**, 13095–13102 (2011).

[63] Cui, H., Cheetham, A. G., Pashuck, E. T. & Stupp, S. I. Amino acid sequence in constitutionally isomeric tetrapeptide amphiphiles dictates architecture of one-dimensional nanostructures. *J. Am. Chem. Soc.* **136**, 12461–12468 (2014).

[64] Xu, X. D., Jin, Y., Liu, Y., Zhang, X. Z. & Zhuo, R. X. Self-assembly behavior of peptide amphiphiles (PAs) with different length of hydrophobic alkyl tails. *Colloids Surf. B Biointerfaces* **81**, 329–335 (2010).

[65] Kuang, Y., Gao, Y. & Xu, B. Supramolecular hydrogelators of N-terminated dipeptides selectively inhibit cancer cells. *Chem. Commun. (Camb).* **47**, 12625–12627 (2011).

[66] Marchesan, S., Qu, Y., Waddington, L. J., *et al.* Self-assembly of ciprofloxacin and a tripeptide into an antimicrobial nanostructured hydrogel. *Biomaterials* **34**, 3678–3687 (2013).

[67] Vemula, P. K., Boilard, E., Syed, A., *et al.* On-demand drug delivery from self-assembled nanofibrous gels: a new approach for treatment of proteolytic disease. *J. Biomed. Mater. Res. A* **97**, 103–110 (2011).

[68] Vemula, P. K., Wiradharma, N., Ankrum, J. A., Miranda, O. R., John, G. & Karp, J. M. Prodrugs as self-assembled hydrogels: a new paradigm for biomaterials. *Curr. Opin. Biotechnol.* **24**, 1174–1182 (2013).

[69] Majumdar, S. & Siahaan, T. J. Peptide-mediated targeted drug delivery. *Med. Res. Rev.* **3**, 632–658 (2012).

[70] Pratt, R. L. & Kinch, M. S. Ligand binding up-regulates EphA2 messenger RNA through the mitogen-activated protein/extracellular signal-regulated kinase pathway. *Mol. Cancer Res.* **1**, 1070–1076 (2003).

[71] Chograni, M., Chaabouni, M., Maazoul, F. & Bouhamed, H. Investigation of four genes responsible for autosomal recessive congenital cataract and highly expressed in the brain in four unrelated Tunisian families. *Open J. Ophthalmol.* **02**, 64–70 (2012).

[72] Wang, S., Placzek, W. J., Stebbins, J. L. *et al.* Novel targeted system to deliver chemotherapeutic drugs to EphA2-expressing cancer cells. *J. Med. Chem.* **55**, 2427–2436 (2012).

[73] Wang, H., Yang, C., Wang, L., Kong, D., Zhang, Y. & Yang, Z., Self-assembled nanospheres as a novel delivery system for taxol: a molecular hydrogel with nanosphere morphology. *Chem. Comm. (Camb).* **47**, 4439–4441 (2011).

[74] Webber, M. J., Kessler, J. A. & Stupp, S. I. Emerging peptide nanomedicine to regenerate tissues and organs. *J. Intern. Med.* **267**, 71–88 (2010).

[75] Nune, M., Kumaraswamy, P., Maheswari Krishnan, U. & Sethuraman, S. Self-assembling peptide nanofibrous scaffolds for tissue engineering: novel approaches and strategies for effective functional regeneration. *Curr. Protein Pept. Sci.* **14**, 70–84 (2013).

[76] Kisiday, J., Jin, M., Kurz, B., *et al.* Self-assembling peptide hydrogel fosters chondrocyte extracellular matrix production and cell division: implications for cartilage tissue repair. *Proc. Natl. Acad. Sci. U. S. A.* **99**, 9996–10001 (2002).

[77] Ustun, S., Tombuloglu, A., Kilinc, M., Guler, M. O. & Tekinay, A. B. Growth and differentiation of prechondrogenic cells on bioactive self-assembled peptide nanofibers. *Biomacromolecules* **14**, 17–26 (2013).

[78] Lee, S. S., Hsu, E. L., Mendoza, M., *et al.* Gel scaffolds of BMP-2-binding peptide amphiphile nanofibers for spinal arthrodesis. *Adv. Healthc. Mater.* **4**, 131–141 (2015).

[79] Kocabey, S., Ceylan, H., Tekinay, A. B. & Guler, M. O. Glycosaminoglycan mimetic peptide nanofibers promote mineralization by osteogenic cells. *Acta Biomater.* **9**, 9075–9085 (2013).

[80] Ceylan, H., Kocabey, S., Gulsuner, H., Balcik, O. S, Guler, M. O. & Tekinay, A. B. Bone-like mineral nucleating peptide nanofibers induce differentiation of human mesenchymal stem cells into mature osteoblasts. *Biomacromolecules* **15**, 2407–2418 (2014).

[81] Li, A., Hokugo, A., Yalom, A., *et al.* A bioengineered peripheral nerve construct using aligned peptide amphiphile nanofibers. *Biomaterials* **35**, 8780–8790 (2014).

[82] Galler, K. M., Hartgerink, J. D., Cavender, A. C., Schmalz, G. & D'Souza, R. N. A customized self-assembling peptide hydrogel for dental pulp tissue engineering. *Tissue Eng. Part A* **18**, 176–184 (2012).

[83] Hsieh, P. C. H., Davis, M. E., Gannon, J., MacGillivray, C. & Lee, R. T. Controlled delivery of PDGF-BB for myocardial protection using injectable self-assembling peptide nanofibers. *J. Clin. Invest.* **116**, 237–248 (2006).

Part III

Nanomaterials for bacterial infections and their toxicity

Chapter 7
Antimicrobial properties of nanomaterials

Nhu-Y. Nguyen[1] *and Hiunan Liu*[1,2,3]

Abstract

Medical device related infections account for almost 64% of all hospital acquired infections. Even with sterile conditions and handling, infection occurs when the bacteria encounters the patients' protein, and usually leads to biofilm formation. The major infection players are *Staphylococcus aureus*, *Staphylococcus epidermidis*, *Escherichia coli*, and *Pseudomonas aeruginosa*. Some of the bacteria had become resistant to certain widely used antibiotics, but there is also a danger of them becoming multidrug resistant organisms, especially *S. aureus*. One way of combating these infections is using nanoparticles (NPs) in medical devices. Currently, NPs exhibit promising applications in medical devices to inhibit or prevent these bacterial infections. Some of the NPs are silver bromide (AgBr), metal oxides that include titanium dioxide (TiO_2), zinc oxide (ZnO), Magnetite (Fe_3O_4), and more recently magnesium oxide (MgO). All of these NPs have shown antimicrobial properties against gram-positive and gram-negative bacteria, while TiO_2 and ZnO also exhibit antifungal properties. Even though some NPs (i.e. TiO_2 and ZnO) have toxicity mechanisms against mammalian cells in vitro, they are still promising to be used as coating materials or as components in composites for medical devices to fight infections. This chapter will discuss the properties, characterization, and applications of these NPs for biomedical applications.

7.1 Introduction: biofilm formation results from bacteria adherence to medical devices

Medical devices made from synthetic materials for temporary or permanent implantation have been one of the major advances in the medical field and have improved many patients' lives [1]. These medical devices include but are not limited to heart valves, artificial veins, prosthetics, urinary catheter, and intravenous

[1]Department of Microbiology, University of California-Riverside, California, USA
[2]Department of Bioengineering, University of California-Riverside, California, USA
[3]Materials Science and Engineering Program, University of California-Riverside, California, USA

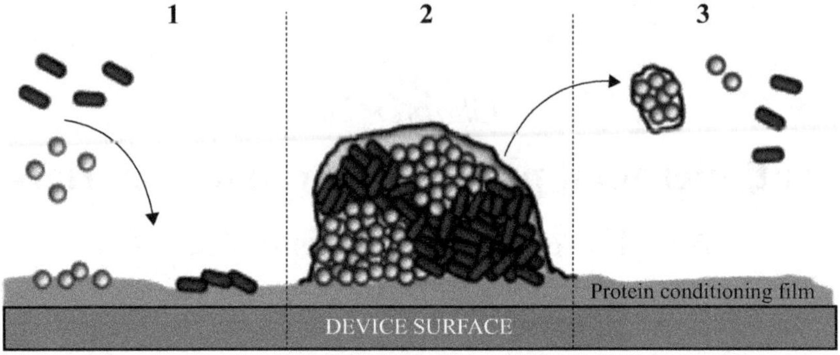

Figure 7.1 Schematic representation of biofilm formation. Stage 1: microorganism attached to the surface, Stage 2: production of exopolysaccharides and biofilm formed, and Stage 3: detachment. © 2010 Wiley. Reprinted with permission from Reference 6

catheter assisted health restoration, but they also face bacterial infection complications [1–3]. Bacterial infection directly contributes to implant failure, which develop immediately after implantation or years later [1]. Some of the most frequently isolated microbes are gram-positive bacteria, gram-negative bacteria, and yeast isolated from the indwelling medical devices (i.e. *S. aureus*, *S. epidermidis*, *E. coli*, *P. aeruginosa*, and *Candida* species) [4].

When foreign devices are implanted, the body reacts by covering it with a film of proteins like fibronectin, laminin, fibrin, collagen, and immunoglobulins, which some serve as receptors to bacteria. Bacteria causing infections can be inherent to the implant sites or introduced during surgical procedures from the patient and surgeons [2,5]. These infections can cause tissue destruction, dysfunction of a prosthetic device, systemic dissemination of the pathogen, and biofilm formation from bacteria attachment on the devices that lead to device removal [2,5–7].

Bacteria adherence will lead to microbial biofilms that could result in single or polymicrobial communities with different bacteria and fungus co-existing together [6]. As illustrated in Figure 7.1, bacteria need to adhere, produce exopolysaccharide layers, detach, and start over at another site. These microbial biofilms are an ideal niche for drug resistance to occur naturally through mutations or induced with antibiotic plasmid exchange. Drugs cannot penetrate these biofilms easily and the slow bacterial metabolism makes them inaccessible to many antibiotics [6].

As shown in Figure 7.2, the prevention of bacteria adherence is crucial in preventing biofilm formation and bacterial colonization. One of the emerging novel antimicrobial agents is nanoscale materials [8] due to their high surface to volume ratio with unique chemical and physical properties [9]. Nanoparticles are clusters of particles in the size range of 1–100 nm. For example, silver nanoparticles (nAg) [10], magnesium, copper, zinc, and titanium [11] are promising because they show antibacterial properties against a broad spectrum of bacteria, fungi, and viruses [12].

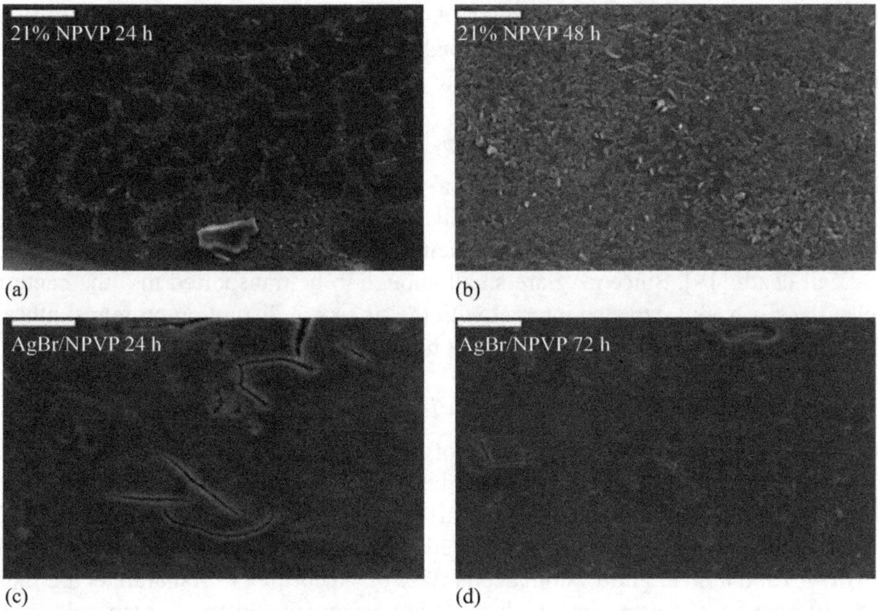

Figure 7.2 *SEM images of* P. aeruginosa *at 24 and 72 h on the nanocomposite. Dense bacteria were seen in the NPVP control but much less in the AgBr/NPVP composite. Scale bar = 10 μm. © 2006 ACS. Reprinted with permission from Reference 10*

Bacteria adhering to the surfaces of implants had been an ongoing problem; therefore, many different research groups have investigated different ways to combat infection by preventing bacteria attachment. For this reason, nanoparticles had been studied extensively to be used as antimicrobial agent for biomedical applications. The following sections represent some of the current nanoparticles being researched and explored for variety of biomedical applications.

7.2 Silver bromide nanoparticles

For centuries, silver has been used for treatment of different diseases due to its antimicrobial nature. Hippocrates treated ulcers with silver and Credeeven treated gonococcal infections in newborns [13]. In more recent times, sliver has been used in different forms for the treatment of a wide variety of infections, such as conjunctivitis and burns [13,14]. Through the years, a silver complex with other compounds was synthesized to treat different diseases. An example is silver nitrate, which was used to treat verneral disease, bone, and perianal abscesses [15,16]. For medical devices, silver has been used as a coating material to prevent bacteria attachment. Silver sulfadiazine is a combination of silver and sulfadiazine, which is used especially for burn wounds treatment [17]. Silver zeolite is a complex of alkaline earth metal with silver ions for food preservation and disinfection of medical devices [9]. Even though nAg has proven to be antimicrobial, its use is limited and careful consideration is needed in any application due to toxicity issues for mammalian cells.

7.2.1 Antimicrobial properties of nAg

The antimicrobial mechanism of nAg has been widely accepted to be partial oxidation and release of silver ions [8,18]. Silver ions interact with the peptidoglycan in the cell wall and the lipids in the cell membrane leading to increased permeability and cell death [19]. Since nAg are small enough to be transported into the bacteria cells, the silver ions can also interact with the ribosome disrupting protein synthesis and preventing bacteria from replicating by binding to the DNA bases [20,21].

7.2.2 Cytotoxicity and genotoxicity of nAg to mammalian cells

Similarly, the antimicrobial properties of nAg have been studied extensively; the cytotoxicity and genotoxicity have also been given a lot more attention in the last decade or more due to failure of implants with nAg coatings. Asharani *et al.* [22] published a paper examining starch coated nAg against human lung fibroblast cells (IMR-90) and human glioblastoma cells (U251). According to Asharani et al., there are different mechanisms in which nAg causes toxicity in the human cells [22]. First, the nanoparticles increased the production of reactive oxygen species (ROS) and mitochondria damage, which leads to DNA damage and causes cell cycle arrest at the G_2 phase of mitosis. This research suggests future studies to examine cell toxicity for different applications rather than just embracing the antimicrobial properties of these nAg.

7.2.3 Silver nanocomposites as a new approach to reduce toxicity

Due to nAg potential long-term toxicity in mammalian cells, debates ensue regarding whether or not it should be used for implantable nanobiomaterials. Newer studies have begun to combine nAg with polymers. A well-dispersed polymer matrix can also control the amount of bare nAg being released to the surrounding cells. The higher the concentrations of nAg, the greater the cytotoxicity. Thus, if some small amounts are being released overtime, then the detrimental effect on the local and systemic cells would be reduced. One example is a nanocomposite of nAgBr and poly(4-vinyl-*N*-hexylpyridinium bromide)-NPVP synthesized by Sambhy *et al.* for biomedical applications. This NPVP cationic polymer is known to not be toxic for mammalian cells, and the nanocomposite was made by precipitation chemistry [10]. Transmission electron microscopy (TEM) confirmed spherical nanoparticles embedded inside the solid polymer and the size was around 10 nm. The researchers tested their nanocomposite with both gram-positive (*Bacillus cereus* and *S. aureus*) and gram-negative (*E. coli* and *P. aeruginosa*) bacteria. The nanocomposite was able to prevent bacteria adherence onto the polymers at 24 and 48 h. As shown in Figure 7.2, there were much less bacteria adhered to the surface of the AgBr/NPVP nanocomposite compared to the NPVP polymer alone even at the 48 h incubation period. In this experiment, it was important to show that nAg retain their antimicrobial properties and yet they were not free floating. No in vitro or in vivo model was done for this nanocomposite; thus the toxicity was not evaluated.

More recently, nAg were combined with bioactive chitosan in a 3D hydrogel and was shown to maintain antimicrobial properties and exhibit no cytotoxicity to mammalian cells [23]. According to Travan *et al.*, the nanocomposite was made with a genipin crosslinked chitosan (GC) film with nAg embedded inside. nAg acts as a structural reinforcement and also adds antimicrobial properties to the polymer. The nanocomposite was characterized with UV–vis which showed that the nAg were well dispersed and stable. TEM images showed the GC/nAg matrix had no aggregation. To study the cytocompatibility, the GC/nAg matrix was cultured with L929 (rat fibroblast) cells. The cells were able to attach and proliferate on the nanocomposite without any sign of cytotoxicity when compared to the control. As for the antimicrobial properties, the GC/nAg matrix was tested against *E. coli*, which exhibited a high percentage of bacteria killed and also no bacteria were seen adhered to the surfaces of the matrix.

As described above, nAg when combined with other polymers could still maintain their antimicrobial properties and reduce cytotoxicity to mammalian cells. It could be that the nAg is not in free form, but they are confined in the polymer. This would prevent the nAg uptake by cells and reduce cytotoxicity or genotoxicity as described in the above section.

7.3 Titanium dioxide nanoparticles (nTiO_2)

nTiO_2 are well established photocatalytically activated antimicrobial properties [24]. nTiO_2 have been shown to be lethal to a broad spectrum of microorganisms, including gram-negative and gram-positive bacteria, bacteriophages, protozoa, and viruses [25,26]. TiO_2 primarily exhibits antimicrobial activity when photo-activated, but is also capable of antimicrobial activity in the absence of light [24]. The level of antimicrobial activity activated by visible light is still debatable [24]. Along with antimicrobial properties, TiO_2 had also been shown to cause cytotoxicity and genotoxicity in mammalian cells [27,28].

7.3.1 Photoactivation of TiO_2

TiO_2 is a semi-conductor which enables nTiO_2 to be responsive to light [24]. During photoactivation, some products are generated like ROS, superoxide anion (O_2^-) and hydroxyl radicals [24,25]. Figure 7.3 is a diagram showing the process of photocatalysis [29]. When nTiO_2 are illuminated by light with energy higher than the band gap, the electrons are promoted to the conduction band and creating a hole in the valence band [28,29]. There is high oxidizing power from the holes in the valence band and the conduction band has high reducing power, which allows electrons to react with other oxygen molecules to generate superoxide anions and hydroxyl radicals [25].

The efficiency of photocatalysis is dependent on the polymorphic form of TiO_2 [24]. There are three forms of TiO_2: anatase, brookite, and rutile. Anatase is

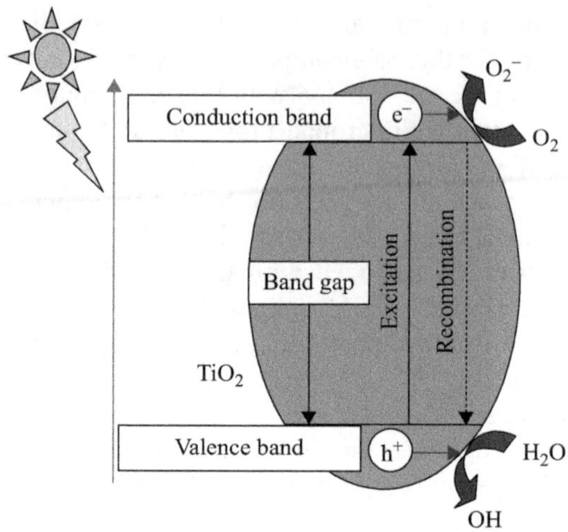

Figure 7.3 Photocatalytic reaction. © 2013 RSC. Reprinted with permission from Reference 29

reported to be the most efficient form, although a combination of anatase and brookite or rutile exhibits even higher efficiency than anatase alone [30,31]. The ROS generated during photoactivation play a critical role in the antimicrobial properties for microorganisms and cytotoxicity in mammalian cells [25,27].

7.3.2 Antimicrobial mechanism of TiO_2

Photocatalysis initiates the antimicrobial activity of TiO_2. As mentioned above, when $nTiO_2$ are irradiated with light, ROS are produced (superoxide anion and hydroxyl radicals) [24]. These ROS species can damage all types of organic biomolecules like carbohydrate, lipids, proteins and DNA [25,32]. The ROS causes damage to the bacteria membrane, which leads to potassium ion leakage, an indicator of cell death [33,34]. The sensitivity levels are different when comparing gram-negative and gram-positive bacteria, and these differences are linked to their cell wall structures [31,35]. In the cell wall of gram-positive bacteria, there is a much thicker peptidoglycan layer than gram-negative bacteria. SEM and TEM studies of *P. aeruginosa* conducted by Amezaga-Madrid *et al.* revealed significant disruption of membrane integrity [36]. When the cell wall and cell membrane are damaged, this also caused morphological changes to the bacteria. A study conducted by Park *et al.* showed the changed in size and shape of *E. coli* when they were exposed to $nTiO_2$. Most studies conclude that gram-negative bacteria are more sensitive to TiO_2. However, in some instance gram-positive bacteria unexpectedly have elevated sensitivity as compared to gram-negative bacteria. This is suggested to be a result from increased contact between the microorganism and TiO_2 catalyst [24].

There are several factors that can enhance the antimicrobials properties of $nTiO_2$. One factor is the size of the nanoparticles, the smaller the size the more lethal it is to bacteria [26]. Contact and the time of exposure between the microorganism and TiO_2 is also a factor in contributing to the degree of antimicrobial efficiency [26]. Longer exposure time of the nanoparticles to the bacteria will also increase bacteria killing percentage.

7.3.3 Cytotoxicity and genotoxicity of nTiO_2 in mammalian cells

Due to the increased use of nanomaterials for biomedical applications, it is important to identify any potential health risks [37]. Similar to nAg above, there have been many reports on the cytotoxicity and genotoxicity of nTiO_2. The studies have been conducted in vitro and then in vivo. The reason for this is because sometimes certain properties can change once the model changes from in vitro to in vivo models. In vivo models also give a better representation of how the nanoparticles will react in the human body.

Cytotoxicity and genotoxicity had been shown in many different cells line: human epidermal cells, human HaCat keratinocytes, and human fibroblast cells [27,37,38]. In human epidermal cells, Shukla *et al.* [37] studied cell cytotoxicity by looking at the mitochondrial activity using MTT assay kit and the cells exposed to nTiO_2 had 73% reduction compared to control. In genotoxicity assessment, Shukla *et al.* found DNA damage, micronucleus induction and also an increase of intracellular ROS production. Jaeger *et al.* [27] found similar results on human HaCat keratinocytes. Jaeger *et al.* show the agglomeration in the intracellular and extracellular space, which means the nanoparticles could be endocytosed by the cells. Further, it reveals that nTiO_2 have access to other cellular organelles causing genotoxicity and cytotoxicity.

In addition to the in vitro studies described above, Sycheva *et al.* [28] did an in vivo model using male CBAxB6 mice to study the cytotoxicity and genotoxicity of nTiO_2. The mice received seven daily doses of treatment. After 24 h from the last dose, organs including bone marrow, fore stomach, colon, testis, liver and brain were collected and tested. Genotoxicity was examined for all of the organs. The results showed that there was DNA damage and that micronuclei in the bone marrow cells and the liver cells had DNA damage. For the other organs, like the testis, the spermatids had two or more nuclei while there was higher apoptosis in the fore stomach. Inflammation and oxidative stress were also found in some of the organs, indicating there was a systemic problem [28].

7.3.4 nTiO_2 nanocomposite with polymer reducing toxicity

There are risks in using this nanoparticle due to its toxicity in mammalian cells, which means it should not be used in its bare state. Archana *et al.* [39] used nTiO_2 with chitosan and PVP to make a nanocomposite in order to achieve a reduction in toxicity and also maintain the antimicrobial properties of nTiO_2. To make the nanocomposite, chitosan was dissolved in 1% acetic solution, after the dissolved PVP solution was added, and finally 10 mg of nTiO_2 was put into to the mixture. FTIR showed a –OH band at 3,422 and 3,405 cm^{-1} indicative of chitosan and chitosan in the nanocomposite, respectively. TEM showed spherical particles with a polycrystalline nature as analyzed by the electron diffraction pattern.

In order to study the antimicrobial properties and cytotoxicity of chitosan–PVP–$nTiO_2$, the nanocomposite was cultured with bacteria strains of *S. aureus*, *P. aeruginosa*, *Bacillus subtilis*, and also against mouse fibroblast and embryonic cells. The nanocomposite was able to inhibit the growth of all the tested bacteria strains. Gram-positive bacteria showed more sensitivity to the nanocomposite compared to the gram-negative bacteria. The viability of the cells after culturing 3 days and 7 days was evaluated with Alamar Blue, and there was over 97% viability even after 7 days of culture. In addition, an in vivo model was done with an albino rat model to see the wound closure rate. The wounds treated with the chitosan–PVP–$nTiO_2$ healed significantly faster than the chitosan alone and the controls groups. The nanocomposites were able to reduce the toxicity compared to the bare $nTiO_2$ and still maintain its antimicrobial properties against gram-positive and gram-negative bacteria [39].

7.4 Zinc oxide nanoparticles (nZnO)

nZnO have a wide range of applications in different industries (cosmetics, paint, electronic, building materials, food, and medicine) due to their unique properties such as: catalytic capacity, optoelectronic, photocatalytic, electrical and anti-microbial [40,41]. In the year 2003/2004, the global production of nZnO and $nTiO_2$ was around 1,000 tons in sunscreen [42,43]. ZnO is also a semiconductor like TiO_2, but they have a larger band gap of 3.37 eV and do not need ultraviolet (UV) light to become catalytically active [44,45].

7.4.1 nZnO use in cosmetics, textiles, and food industries

With their unique UV blocking properties, nZnO are used in cosmetics and the textiles industries. nZnO are added in sunscreen to protect the skin from adverse effects of UVB and UVA radiation [46,47]. In textiles, nZnO are used as UV-absorbers to further protect human against solar radiation [48]. Nanomaterials have high surface area to volume ratio, which increases the effectiveness in blocking UV radiation compare to bulk material [48]. nZnO are inorganic materials that are preferred over organic materials because they are chemically stable under harsh production conditions of high temperature and pressure [49].

Textiles industries take advantage of nZnO antimicrobial properties to enhance the quality of their fabrics [48]. The use of nanotechnology has increased because conventional methods do not have permanent effects and lose function after laundering or wearing [48]. Nanoparticles provide higher durability and ensure better affinity for fabric [48]. There are three problems that microorganisms pose on fabrics: degradation phenomena of color, unpleasant odors from the microorganism byproducts, and increased health risks [50]. Conventional fibers and polymers are prone to accumulation, multiplication, and proliferation of microorganisms into surrounding environment. Several factors such as temperature, humidity, present of dust, soil, dead cells, and sweats from skin can make textiles optimal enrichment cultures for the rapid multiplication of organisms [50].

Annually, the food sector is worth over 4 trillion globally, and food related applications for nanotechnologies offered a wide range of benefits, notably antimicrobial properties [51]. nZnO are used in food packaging applications, nanocoatings on food contact surfaces to prevent bacterial growth during processing [51]. nZnO had also been incorporated into linings of can foods for meat, fish, corn, and peas to preserve color and prevent spoilage [52].

7.4.2 Antimicrobial mechanism of nZnO

nZnO are effective against both a broad spectrum of bacteria to fungi. The following microorganism are known to be inhibited by nZnO: foodborne pathogens – *E. coli* O157:H7 [49,53], *Listeria monocytogenes* [54], *B. cereus* [54], *Campylobacter jejuni* [52], *Salmonella typhimurium* [55]; other pathogenic bacteria – *Streptococcus mutans* [56], *S. arueus* [57], *B. subtilis* [58], *E. coli*, MRSA, *Streptococcus pyogenes* and *Enterococcus faecalis* [59]; and also fungi – *Botrytis cinerea, Penicillium expansum* [60].

The antimicrobial mechanism of nZnO includes the production of ROS (hydrogen peroxide), intracellular content interaction with nZnO and cell membrane damage [49,52,57,59]. Liu *et al.* showed the membrane of *E. coli* O157:H7 was damaged and deformed with disorganized intracellular structures after 12 h of incubation with 12 mmol/L compared to nZnO-free media group (Figure 7.5) [49]. An earlier experiment done by Brayner *et al.* also indicated damaged and disorganization of the cell wall, a change in *E. coli* morphology and the internalization of the nanoparticles (Figure 7.4) [44]. Xie *et al.* showed nZnO had direct interaction with the surfaces of *C. jejuni* and changed their morphology from spiral to coccoid form. In addition, Xie *et al.* reported nZnO induced oxidative gene (*kat*A, *ahp*C, and *dna*K) expression up to a 52-fold increase [52]. nZnO also showed antifungal toward fruit fungus like *B. cinerea* and *P. expansum* [60]. The nanoparticles inhibited the growth of both fungus and caused morphological changed. After exposure to nZnO, *B. cinerea* hyphae were rough with unusual bulges on the surface and *P. expansum* no longer able to produce conidia [60]. The mechanism of nZnO toward the fungus was interfering with cell functions.

Both bacteria and fungus showed morphological changes after exposure to nZnO. These morphology changes can be through the direct contact of the nZnO to the cell wall of the bacteria and fungi. Production of ROS also plays a role in the death of these microorganisms.

7.4.3 Toxicity of nZnO for mammalian cells

The production of nZnO has increased greatly in the last decades; however, attention to toxicity evaluation was only recently. nZnO were "generally consider as safe" by the Food and Drug Administration; therefore toxicity was not evaluated until recently with the rise and development of nanotoxicology [41]. The toxicology studies of nZnO are far behind the nZnO application development [41]. Human contact with nZnO increases due to direct skin exposure through sunscreen, nanoparticles dust in factories, and uncontrolled released into the environment like the

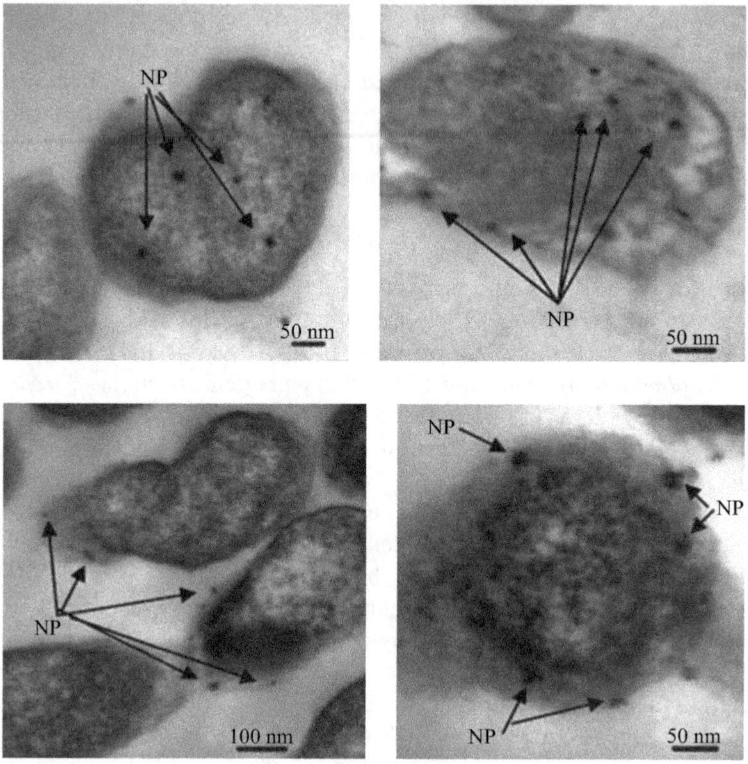

Figure 7.4 TEM images of E. coli *internalizing the nZnO. NP: nanoparticles. © 2006 ACS. Reprinted with permission from Reference 44*

aquatic ecosystem [40,61]. The toxicity of nZnO is still being investigated because there are debates and contradictory findings due to the different types of nZnO and their synthesis conditions such as temperature, pH, and light exposure [40,41].

The toxicity of some nanoparticles has been associated with production of ROS. Song *et al.* showed that Zn^{2+} played an important role in the ZnO toxicity effect by the induction of intracellular ROS production when Zn^{2+} enters the macrophage cells. According to Song *et al.*, the toxicity of nZnO correlates with dose but is not size dependent [41]. Additionally, Sharma *et al.* reported the genotoxicity with DNA damage caused by oxidative stress and the production of ROS leading to mitochondria mediated cell apoptosis in liver cells (HepG2) [62]. Furthermore, nZnO posed toxicity concerns for the aquatic environment [40]. When Tilapia fish were exposed to 1 and 10 mg/L of small (10–30 nm) and larger size (100 nm) ZnO for 14 days, various pathologies to the kidney (tubular deromation, necrosis), liver (edema, mononuclear cell in filtration), gill (hyperplasia, epithelial lifting), and intestine (hyperplasia, swelled of goblet cells and villus deformations) were found. Kaya *et al.* also reported osmoregulatory changes and immune disorders in Tilapia fish.

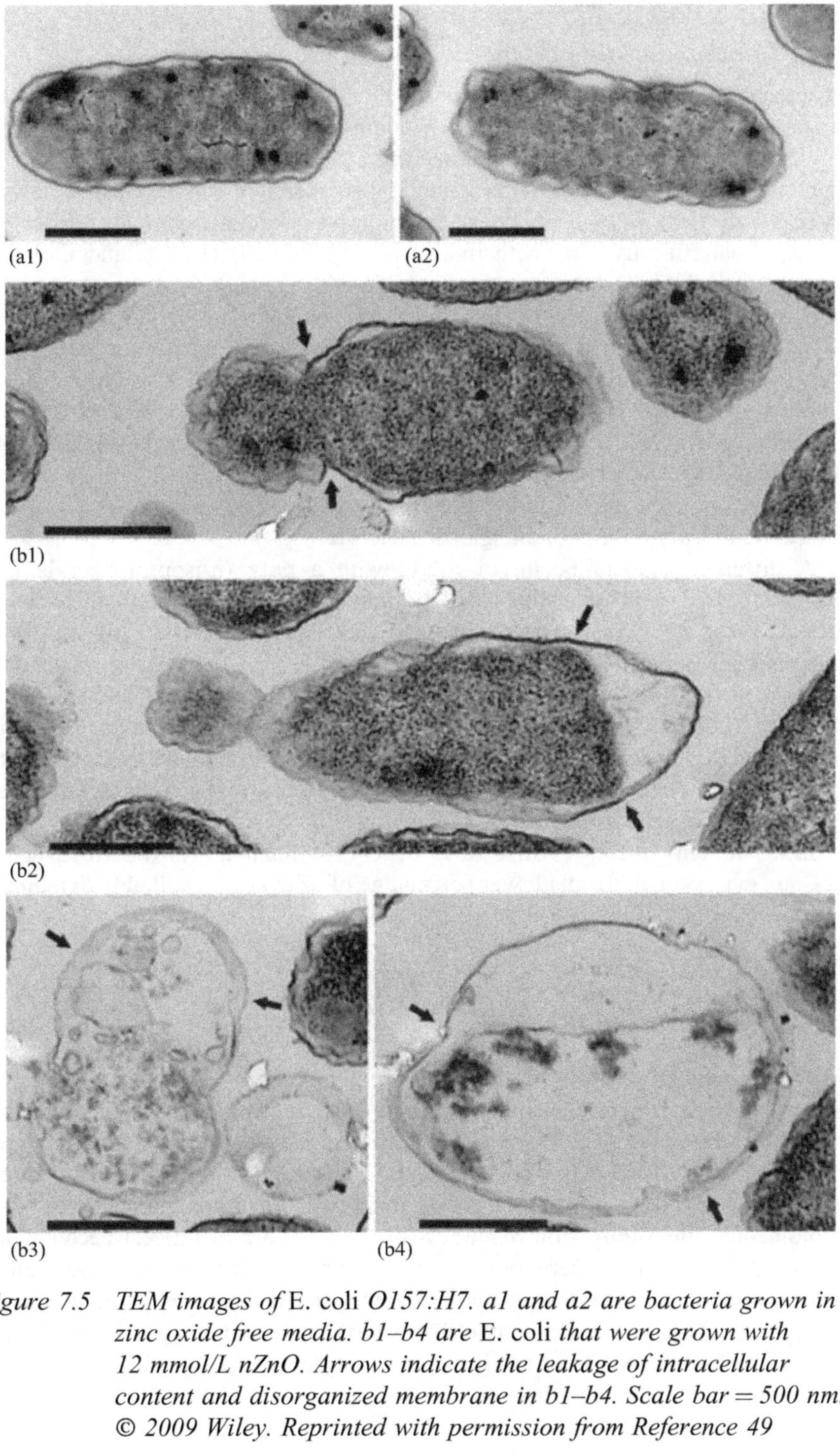

Figure 7.5 *TEM images of* E. coli *O157:H7. a1 and a2 are bacteria grown in zinc oxide free media. b1–b4 are* E. coli *that were grown with 12 mmol/L nZnO. Arrows indicate the leakage of intracellular content and disorganized membrane in b1–b4. Scale bar = 500 nm. © 2009 Wiley. Reprinted with permission from Reference 49*

7.4.4 Reducing nZnO toxicity by combining with other biocompatible polymers

ZnO are combined with other biocompatible materials to reduce nZnO toxicity and maintain their antimicrobial properties. Shalumon *et al.* combined nZnO with poly (vinyl alcohol) and sodium alginate making a Sa/PVA/ZnO nanofibrous mat for wound dressing applications. This material demonstrated inhibition of *S. aureus* and *E. coli* through diffusion disk tests. In addition, Shalumon *et al.* also tested SA/PVA/ZnO nanofibrous mats with mouse fibroblasts cells (L929), and the blend with 0.5% of nZnO oxide was cytocompatible with the cells [63].

Another study done by Munchow *et al.* was to synthesize poly(e-caprolactone) (PCL) and PCL/gelatin membranes loaded with 0–30% nZnO for periodontal regeneration. These nZnO loaded membranes were cultured with human dental pulp stem cells (hDPSCs) for 48 h, membranes loaded up to 15% showed cytocompatibility with hDPSCs and the 30% nZnO content had significantly decreased cell viability. However, the PCL or PCL/gelatin membrane loaded with 15% nZnO showed a clear zone of inhibition against *Porphyromonas gingivalis* and *Fusobacterium nucleatum* through an agar diffusion test [64].

A different nanocomposite of nZnO with a poly (n-isopropylacrylamide) (PNIPAAm) hydrogel was synthesized by Schwartz *et al.* for a biomedical device coating. One to 10% of nZnO were blended with PNIPAAm. The nanocomposites were made into thin (60–85 nm) and thick (230–400 nm) films. After exposure to *E. coli*, both thin and thick hydrogel nanocomposite films were able to reduce bacteria growth compared to control or 1% nZnO. In the 5% nZnO blend, the thick film showed higher efficacy of inhibiting bacteria compared to the thin film. At the same 5% nZnO blend, the hydrogel nanocomposite film was non-cytotoxic toward mouse fibroblast cells following 24 h co-culture [65].

nZnO toxicity increases in a dose dependent manner. As described above, nZnO are cytocompatible at a lower percentage of nZnO but is still able to maintain their antimicrobial effects.

7.5 Magnetite nanoparticles (nFe_3O_4)

nFe_3O_4 or magnetite has been used for drug delivery, tissue repair, magnetic resonance imaging, treatment of cancer and more recently for use in decreasing implant infections [66,67]. The pure metal iron possesses higher saturation magnetization, but it is also toxic and sensitive to oxidation. Iron oxide or magnetite is less sensitive to oxidation and is also less toxic which is more applicable to use in the biomedical field [66]. Iron oxide less than 15 nm are of interest because they show biocompatibility, superparamagnetic properties, and also avoid filtration by the reticulo-endothelial system which allows longer circulation in the body [68,69]. Superparamagnetic properties enable the material to lose its magnetic capabilities after the initial magnetic field has been removed [66].

7.5.1 Antimicrobial and biocompatibility properties of nFe_3O_4

Due to the nature of the biofilm formation on medical implants, it is difficult for antibiotics and antifungals to penetrate and kill microorganism and sometimes can lead to drug resistance [70]. Even some current nanoparticles cannot penetrate this biofilm; thus they are used to prevent initial microorganism adhesion before the formation of the biofilm. However, nFe_3O_4 have been proposed to be used to treat biofilm formation because the magnetic field can facilitate deep penetration into the biofilm and kill the microorganism inside [1,70].

7.5.1.1 Superparamagnetic iron oxide nanoparticles (SPIONs) antibacterial against staphylococci

Many studies had been done to evaluate SPIONs in preventing infections and also treat grown biofilm. One of the many recent experiments evaluating SPIONs antimicrobial properties was conducted by Subbiahdoss *et al.* [1]. The researchers modified nFe_3O_4 by surface grafting functional groups of carboxyethylsilanetriol (CES), aminopropyltriethoxysilane (APTES), or PEGylated (PEG). All of the three different functional grafted groups had sizes ranging from 13 to 20 nm using TEM, insinuating the size variation was small.

In order to study the antimicrobial properties of SPIONs grafted functional groups, gram-positive *S. epidermidis* gentamycin resistant strain was used. The *S. epidermidis* was allowed to grow for 24 h without the CES grafted SPIONs or other experimental materials. After 24 h cultures, the CES-grafted SPIONs group was added to culture and this was allowed to grow for another 24 h. With the targeted CES-grafted and APTES-grafted SPIONs, there were more dead bacteria observed on the bottom of the biofilm compared to the control group. There are several antimicrobial mechanisms proposed for SPIONs. One of the more well known is the production of ROS that generates superoxide radicals, hydroxyl radicals, hydrogen perroxides and singlet oxygen that cause DNA and protein damage [71,72].

An earlier study done by Taylor *et al.* [70] also looked at just SPIONs bare particles with *S. epidermidis*. As concentrations of SPIONs increased, there was more bacteria death in solution compared to the control group which had no SPIONs. This study showed an inhibition of biofilm formation, which is similar to many other nanoparticles.

7.5.2 Biocompatibility of magnetic particles to mammalian cells

Biocompatibility is an important consideration for all nanoparticles. However, the biocompatibility of magnetic particles has not been studied as much compared to other nanoparticles, and there is no clear conclusion whether the nanoparticles are really toxic or not.

Subbiahdoss *et al.* [1] looked into the biocompatibility of the different SPIONs grafted functional groups against U2OS osteosarcoma cells. The cells were cultured for 12 h before the addition of bare SPIONs and the different functionalized groups.

There were no differences in the growth of the cells at 24 h between all groups compared to the control (cells only). Even the bare SPIONs group had similar results to the control. According to Subbiahdoss *et al.*, there was no toxicity found in any of the SPIONs functionalized or bare particle groups.

On other hand, another study done by Gupta and Gupta [73] showed there was toxicity in the bare SPIONs as compared to the coated Pullulan (Pn) SPIONs. Human fibroblasts were cultured without any of the materials for 24 h. Post 24 h culture, media was changed, and 0–2.0 mg/mL SPIONs bare or Pn coated were added into the cultured plates and let the cells grow for an additional 24 h. The results showed that the bare SPIONs group had a reduction of 80% cell viability compared to the control group even at the lower concentration of 0.5 mg/mL. While the coated Pn SPIONs had more than 92% viability compared to the control group even at the highest concentration of 2.0 mg/mL.

There are some other studies that showed the magnetic nanoparticles are not toxic or have very little acute toxicity. For example, Cengelli *et al.* [74] reported no toxicity found in brain-derived endothelial cells and murine microglial cells when they were cultured with iron oxide nanoparticles coated with dextran. Another experiment done by Muller *et al.* [75] showed no cytotoxic effects on human macrophages cultured with iron oxide nanoparticles at 1 mg/mL and acute toxicity at 10 mg/mL. Most recently, in 2014, Peter *et al.* showed that there was no evidence of toxicity with brain cells including: astrocytes, microglia, oligodendrocytes, and neurons [76].

More studies with in vivo models and also longer time points will help to further elucidate toxicity problems. There is one clear point, at low concentrations there is very low toxicity. However, another point to consider is whether or not the low concentrations of iron oxide can still maintain its antimicrobial properties in order to be used in certain biomedical applications.

7.6 Magnesium oxide nanoparticles (nMgO)

Magnesium oxide is one of the newer metal oxide nanoparticles with an increase of research being performed over the past decade. Since it is still fairly new, the mechanism for killing bacteria is still not fully elucidated. In addition, there are very few studies done to assess the biocompatibility of nMgO with mammalian cells.

7.6.1 Unique properties of nMgO

As an inorganic ionic compound, nMgO are stable and able to endure harsh processing [77,78]. MgO is able to withstand high pressure and temperature conditions (i.e. during calcination, a thermal decomposition procedure used in the synthesis of nMgO, temperatures of 600–800°C are reached) [77–79]. nMgO have a high specific surface area which provides a large reaction platform. In addition, nMgO possess a high concentration of low coordinated sites and crystal defects on its surface [79]. Low coordinated sites and crystal defects translate to the presence of atoms with unsatisfied valancies, which can move toward the surface of the particle and exhibit varying reactivity toward adhered chemicals [80]. As a result, nMgO possess the unique ability to absorb various gases such as halogens [77].

MgO can take on the form of powder or liquid slurries. MgO can be synthesized into the following morphology: one-dimensional wire, rod, fiber, or whisker. These specific morphologies have gained interest for a variety of applications [78].

7.6.2 Antimicrobial properties of nMgO

nMgO are classified as a bactericide against many bacteria and spores [81]. nMgO also exhibit bacteriostatic and bactericidal activity. Bactericidal activity is the capability to kill the bacteria, while bacteriostatic refers to the ability to inhibit any further growth.

It had been well received that the antimicrobial mechanism is via the production of a superoxide ion (O_2^-). Superoxide O_2^- has previously been detected on MgO on the surface of the MgO powder slurry (0.5 mg/mL concentration of nMgO) through chemiluminescence [82]. nMgO are easily hydrated in aqueous solution and forms $Mg(OH)_2$. Hydroxide ions dissociate from $Mg(OH)_2$ creating an alkaline environment of approximately pH 10. In the presence of dissolved oxygen, oxygen can be reduced to form O_2^- radicals and are stabilized by the alkaline environment. The O_2^- radical performs a nucleophilic attack on the carbonyl carbon ($C^{\delta+}$) within the peptide linkages of proteins in the cell coat and wall of spores, and bacteria, respectively [77,83,84].

However, a more recent paper has shown that the mechanism of nMgO against *E. coli* was non-ROS mediated and the bacteria death is due more to the damage of the cell wall and cell membrane [85]. Earlier reports also supported the theory of cell wall and cell membrane damage [77]. MgO has a positive zeta potential which is essential in making the initial contact to the negatively charged bacterial cell wall and membrane components [77,83]. Leung *et al.* [85] showed that there was no peak at the position that is indicative of ROS compared to a control of H_2O_2in the electron spin resonance spectra. TEM of the bacteria cultured with nMgO showed a change in the bacteria morphology and cell membrane damage. The researchers also showed proteomics data with the absence of oxidative stress proteins and a decrease in outer membrane protein components within the 20 min incubations of bacteria and the nMgO [85].

The general consensus among literature is that the smaller particle size enhances bactericidal activity. Again, this can be ascribed to the fact that at smaller particle sizes, there is higher surface area to interact with the microorganism. To elucidate this trend Makhluf *et al.* varied MgO nanoparticle sizes (8, 11, 15, 18, and 23 nm) and tested them against *E. coli* and *S. aureus* at 1 mg/mL concentrations. The 8 $\pm$ 1 nm particle size was most effective at killing both bacterial species after 4 h. *Escherichia coli* was inactivated to a greater extent than its gram positive counterpart *S. aureus* (95% inactivation for *S. aureus* and $>$ 99.99% for *E. coli*) [79]. This is consistent with the Al-Hazmi *et al.* [77], in which the reduced sensitivity is ascribed to the MgO particles becoming trapped in the thicker peptidoglycan layer of gram-positive species.

7.6.3 Biocompatibility of nMgO

Even though the antibacterial activity of MgO has been reported and several mechanisms hypothesized, few cytotoxicity reports of mammalian cells have been reported. The 2012 paper by Horie *et al.* analyzed the cytotoxicity of HaCat cells and human lung carcinoma A549 cells when exposed to nMgO for 24 h. There were less viable cells when they were exposed to 1 mg/mL nMgO compared to the control, but the cell viability remained high at 0.1 mg/mL nMgO [86].

Conversely, a recent report from Monzavi *et al.* [87] showed that there was no cytotoxicity found in HeLa cells by measuring the lactate dehydrogenase (LDH) levels. The cells were incubated with 2.5–10 mg/L nMgO for 24 h and LDH levels were the same in all nMgO treatments compared to the control with no treatment. Although, Monzavi *et al.* group did suggest to have a longer time point study to confirm and ensure that there is actually no toxicity [87].

7.7 Conclusion

Throughout this chapter, several nanoparticles possessing antimicrobial properties have been described and taken into account their cytotoxic and/or genotoxic nature. The cytotoxicity was reduced when the nanoparticles were coated or embedded into a nanocomposite with a biocompatible polymer. Such nanocomposites are being proposed for applications like wound dressing to enhance wound healing and also to fight infections by using chitosan–PVP–nTiO_2. Other potential applications that were proposed include but were not limited to neuronal applications, treating an established biofilm that is resistant to drugs using nFe_3O_4, and other orthopedic implants to prevent the initial adherence of bacteria. These nanomaterials, while being coated or embedded in a nanocomposite, still maintain their antimicrobial properties while limiting cytotoxic effects. If these nanocomposites or coated nanomaterials are made and used for the right application; they could positively aid in the prevention of implant related infections.

References

[1] Subbiahdoss, G., S. Sharifi, D. Frijpma, *et al.*, Magnetic targeting of surface-modified superparamagnetic iron oxide nanoparticles yields antibacterial efficacy against biofilms of gentamicin-resistant staphylococci. *Acta Biomaterialia*, 2012. **8**(6): 2047–2055.

[2] Dickinson, G.M. and A.L. Bisno, Infections associated with indwelling devices: concepts of pathogenesis; infections associated with intravascular devices. *Antimicrobial Agents and Chemotherapy*, 1989. **33**(5): 597.

[3] Reid, G., Biofilms in infectious disease and on medical devices. *International Journal of Antimicrobial Agents*, 1999. **11**(3): 223–226.

[4] Donlan, R.M., Biofilms and device-associated infections. *Emerging Infectious Diseases*, 2001. **7**(2): 277.

[5] Weinstein, R.A. and R.O. Darouiche, Device-associated infections: a macroproblem that starts with microadherence. *Clinical Infectious Diseases*, 2001. **33**(9): 1567–1572.

[6] Francolini, I. and G. Donelli, Prevention and control of biofilm-based medical-device-related infections. *FEMS Immunology & Medical Microbiology*, 2010. **59**(3): 227–238.

[7] Zimmerli, W., A. Trampuz, and P.E. Ochsner, Prosthetic-joint infections. *New England Journal of Medicine*, 2004. **351**(16): 1645–1654.

[8] Morones, J.R., J.L. Elechiguerra, A. Camacho, *et al.*, The bactericidal effect of silver nanoparticles. *Nanotechnology*, 2005. **16**(10): 2346.

[9] Rai, M., A. Yadav, and A. Gade, Silver nanoparticles as a new generation of antimicrobials. *Biotechnology Advances*, 2009. **27**(1): 76–83.

[10] Sambhy, V., M. MacBride, B. Peterson, and A. Sen, Silver bromide nanoparticle/polymer composites: dual action tunable antimicrobial materials. *Journal of the American Chemical Society*, 2006. **128**(30): 9798–9808.

[11] Schabes-Retchkiman, P., G. Canizal, R. Herrera-Becerra, C. Zorilla, H.B. Liu, and J.A. Ascencio, Biosynthesis and characterization of Ti/Ni bimetallic nanoparticles. *Optical Materials*, 2006. **29**(1): 95–99.

[12] Gong, P., H. Li, X. He, *et al.*, Preparation and antibacterial activity of Fe_3O_4@Ag nanoparticles. *Nanotechnology*, 2007. **18**(28): 285604.

[13] Chaloupka, K., Y. Malam, and A.M. Seifalian, Nanosilver as a new generation of nanoproduct in biomedical applications. *Trends in Biotechnology*, 2010. **28**(11): 580–588.

[14] Huang, Y., X. Li, Z. Liao, *et al.*, A randomized comparative trial between Acticoat and SD-Ag in the treatment of residual burn wounds, including safety analysis. *Burns*, 2007. **33**(2): 161–166.

[15] Klasen, H., Historical review of the use of silver in the treatment of burns. I. Early uses. *Burns*, 2000. **26**(2): 117–130.

[16] Lansdown, A., Silver I: its antibacterial properties and mechanism of action. *Journal of Wound Care*, 2002. **11**(4): 125–130.

[17] Atiyeh, B.S., M. Costagliola, S.N. Hayek, and S.A. Dibo, Effect of silver on burn wound infection control and healing: review of the literature. *Burns*, 2007. **33**(2): 139–148.

[18] Lok, C.-N., C.M. Ho, R. Chen, *et al.*, Silver nanoparticles: partial oxidation and antibacterial activities. *JBIC Journal of Biological Inorganic Chemistry*, 2007. **12**(4): 527–534.

[19] Sondi, I. and B. Salopek-Sondi, Silver nanoparticles as antimicrobial agent: a case study on *E. coli* as a model for Gram-negative bacteria. *Journal of Colloid and Interface Science*, 2004. **275**(1): 177–182.

[20] Yang, W., C. Shen, Q. Ji, *et al.*, Food storage material silver nanoparticles interfere with DNA replication fidelity and bind with DNA. *Nanotechnology*, 2009. **20**(8): 085102.

[21] Yamanaka, M., K. Hara, and J. Kudo, Bactericidal actions of a silver ion solution on *Escherichia coli*, studied by energy-filtering transmission

electron microscopy and proteomic analysis. *Applied and Environmental Microbiology*, 2005. **71**(11): 7589–7593.

[22] Asharani, P., Y.L. Wu, Z. Gong, and S. Valiyaveettil, Toxicity of silver nanoparticles in zebrafish models. *Nanotechnology*, 2008. **19**(25): 255102.

[23] Travan, A., C. Pelillo, I. Donati, *et al.*, Non-cytotoxic silver nanoparticle-polysaccharide nanocomposites with antimicrobial activity. *Biomacromolecules*, 2009. **10**(6): 1429–1435.

[24] Foster, H.A., I. Ditta, S. Varghese, and A. Steele, Photocatalytic disinfection using titanium dioxide: spectrum and mechanism of antimicrobial activity. *Applied Microbiology and Biotechnology*, 2011. **90**(6): 1847–1868.

[25] Li, Y., W. Zhang, J. Niu, and Y. Chen, Mechanism of photogenerated reactive oxygen species and correlation with the antibacterial properties of engineered metal-oxide nanoparticles. *ACS Nano*, 2012. **6**(6): 5164–5173.

[26] Park, S., S. Lee, B. Kim, *et al.*, Toxic effects of titanium dioxide nanoparticles on microbial activity and metabolic flux. *Biotechnology and Bioprocess Engineering*, 2012. **17**(2): 276–282.

[27] Jaeger, A., D.G. Weiss, L. Jonas, and R. Kriehuber, Oxidative stress-induced cytotoxic and genotoxic effects of nano-sized titanium dioxide particles in human HaCaT keratinocytes. *Toxicology*, 2012. **296**(1): 27–36.

[28] Sycheva, L.P., V.S. Zhurkovet, V.V. Iurchenko, *et al.*, Investigation of genotoxic and cytotoxic effects of micro-and nanosized titanium dioxide in six organs of mice in vivo. *Mutation Research/Genetic Toxicology and Environmental Mutagenesis*, 2011. **726**(1): 8–14.

[29] Yin, Z.F., L. Wu, W.G. Yang, and Y.H. Su, Recent progress in biomedical applications of titanium dioxide. *Physical Chemistry Chemical Physics*, 2013. **15**(14): 4844–4858.

[30] Miyagi, T., M. Kamei, T. Mitsuhashi, T. Ishigaki, and A. Yamazaki, Charge separation at the rutile/anatase interface: a dominant factor of photocatalytic activity. *Chemical Physics Letters*, 2004. **390**(4): 399–402.

[31] Shah, R.R., S. Kaewgum, B. Lee, and T.R. Tzeng, The antibacterial effects of biphasic brookite-anatase titanium dioxide nanoparticles on multiple-drug-resistant *Staphylococcus aureus*. *Journal of Biomedical Nanotechnology*, 2008. **4**(3): 339–348.

[32] Visai, L., L. De Nardo, C. Punta, *et al.*, Titanium oxide antibacterial surfaces in biomedical devices. *International Journal of Artificial Organs*, 2011. (**34**): 929–946.

[33] Hu, C., J. Guo, J. Qu, and X. Hu, Photocatalytic degradation of pathogenic bacteria with AgI/TiO_2 under visible light irradiation. *Langmuir*, 2007. **23**(9): 4982–4987.

[34] Kambala, V.S.R. and R. Naidu, Disinfection studies on TiO_2 thin films prepared by a sol–gel method. *Journal of Biomedical Nanotechnology*, 2009. **5**(1): 121–129.

[35] Kim, B., D. Kim, D. Cho, and S. Cho, Bactericidal effect of TiO_2 photocatalyst on selected food-borne pathogenic bacteria. *Chemosphere*, 2003. **52**(1): 277–281.

[36] Amezaga-Madrid, P., R. Silveyra-Morales, L. Córdoba-Fierro, *et al.*, TEM evidence of ultrastructural alteration on Pseudomonas aeruginosa by photocatalytic TiO2 thin films. *Journal of Photochemistry and Photobiology B: Biology*, 2003. **70**(1): 45–50.

[37] Shukla, R.K., V. Sharma, A. Pandey, S. Singh, S. Sultana, and A. Dhawan, ROS-mediated genotoxicity induced by titanium dioxide nanoparticles in human epidermal cells. *Toxicology In Vitro*, 2011. **25**(1): 231–241.

[38] Setyawati, M.I., P.K.T. Khoo, B.H. Eng, *et al.*, Cytotoxic and genotoxic characterization of titanium dioxide, gadolinium oxide, and poly (lactic-co-glycolic acid) nanoparticles in human fibroblasts. *Journal of Biomedical Materials Research Part A*, 2013. **101**(3): 633–640.

[39] Archana, D., B. Singh, J. Dutta, and P.K. Dutta, In vivo evaluation of chitosan–PVP–titanium dioxide nanocomposite as wound dressing material. *Carbohydrate Polymers*, 2013. **95**(1): 530–539.

[40] Kaya, H., F. Aydin, M. Gürkan, *et al.*, A comparative toxicity study between small and large size zinc oxide nanoparticles in tilapia (*Oreochromis niloticus*): organ pathologies, osmoregulatory responses and immunological parameters. *Chemosphere*, 2016. **144**: 571–582.

[41] Song, W., J. Zhang, J. Guo, *et al.*, Role of the dissolved zinc ion and reactive oxygen species in cytotoxicity of ZnO nanoparticles. *Toxicology Letters*, 2010. **199**(3): 389–397.

[42] Piccinno, F., F. Gottschalk, S. Seeger, and B. Nowack, Industrial production quantities and uses of ten engineered nanomaterials in Europe and the world. *Journal of Nanoparticle Research*, 2012. **14**(9): 1–11.

[43] Ma, H., P.L. Williams, and S.A. Diamond, Ecotoxicity of manufactured ZnO nanoparticles–a review. *Environmental Pollution*, 2013. **172**: 76–85.

[44] Brayner, R., R. Ferrari-Iliou, N. Brivois, S. Djediat, M. Benedetti, and F. Fiévet, Toxicological impact studies based on *Escherichia coli* bacteria in ultrafine ZnO nanoparticles colloidal medium. *Nano Letters*, 2006. **6**(4): 866–870.

[45] Rajendra, R., C. Balakumar, H.A.M. Ahmmed, S. Jayakumar, K. Vaideki, and E. Rajesh, Use of zinc oxide nano particles for production of antimicrobial textiles. *International Journal of Engineering, Science and Technology*, 2010. **2**(1): 202–208.

[46] Smijs, T.G. and S. Pavel, Titanium dioxide and zinc oxide nanoparticles in sunscreens: focus on their safety and effectiveness. *Nanotechnology, Science and Applications*, 2011. **4**: 95.

[47] Cuddy, M.F., A.R. Poda, R.D. Moser, C.A. Weiss, C. Cairns, and J.A. Steevens, A weight-of-evidence approach to identify nanomaterials in consumer products: a case study of nanoparticles in commercial sunscreens. *Journal of Exposure Science and Environmental Epidemiology*, 2015.

[48] Becheri, A., M. Durr, P.L. Nostro, and P. Baglioni, Synthesis and characterization of zinc oxide nanoparticles: application to textiles as UV-absorbers. *Journal of Nanoparticle Research*, 2008. **10**(4): 679–689.

[49] Liu, Y., L. He, A. Mustapha, H. Li, Z.Q. Hu, and M. Lin, Antibacterial activities of zinc oxide nanoparticles against *Escherichia coli* O157: H7. *Journal of Applied Microbiology*, 2009. **107**(4): 1193–1201.

[50] Dastjerdi, R. and M. Montazer, A review on the application of inorganic nano-structured materials in the modification of textiles: focus on anti-microbial properties. *Colloids and Surfaces B: Biointerfaces*, 2010. **79**(1): 5–18.

[51] Chaudhry, Q. and L. Castle, Food applications of nanotechnologies: an overview of opportunities and challenges for developing countries. *Trends in Food Science & Technology*, 2011. **22**(11): 595–603.

[52] Xie, Y., Y. He, P. Irwin, T. Jin, and X. Shi, Antibacterial activity and mechanism of action of zinc oxide nanoparticles against *Campylobacter jejuni*. *Applied and Environmental Microbiology*, 2011. **77**(7): 2325–2331.

[53] Sharifalhoseini, Z., M.H. Entezari, and R. Jalal, Direct and indirect sonication affect differently the microstructure and the morphology of ZnO nanoparticles: optical behavior and its antibacterial activity. *Ultrasonics Sonochemistry*, 2015. **27**: 466–473.

[54] Mirhosseini, M. and F. Barzegari Firouzabadi, Reduction of listeria monocytogenes and *Bacillus cereus* in milk by zinc oxide nanoparticles. *Iranian Journal of Pathology*, 2015. **10**(2): 97–104.

[55] Tayel, A.A., W.F. ElTras, S. Moussa, *et al.*, Antibacterial action of zinc oxide nanoparticles against foodborne pathogens. *Journal of Food Safety*, 2011. **31**(2): 211–218.

[56] Hernández-Sierra, J.F., F. Ruiz, D.C.C. Pena, *et al.*, The antimicrobial sensitivity of Streptococcus mutans to nanoparticles of silver, zinc oxide, and gold. *Nanomedicine: Nanotechnology, Biology and Medicine*, 2008. **4**(3): 237–240.

[57] Sawai, J. and T. Yoshikawa, Quantitative evaluation of antifungal activity of metallic oxide powders (MgO, CaO and ZnO) by an indirect conductimetric assay. *Journal of Applied Microbiology*, 2004. **96**(4): 803–809.

[58] Meruvu, H., M. Vangalapati, S.C. Chippada, and S.R. Bammidi, Synthesis and characterization of zinc oxide nanoparticles and its antimicrobial activity against *Bacillus subtilis and Escherichia coli. Chem*, 2011. **4**(1): 217–222.

[59] Jones, N., B. Ray, K. Ranjit, and A. Manna, Antibacterial activity of ZnO nanoparticle suspensions on a broad spectrum of microorganisms. *FEMS Microbiology Letters*, 2008. **279**(1): 71–76.

[60] He, L., Y. Liu, A. Mustapha, and M. Lin, Antifungal activity of zinc oxide nanoparticles against *Botrytis cinerea* and *Penicillium expansum*. *Microbiological Research*, 2011. **166**(3): 207–215.

[61] Vandebriel, R.J. and W.H. De Jong, A review of mammalian toxicity of ZnO nanoparticles. *Nanotechnology, Science and Applications*, 2012. **5**: 61.

[62] Sharma, V., D. Anderson, and A. Dhawan, Zinc oxide nanoparticles induce oxidative DNA damage and ROS-triggered mitochondria mediated apoptosis in human liver cells (HepG2). *Apoptosis*, 2012. **17**(8): 852–870.

[63] Shalumon, K., K.H. Anulekha, S. Nair, S.V. Nair, K.P. Chennazhi, and R. Jayakumar, Sodium alginate/poly (vinyl alcohol)/nano ZnO composite nanofibers for antibacterial wound dressings. *International Journal of Biological Macromolecules*, 2011. **49**(3): 247–254.

[64] Münchow, E.A., M.T.P. Albuquerque, B. Zero, *et al.*, Development and characterization of novel ZnO-loaded electrospun membranes for periodontal regeneration. *Dental Materials*, 2015. **31**(9): 1038–1051.

[65] Schwartz, V.B., F. Thétiot, S. Ritz, *et al.*, Antibacterial surface coatings from zinc oxide nanoparticles embedded in poly (N-isopropylacrylamide) hydrogel surface layers. *Advanced Functional Materials*, 2012. **22**(11): 2376–2386.

[66] Tran, N. and T.J. Webster, Magnetic nanoparticles: biomedical applications and challenges. *Journal of Materials Chemistry*, 2010. **20**(40): 8760–8767.

[67] Abe, S., I. Kida, M. Esaki, *et al.*, Biodistribution imaging of magnetic particles in mice: X-ray scanning analytical microscopy and magnetic resonance imaging. *Bio-Medical Materials and Engineering*, 2009. **19**(2): 213–220.

[68] Gupta, A.K. and M. Gupta, Synthesis and surface engineering of iron oxide nanoparticles for biomedical applications. *Biomaterials*, 2005. **26**(18): 3995–4021.

[69] Song, S., G. Bohuslav, A. Capitano, *et al.*, Experimental characterization of electrochemical synthesized Fe nanowires for biomedical applications. *Journal of Applied Physics*, 2012. **111**(5): 056103.

[70] Taylor, E.N. and T.J. Webster, The use of superparamagnetic nanoparticles for prosthetic biofilm prevention. *International Journal of Nanomedicine*, 2009. **4**: 145.

[71] Kim, J.S., E. Kuk, K.N. Yu, *et al.*, Antimicrobial effects of silver nanoparticles. *Nanomedicine: Nanotechnology, Biology and Medicine*, 2007. **3**(1): 95–101.

[72] Li, B. and B.E. Logan, Bacterial adhesion to glass and metal-oxide surfaces. *Colloids and Surfaces B: Biointerfaces*, 2004. **36**(2): 81–90.

[73] Gupta, A.K. and M. Gupta, Cytotoxicity suppression and cellular uptake enhancement of surface modified magnetic nanoparticles. *Biomaterials*, 2005. **26**(13): 1565–1573.

[74] Cengelli, F., D. Maysinger, F. Tschudi-Monnet, *et al.*, Interaction of functionalized superparamagnetic iron oxide nanoparticles with brain structures. *Journal of Pharmacology and Experimental Therapeutics*, 2006. **318**(1): 108–116.

[75] Müller, K., J.N. Skepper, M. Posfai, *et al.*, Effect of ultrasmall superparamagnetic iron oxide nanoparticles (Ferumoxtran-10) on human monocyte-macrophages in vitro. *Biomaterials*, 2007. **28**(9): 1629–1642.

[76] Petters, C., E. Irrsack, M. Koch, and R. Dringen, Uptake and metabolism of iron oxide nanoparticles in brain cells. *Neurochemical Research*, 2014. **39**(9): 1648–1660.

[77] Al-Hazmi, F., F. Alnowaiser, A.A. Al-Ghamdi, *et al.*, A new large-scale synthesis of magnesium oxide nanowires: structural and antibacterial properties. *Superlattices and Microstructures*, 2012. **52**(2): 200–209.

[78] Comini, E., G. Faglia, M. Ferroni, A. Ponzoni, A. Vomiero, and G. Sberveglieri, Metal oxide nanowires: preparation and application in gas sensing. *Journal of Molecular Catalysis A: Chemical*, 2009. **305**(1): 170–177.

[79] Makhluf, S., R. Dror, Y. Nitzan, Y. Abramovich, R. Jelinek, and A. Gedanken, Microwave-assisted synthesis of nanocrystalline MgO and its use as a bacteriocide. *Advanced Functional Materials*, 2005. **15**(10): 1708–1715.

[80] Mishakov, I.V., A.F. Bedilo, R.M. Richards, *et al.*, Nanocrystalline MgO as a dehydrohalogenation catalyst. *Journal of Catalysis*, 2002. **206**(1): 40–48.

[81] Koper, O.B., J. Klabunde, G. Marchin, K. Klabunde, P. Stoimenov, and L. Bohra, Nanoscale powders and formulations with biocidal activity toward spores and vegetative cells of *Bacillus* species, viruses, and toxins. *Current Microbiology*, 2002. **44**(1): 49–55.

[82] Sawai, J., H. Kohima, H. Igarashi, *et al.*, Antibacterial characteristics of magnesium oxide powder. *World Journal of Microbiology and Biotechnology*, 2000. **16**(2): 187–194.

[83] Huang, L., D.-Q. Li, Y.-J. Lin, M. Wei, D. Evans, and X. Duan, Controllable preparation of Nano-MgO and investigation of its bactericidal properties. *Journal of Inorganic Biochemistry*, 2005. **99**(5): 986–993.

[84] Seil, J.T. and T.J. Webster, Antimicrobial applications of nanotechnology: methods and literature. *International Journal of Nanomedicine*, 2012. **7**: 2767.

[85] Leung, Y.H., A. Ng, X. Xu, *et al.*, Mechanisms of Antibacterial activity of MgO: non-ROS mediated toxicity of MgO nanoparticles towards *Escherichia coli. Small*, 2014. **10**(6): 1171–1183.

[86] Horie, M., K. Fujita, H. Kato, *et al.*, Association of the physical and chemical properties and the cytotoxicity of metal oxide nanoparticles: metal ion release, adsorption ability and specific surface area. *Metallomics*, 2012. **4**(4): 350–360.

[87] Monzavi, A., S. Eshraghi, R. Hashemian, and F. Momen-Heravi, In vitro and ex vivo antimicrobial efficacy of nano-MgO in the elimination of endodontic pathogens. *Clinical Oral Investigations*, 2015. **19**(2): 349–356.

Chapter 8

Nanoparticle-enhanced radiation killing of bacteria

Yang Luo[1,2], Yong Qiao[1], Sichao Hou[1] and Ming Su[1]

Abstract

Multidrug-resistant (MDR) bacterial infections of deep wounds are hard to be treated. Physical antimicrobial agents can be used to remove bacteria from superficial wounds, but are less effective in removing bacteria deeply buried underneath the dermis. This chapter describes a nanoparticle enhanced X-ray radiation technique that can be used to kill MDR bacteria. Taking *Pseudomonas aeruginosa* as an example, polyclonal antibody modified bismuth nanoparticles are introduced into bacterial culture to specifically target *P. aeruginosa*. After washing off uncombined bismuth nanoparticles, the bacteria are irradiated with X-rays, using a setup that mimics a deeply buried wound in humans. Results show that up to 90% of *P. aeruginosa* are killed in the presence of 200 mg/mL bismuth nanoparticles, whereas only 6% are killed in the absence of bismuth nanoparticles when exposed to 40 kV X-rays for 10 min. The 200 mg/mL bismuth nanoparticles can enhance localized X-ray dose by 35 times higher than the control with no nanoparticles. In addition, no significant harmful effects on human cell lines have been observed with 200 mg/mL bismuth nanoparticles and 10 min of 40 kV X-ray irradiation exposure.

8.1 Introduction

The worldwide misuse and overuse of antibiotics have generated strains with MDR [1]. MDR bacteria induced wound infections can be life threatening if not properly treated. In the United States, MDR bacteria cause a majority of nosocomial infection related death [2]. The discovery of antibiotics targeting new biochemical pathways is a serious health challenge with the utmost importance and urgency.

[1]Department of Chemical Engineering, Northeastern University, Boston, MA, USA
[2]Medical Research Center, Southwest Hospital, Third Military Medical University, Chongqing, China

But the pipeline for new antibiotics has slowed: New drug approvals for antibiotics have decreased for every 5-year period over the last 25 years with only four approved during 2003–2007, and two new classes of antibiotics have been introduced in the past 40 years [3–5]. Although the combination of antimicrobial therapy can deal with MDR induced infections, the misuse of multiple antibiotics can potentially lead to pan-drug resistant bacteria, which will pose an even more serious threat on the human health [6–7].

Given that pathogenic bacteria have evolved drug resistance, physical methods that kill bacteria without triggering new resistance have received a lot of attention [8]. Although a variety of physical or chemical methods has been applied for sterilization, most of them are only applicable to non-living objects [9]. Autoclaving is the most effective way of sterilization, but cannot be used in vivo due to high temperature [10]. Ultraviolet irradiation can prevent bacteria growth by causing adjacent thymine on DNA to dimerize, thereby inhibiting DNA replication. For in vivo removal of bacterial from the wound, silver nanoparticles have been added into bandages to prevent infection and help wounds heal [11]. Gold nanoparticles have been used to enhance the photo-thermal killing [12–14]. Aseptic debridement or intensive cleaning can remove MDR bacteria on superficial wounds; they cannot remove bacteria including MDR ones deeply buried underneath the dermis.

Ionizing radiations such as X-ray and Gamma ray can kill bacteria by causing DNA damage. Rather than causing contamination as radioactive materials, high energy X-ray radiation can easily penetrate most of tissues with excellent transmission efficiency. The Food and Drug Administration (FDA) has approved the use of X-ray irradiation to remove pathogens from foods such as fruits, vegetables, and grains. More recently, X-ray radiation has been developed to eliminate bacteria in living oysters without causing the death of oysters. Many gram-negative bacteria such as *P. aeruginosa*, *Escherichia coli*, and *Salmonella* species can be effectively controlled in vitro by X-rays, but in vivo use of X-ray radiation on the human body is rare due to concerns related to the carcinogenic nature of ionizing radiation. Reducing X-ray dose can lower radiation damage, but the bactericide activity is also decreased. Metallic nanoparticles can be used as radiosensitizers due to a large cross-section of X-ray absorption and secondary electron generation. The electrons will cause water radiolysis and produce free radicals (hydroxyl radicals) to damage bacterial DNA. If damaged DNAs cannot be repaired by enzymes in bacteria, damage can be accumulated, eventually causing bacterial death.

This chapter describes a new bactericide method that is based on nanoparticle enhanced X-ray radiation by taking MDR *P. aeruginosa* as an example. Bismuth nanoparticles are mixed with bacteria or modified with polyclonal antibodies against bacteria. After removing excess nanoparticles, low dose X-ray is used to irradiate bacteria (Figure 8.1(a)). The effect of the combined use of nanoparticles and X-rays for in vitro bacteria killing has been confirmed with a variety of techniques such as biochemical methods, and bacterial culture and counting.

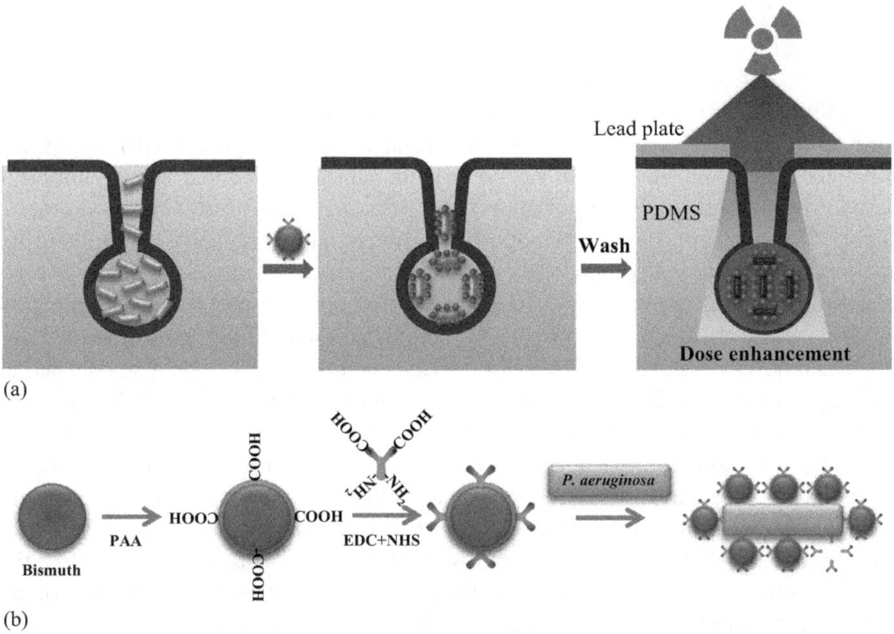

Figure 8.1 *Nanoparticle enhanced X-ray radiation killing of MDR bacteria in deep tissue (a), where bacteria are grown underneath a 2 cm thick PMMA block and are irradiated vertically. (b) The scheme of antibody-conjugated nanoparticles and bacteria*

8.2 Experimental

Bismuth nanoparticles are made as follows: 0.1 mmol of bismuth nitrate ($Bi(NO_3)_3 \cdot 5H_2O$) and 0.5 mmol polyvinylpyrrolidone (PVP) are dissolved in 10 mL of dimethylformamide (DMF). The mixture is degassed with argon under stirring for 15 min. 300 μl of 1 M $NaBH_4$ in water is mixed with 10 mL of DMF and added in the mixture of $Bi(NO_3)_3$ and PVP under vigorous stirring and argon flow for 5 min. The nanoparticles are precipitated by adding acetone, followed by centrifugation, washing with acetone and drying in vacuum. The morphology of bismuth nanoparticles is derived using a JEOL 1011 transmission electron microscope (TEM) operated at 100 kV. An X-ray spectrometer (Amptek X-123) with Si-PIN photodiode is used to analyze X-ray fluorescence (XRF) emissions from nanoparticles in the transmission mode. 25 μm thick silver and 250 μm thick aluminum filters are used to reduce background and improve signal-to-noise ratio in the low energy (0–15 keV) region of spectrum.

Carboxyl acid group terminated bismuth nanoparticles are obtained by mixing nanoparticles with a 0.5 M polyacrylic acid (PAA) solution and stirring for 24 h and washed with distilled water. The carboxyl acid modified nanoparticles are dispersed in phosphate buffer saline (PBS, pH 7.4). To conjugate the antibody, the carboxylic-terminated nanoparticles are allowed to react with the amine groups of *P. aeruginosa* antibodies for 3 h in the presence of 1-ethyl-3-(3-dimethylamino-propyl) carbodiimide (EDC), a water-soluble carbodiimide that promotes amide bond formation between carboxylic acid and primary amine. Figure 8.1(b) shows the scheme of the coupling procedure. The unbound antibody in supernatant solution is derived by measuring the absorbance at 280 nm and compared with a calibration curve. The amount of antibody bound to bismuth nanoparticles is determined by subtracting the amount of unbound antibody from the total antibody added. Fourier transform infrared spectroscopy (FTIR) measurements are performed in reflection mode to verify the conjugation between nanoparticle and *P. aeruginosa* antibody.

Bacteria are cultured in Luria Broth (LB) medium at 37°C on a shaker bed at 200 rpm for 10 h and diluted with the fresh broth to a concentration of 10^8 CFU/mL (corresponding to an optical density of 0.1 at 600 nm). 10 μl of bacterial suspension and 100 μl of antibody modified nanoparticle suspension are added into 1.89 mL of broth in test tubes and cultured at 37°C on a shaker bed at 200 rpm. A 2 mL of broth contains the according concentration of bacteria used as the blank control. 100 μl of the mixture is pipetted in a falcon 96-well plate to measure $QD_{600\ nm}$. After the bacteria concentration is determined, 100 μl of the suspensions is inoculated on an agar plate containing LB supplemented with 2% agar. The concentration of plated cells is adjusted to achieve full coverage of the plate once colonies are developed (10^3 cells are spread over the plate which corresponds to 10 cells/cm^2).

A Mini-X X-ray tube from Amptek (Bedford, MA) with a silver anode operating at 40 kV and 100 μA is used to produce primary X-rays irradiation. The tube is fitted with a series of brass and tungsten collimators to reduce outgoing beam size from 2 mm to 400 μm in diameter. For bactericide activity experiment, all types of bacteria grown in the broth and the LB agar plates are placed 5 cm away from X-ray generator and irradiated perpendicularly with the same dose at room temperature. After irradiation, all bacteria cultures are incubated overnight at 37°C. Then the viable cells in the agar plates are counted in form of colonies and the cells grown in the broth are counted by measuring the $QD_{600\ nm}$. For the X-ray penetrating experiment, a polydimethylsiloxane (PDMS) film with thickness ranging from 2 to 10 cm is applied to mimic human tissue. A charge-coupled device (CCD) is used to test photon number that penetrated across the PDMS. As X-rays impinge on the CCD, individual X-ray photons are converted into light photons that appear as bright dots on the screen and can be counted using Image plus Pro 6 software.

8.3 Results and discussion

8.3.1 Nanoparticle synthesis and modification

Bismuth nanoparticles are made by thermal decompositions of organometallic precursor (bismuth acetate). TEM image shows that bismuth nanoparticles have an average diameter of ~20 nm (Figure 8.2(a)). X-ray fluorescence (XRF) spectrum shows the characteristic $L\alpha_1$ and $L\beta_1$ peaks at 10.82 and 13.02 keV,

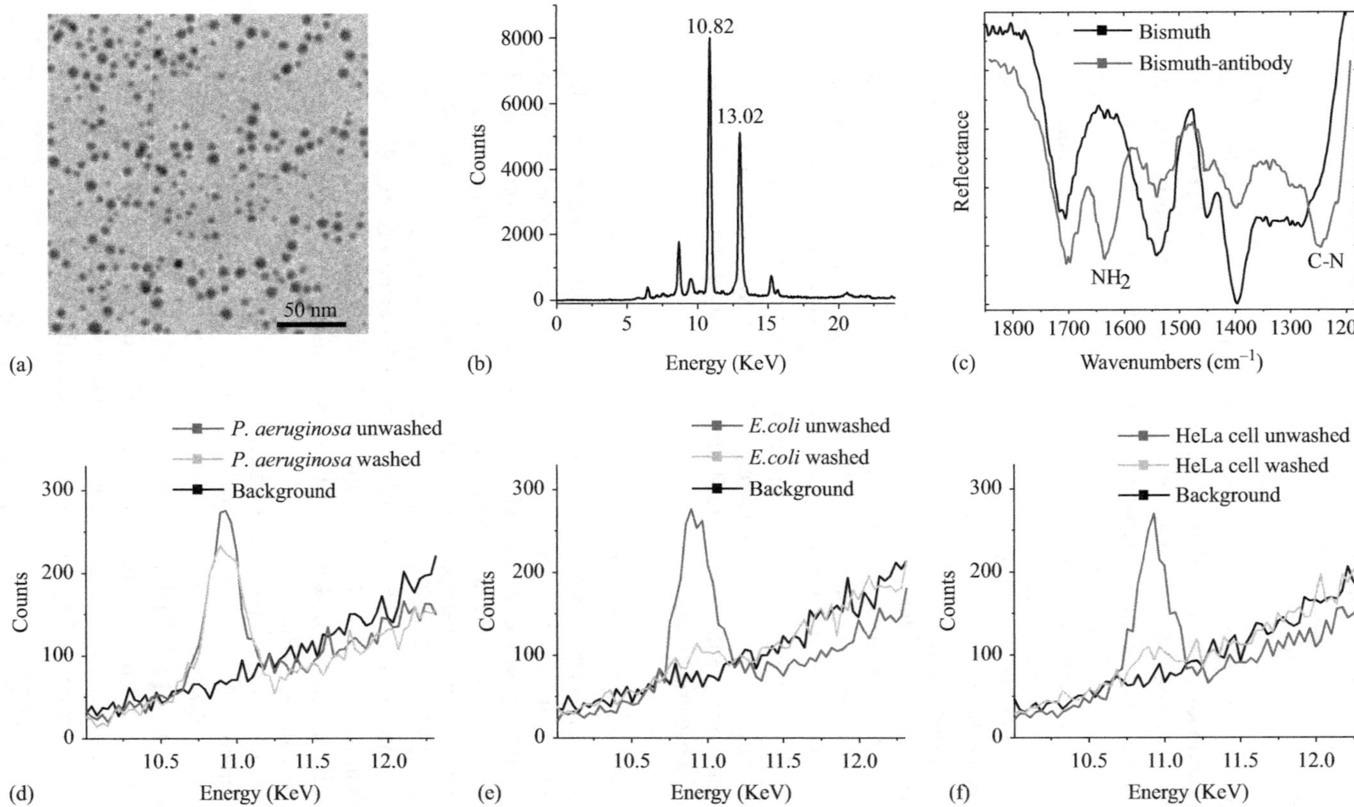

Figure 8.2 Bismuth nanoparticles and their conjugations with antibodies. TEM image (a) and XRF spectrum (b) of bismuth nanoparticles; (c) FTIR spectra of unmodified and antibody-attached bismuth nanoparticles. XRF spectra in the culture of P. aeruginosa *(d), E. coli (e), and HeLa cells (f) after incubating with* P. aeruginosa *antibody-conjugated bismuth nanoparticles for 24 h. Samples were washed intensively prior to XRF measurements. The black, grey and light grey lines are background, unwashed samples, and washed samples, respectively*

respectively (Figure 8.2(b)). The surface of bismuth nanoparticles is slightly oxidized by heating at 120°C in order to be modified with silane derivatives.

The polyclonal *P. aeruginosa* antibody is conjugated to the surface of the bismuth nanoparticle by carboiimide chemistry. The successful conjugation of antibody is verified by FTIR. The presence of C—N (1,000–1,250 cm^{-1}) and NH_2 scissoring (1,550–1,650 cm^{-1}) peaks from amide bonds of free antibody and confirms the attachment of the antibody on nanoparticles (Figure 8.2(c)). The yield of conjugation is determined by measuring the absorbance of the total antibody added in solution, and the unbound antibody in the supernatant after centrifugation. The grafting density of the antibody bound to bismuth nanoparticles is calculated to be ~12.5 per particle by calculating the difference between the initial addition and the amount of unbound antibody determined in the supernatants.

The specificity of the *P. aeruginosa* antibody conjugated to the bismuth nanoparticle is determined by adding the antibody modified nanoparticles in the MDR *P. aeruginosa* suspension and *E. coli* suspension, respectively. A 100 μl of the mixture is inoculated and cultured for 12 h in LB agar plate, followed by an intensive wash to remove excess bismuth nanoparticle. The XRF spectrum shows that the bismuth signal in the *P. aeruginosa* culture plate decreases 20% of the original value after washing (Figure 8.2(d)), while it decreases 90% in the *E. coli* culture plate (Figure 8.2(e)), indicating that the *P. aeruginosa* antibody functionalized bismuth nanoparticle can combine with *P. aeruginosa* with high specificity. The weak XRF signal in washed *E. coli* culture plates may be due to the small amount of non-specifically attached nanoparticles on the bacteria colony.

8.3.2 Nanoparticle cytotoxicity

Beside bacteria control, mammalian cell controls are also performed to test the possibility of nonspecific binding of *P. aeruginosa* antibody coated bismuth nanoparticles. Ovarian cancer HeLa cell and osteosarcoma MG63 cell were adopted in the specificity test, because they are from a different origination. After incubating antibody-conjugated bismuth nanoparticles with cell cultures containing 80% confluent HeLa cell, and MG63 cell for 12 h, cell cultures are intensively washed, and XRF spectra of the plates are recorded. Figure 8.2(f) shows no observable XRF signal of bismuth in the washed HeLa cell culture and in the washed MG63 culture (data not shown), indicating the high specificity of the *P. aeruginosa* antibody functionalized bismuth nanoparticles.

The toxicity of bismuth nanoparticles is tested on both bacteria and HeLa cells. Cell viability is measured after incubating with bare or antibody functionalized bismuth nanoparticles for 2 h, respectively. Figure 8.3(a) shows that bare nanoparticles kill ~4.6% and ~21% of HeLa cells at the concentration of 2 and 200 μg/mL, respectively, while *P. aeruginosa* antibody conjugated nanoparticles kill ~3.2% and ~9% of HeLa cells accordingly. Antibody functionalized nanoparticles kill less HeLa cells than bare nanoparticles, because surface modification reduces the cytotoxicity of bismuth nanoparticles [15].

Apart from cytotoxicity on mammalian cells, nanoparticles show bactericide activity. The growth of *P. aeruginosa* in the presence of bismuth nanoparticles is

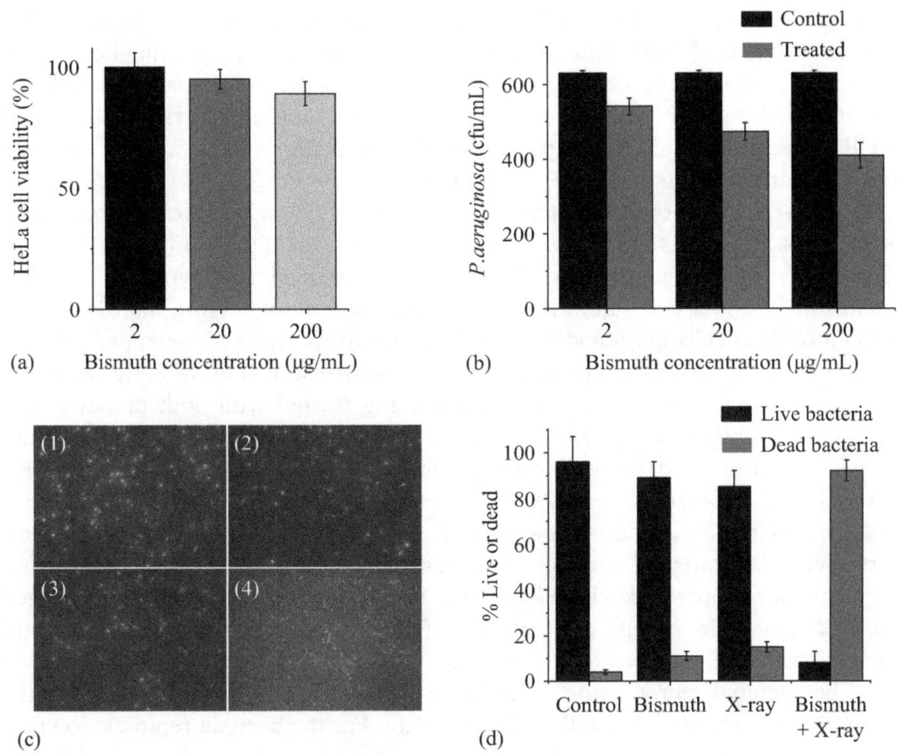

Figure 8.3 (a) HeLa cell viability after exposure to different concentrations of bismuth nanoparticles; (b) P. aeruginosa *cell colony forming units with different concentrations of nanoparticles; (c) fluorescent images of* P. aeruginosa *cells without treatment (1), 40 kV X-ray treated (2), 200 mg/mL nanoparticles treated (3), and both 40 kV radiation and 200 mg/mL nanoparticles treated (4); (d) viability of* P. aeruginosa *after different treatments*

measured by counting the number of colony-forming units (CFUs) in the plate. Figure 8.3(b) shows antibody-conjugated bismuth nanoparticles kill ~14% and ~35% of *P. aeruginosa* at a concentration of 2 and 200 μg/mL, respectively. The antibody conjugated bismuth nanoparticles kill more MDR *P. aeruginosa* than HeLa cells, because the *P. aeruginosa* antibody mediates endocytosis of nanoparticles into bacteria cells.

8.3.3 Bactericidal activity of bismuth nanoparticles enhanced X-ray radiation

The bactericidal activity of nanoparticle enhanced X-ray radiation is tested by determining the instant bacterial killing activity in the LB liquid media and the growing capability of *P. aeruginosa* in the LB agar plate. To evaluate the instant bactericide activity, the LIVE/DEAD bacterial viability assay is conducted to discriminate the dead (stain into red) or live (stain as green) bacteria. Four types of MDR *P. aeruginosa* samples in the LB liquid media are prepared as follows: (1) untreated MDR *P. aeruginosa* as the negative control, (2) radiation treated MDR *P. aeruginosa*, (3) nanoparticle treated MDR *P. aeruginosa*, and (4) radiation and nanoparticle treated MDR *P. aeruginosa*.

All X-ray radiations are carried out on bacteria cell suspension at room temperature for 30 min at the voltage of 40 kV and current of 100 μA. Figure 8.3(c1–c4) shows the fluorescence images of the above-mentioned four types of samples. Almost all bacteria cells are alive in the negative control (Figure 8.3(c1)); small amount of dead cells are found in bismuth nanoparticle treated bacteria suspension (Figure 8.3(c2)) and X-ray irradiated bacteria suspension (Figure 8.3(c3)); while almost all of bacteria cells are dead after being treated with both radiation and nanoparticle (Figure 8.3(c4)). The fluorescence intensities of the samples are shown in Figure 8.3(d), revealing the combination of radiation and nanoparticles kills ~82% *P. aeruginosa* cells, while X-ray radiation kills ~16% *P. aeruginosa* cells, and bismuth nanoparticles kills 10% *P. aeruginosa* cells alone, when compared with the untreated control. These results show that neither X-ray radiation nor bismuth nanoparticles alone have high bactericidal activity, but the combined use of X-ray radiation and antibody modified bismuth nanoparticles can greatly enhance bactericide activity by ~10 times [16].

The bacterial reproduction capability after radiation is tested by counting the MDR *P. aeruginosa* CFUs in the LB agar plate. For the bacteria reproduction test, 100 μl of *P. aeruginosa* suspensions (with/without bismuth nanoparticles) are inoculated on a LB agar plate, followed by radiation at room temperature immediately to ensure that radiation is performed in the lag phase of bacteria growth. The samples are incubated for 16 h after radiation, and the images of bacteria grown on petri dish are taken, and the bacterial CFUs are counted. Figure 8.4(a) shows that the bacteria colonies of untreated control samples distribute uniformly in the LB agar plate. The bacteria treated by either antibody modified nanoparticle (Figure 8.4(b)) or X-ray irradiated alone (Figure 8.4(c)) have fewer colonies than untreated controls. But only a small amount of colony is observed in the sample treated by both nanoparticle and X-ray radiation (Figure 8.4(d)). The colony sizes in the untreated culture control are smaller than that of the other treated samples after culturing for 16 h. This is induced by insufficient nutrient provided for high density of bacteria colonies in the LB petri dish in untreated control, while nutrient is relatively sufficient for decreased colonies with lower bacteria density [17–18]. Figure 8.4(e) shows that the CFUs number reduces 15% and 9% for the radiation group and nanoparticle group, respectively, while the CFUs number reduces ~94% in the dual-treatment (radiation and nanoparticle) group compared to negative control. Thus, the combined use of bismuth nanoparticles and X-ray radiation can significantly reduce the reproduction ability of *P. aeruginosa*. A radial bacteria killing effect is observed in Figure 8.4(d), as there are fewer bacteria colonies in the central of the petri dish than at the edges after radiation. This is because the intensity of X-ray radiation beam is strongest at the center, and

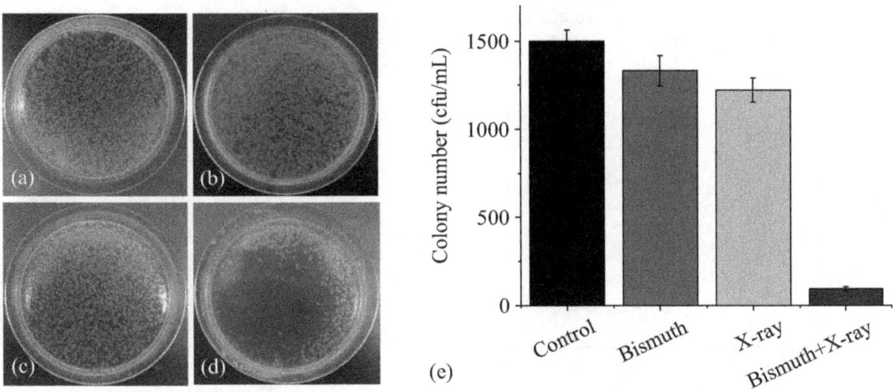

Figure 8.4 P. aeruginosa *cells without treatment (a); treated with nanoparticles (b); treated with radiation (c); and treated with both radiation and nanoparticles (d); colony number after different treatment (e)*

decreases towards the edge. According to the inverse square law, the dose of X-ray radiation is inversely related to the square of the distance from the source, which is, $D_2 = D_1 \times (d_1/d_2)^2$, where D_2 and D_1 are the doses at distances d_2 and d_1 from the X-ray source, respectively [19]. Assuming the intensity of X-ray to be 100% at the center of a petri dish with a radius of 10 cm, the intensity at the edges is about 100 times smaller than that at the center.

8.3.4 Dose-dependent bactericidal activity

To evaluate the dose-dependent bactericidal activity of nanoparticle and radiation, MDR *P. aeruginosa* is irradiated at 40 kV and 100 μA for different times with nanoparticle concentrations of 2, 20, and 200 μg/mL. Figure 8.5(a) shows that bacteria CFUs decrease significantly with the increase of radiation time, which is proportional to the radiation dose. For bacteria treated with 200 μg/mL antibody-conjugated bismuth nanoparticles, the CFUs number reduces ~90% after irradiation for 20 min, and almost all bacteria are killed after irradiation for 40 min. In the meanwhile, for a given radiation dose, increasing the number of nanoparticles enhances bacteria killing (Figure 8.5(b)). Bismuth nanoparticles at a concentration of 2, 20, and 200 μg/mL can kill 15%, 30%, and 90% of bacteria after irradiation for 20 min, respectively. When irradiation time is extended to 40 min, nanoparticles at the three concentrations can kill 92%, 98%, and 100% of bacteria, respectively. Figure 8.5(c) shows the effects of unmodified and antibody modified nanoparticles on *P. aeruginosa* and *E. coli* in the presence of an X-ray, where antibody modified nanoparticles have shown a much higher killing effect on *P. aeruginosa* than on *E. coli*.

To minimize the radiation damage to normal tissue for future use on the human body, a minimal X-ray dose that can trigger the bactericidal activity should be used. X-ray dose is simulated using PCXMC-a PC-based Monte Carlo simulation program. Simulations are done with an X-ray source fixed perpendicularly at 10 cm

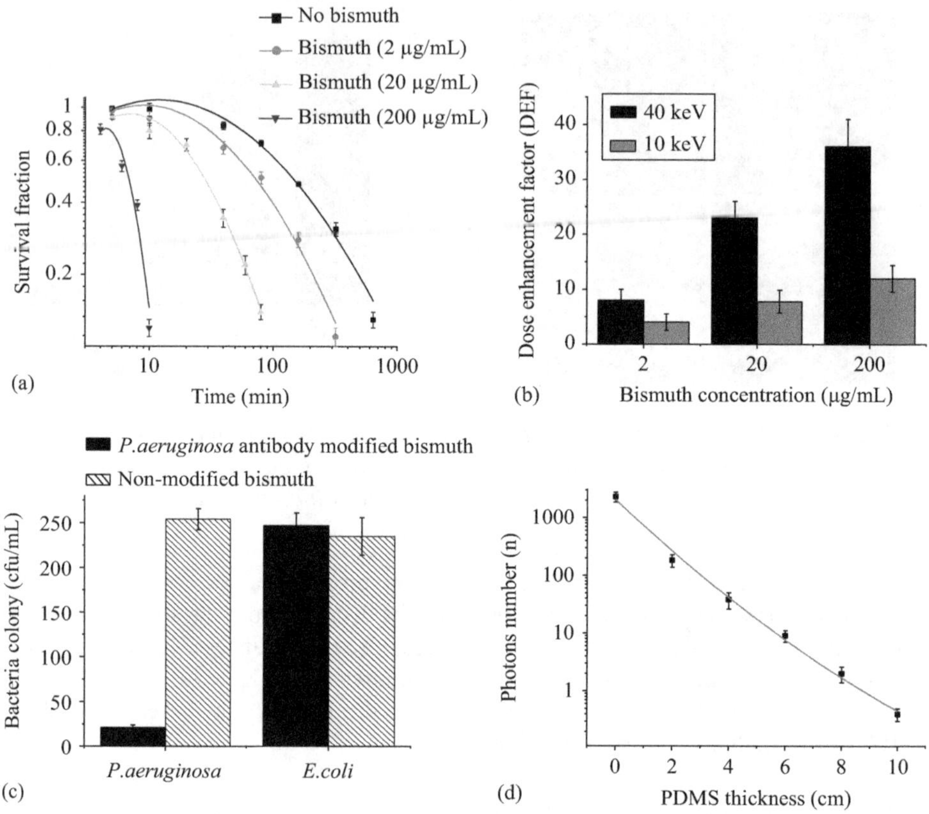

Figure 8.5 Nanoparticle enhanced X-ray killing of bacteria. (a) The survival fraction of irradiated P. aeruginosa *after incubation with different concentrations (2, 20, and 200 mg/mL) of antibody modified nanoparticles; (b) the DEF at different nanoparticle concentration; (c) the antimicrobial activities of unmodified and antibody modified nanoparticles. (d) Penetrating ability of 40 kV X-ray irradiation on a PMMA phantom with a thickness ranging from 0 to 10 cm*

from the center of a wound in the belly of a human phantom. With 40 kV X-ray generated from the tube with an output cone angle of 120°, the distance *r* is 17.32 cm. This covers an area of 943 cm^2 on X-ray exposed surface. At a 100 μA tube current and 120 s exposure time, the simulations result in a surface dose (D_s) of ~1.33 Sv and an effective dose (D_e) of 1.9 mSv, where D_s is the dose at the surface just before the X-ray enters the body and D_e is the dose absorbed by the body. The killing dose (D_k) can be calculated using the equation $D_k = D_s e^{-\mu\rho x}$, where ρ is the tissue density, μ is the mass absorption coefficient and x is the depth of wound. For skin tissue radiated by 40 keV X-rays, ρ is 1 g/cm^3 and μ is 0.27 cm^2/g. Assuming the bacteria is located at a depth of 4 cm from the skin surface, D_k will be 0.45 Sv.

8.3.5 Radiation safety issue

According to the U.S. nuclear regulatory commission, a single D_e is should be less than 0.05 Sv and the sum of the accumulative dose should be lower than 0.5 Sv for the occupational dose limits [20]. The D_e of X-ray radiation used in this work (1.9 mSv) is much lower than the limits and is also lower than the radiation doses used in regular medical procedures such as an abdominal CT scan (10.0 mSv) [21], thus imposing minimal threat on the normal human tissues. On the other hand, a D_k of 0.45 Sv that is posed on the bacteria layer of wound is much higher than the single dose limit and can induce single-strand breaks in bacterial DNA [22]. In addition, the introduced bismuth nanoparticles can significantly enhance DNA damage by generating free radicals [23]. X-ray alone (2 mSV) kills 10% MDR *P. aeruginosa*, while the same X-ray dose can kill 70% of bacteria when 20 μg/mL bismuth nanoparticles are added. The combined use of 20 μg/mL bismuth nanoparticles and 1 mSv X-ray radiation can kill 99.9% bacteria within 10 min. Moreover, for killing localized deeply buried bacteria, a set of collimators can be used to focus the X-ray beam to reduce the irradiation exposure around a wound. For instance, a collimator with an output cone of 8.8° (of 13 mm length and 1 mm radius) can focus a radiation beam within an area of 2 cm^2 at the 10 cm height from the objective.

8.3.6 Penetrating power of X-ray radiation

The penetrating power of X-rays has been tested using PDMS to mimic human tissue due to their similar density [24]. A CCD is used to count the number of X-ray photons penetrating through several PDMS blocks. As X-ray photons impinge on the CCD, they converted to visible light, appearing as bright dots on the screen. The number of bright dots is proportional to the number of X-ray photons penetrating through PDMS. Figure 8.5(d) shows that the number of photons decreases as the PDMS thickness is increased. As for 4 cm thick PDMS, the penetrated photons are ~10% of the original value. According to the above calculation, X-ray radiation for 20 min can kill 50% of bacteria sitting under a 4 cm thick PDMS. In most of cases, the depth of wound is less than 4 cm, which means this method can be used to kill MDR bacteria deeply buried in the tissue. The penetrating power of X-ray increases as its energy. In normal radiation therapy, X-ray with mega voltage energy is used, which leads to penetration into a deeper tissue. In addition, the differential X-ray absorption between bismuth and tissue will increase with X-ray energy; thus the benefit of using nanoparticles to enhance X-ray absorption will be much larger for high energy X-rays.

8.4 Conclusions

A new method is developed to kill MDR *P. aeruginosa* by the combined use of low dose X-ray radiation and antibody functionalized bismuth nanoparticles. The method is able to remove bacteria in deep tissues to prevent wound infection. Compared to antibiotics, the method does not cause MDR or pan-drug resistance and can kill all types of *P. aeruginosa* (including MDR *P. aeruginosa*) using a strain-specific polyclonal antibody. Compared to existing physical and nanoparticle methods, the method allows effective killing of MDR bacteria even if they are located deeply inside tissue. Although MDR *P. aeruginosa* is taken as an example, this method can also be used to kill other bacteria, providing a new strategy in combating MDR bacteria.

References

[1] Tseng, S. H.; Lee, C. M.; Lin, T. Y.; Chang, S. C.; Chang, F. Y. Emergence and spread of multi-drug resistant organisms: think globally and act locally. *J. Microbiol. Immunol. Infect.* **2011**, *44*, 157–165.

[2] Weinstein, R. A. Nosocomial infection update. *Emerg. Infect. Dis.* **1998**, *4*, 416–420.

[3] Kang, C. I.; Kim, S. H.; Kim, H. B.; *et al. Pseudomonas aeruginosa* bacteremia: risk factors for mortality and influence of delayed receipt of effective antimicrobial therapy on clinical outcome. *Clin. Infect. Dis.* **2003**, *37*, 745–751.

[4] Koningsberger, D. C. P., R, *X-Ray Absorption: Principles, Applications, Techniques of Exafs, Sexafs and Xanes*. John Wiley and Sons Inc.: New York, **1987**.

[5] Dincer, A. H.; Baysal, T. Decontamination techniques of pathogen bacteria in meat and poultry. *Crit. Rev. Microbiol.* **2004**, *30*, 197–204.

[6] Diehl, J. F. Food irradiation – past, present and future. *Radiat. Phys. Chem.* **2002**, *63*, 211–215.

[7] Shea, K. M.; Hlth, C. E. Technical report: irradiation of food. *Pediatrics* **2000**, *106*, 1505–1510.

[8] Hu, J. J.; Hong, Y.; Muratore, C.; Su, M.; Voevodin, A. A. In situ transmission electron microscopy of solid-liquid phase transition of silica encapsulated bismuth nanoparticles. *Nanoscale* **2011**, *3*, 3700–3704.

[9] Rabin, O.; Perez, J. M.; Grimm, J.; Wojtkiewicz, G.; Weissleder, R. An X-ray computed tomography imaging agent based on long-circulating bismuth sulphide nanoparticles. *Nat. Mater.* **2006**, *5*, 118–122.

[10] Stevens, K. N. J.; Croes, S.; Boersma, R. S.; *et al.* Hydrophilic surface coatings with embedded biocidal silver nanoparticles and sodium heparin for central venous catheters. *Biomaterials* **2011**, *32*, 1264–1269.

[11] Li, W.R.; Xie, X.B.; Shi, Q.S.; Duan, S.S.; Ouyang, Y.S.; Chen, Y.B. Antibacterial effect of silver nanoparticles on *Staphylococcus aureus*. *Biometals* **2011**, *24*, 135–141.

[12] Norman, R.S.; Stone, J.W.; Gole, A.; Murphy, C.J.; Sabo-Attwood, T.L. Targeted photothermal lysis of the pathogenic bacteria, *Pseudomonas aeruginosa*, with gold nanorods. *Nano Lett.* **2008**, *8*, 302–306.

[13] Huang, W.C.; Tsai, P.J.; Chen, Y.C. Functional gold nanoparticles as photothermal agents for selective-killing of pathogenic bacteria. *Nanomedicine* **2007**, *2*, 777–787.

[14] Zhao, Y.Y.; Tian, Y.; Cui, Y.; Liu, W. W.; Ma, W. S.; Jiang, X. Y. Small molecule-capped gold nanoparticles as potent antibacterial agents that target gram-negative bacteria. *J. Am. Chem. Soc.* **2010**, *132*, 12349–12356.

[15] Hoshino, A.; Hanada, S.; Yamamoto, K. Toxicity of nanocrystal quantum dots: the relevance of surface modifications. *Arch. Toxicol.* **2011**, *85*, 707–720.

[16] Park, S.; Chibli, H.; Wong, J.; Nadeau, J. L. Antimicrobial activity and cellular toxicity of nanoparticle-polymyxin B conjugates. *Nanotechnology* **2011**, *22*, 185101.

[17] Banerjee, M.; Mallick, S.; Paul, A.; Chattopadhyay, A.; Ghosh, S. S. Heightened reactive oxygen species generation in the antimicrobial activity of a three component iodinated chitosan-silver nanoparticle composite. *Langmuir* **2010**, *26*, 5901–5908.

[18] Qiu, Y.; Liu, Y.; Wang, L. M.; *et al.* Surface chemistry and aspect ratio mediated cellular uptake of Au nanorods. *Biomaterials* **2010**, *31*, 7606–7619.

[19] Burlin, T. E. The effect of inverse square law attenuation on the measurement of Grenz-ray exposure with a cavity ionisation chamber. *Br. J. Radiol.* **1964**, *37*, 693–695.

[20] Michaelson, S. M. Standards for protection of personnel against nonionizing radiation. *Am. Ind. Hyg. Assoc. J.* **1974**, *35*, 766–784.

[21] Wall, B. F.; Hart, D. Revised radiation doses for typical X-ray examinations. Report on a recent review of doses to patients from medical X-ray examinations in the UK by NRPB. National Radiological Protection Board. *Br. J. Radiol.* **1997**, *70*, 437–439.

[22] Lehnert, S.; Moroson, H. X-ray-induced single-strand breaks in DNA of *E. coli*-B/R – effect of irradiation and postirradiation conditions. *Radiat. Res.* **1971**, *45*, 299.

[23] Turkez, H.; Geyikoglu, F. The efficiacy of bismuth subnitrate against genotoxicity and oxidative stress induced by aluminum sulphate. *Toxicol. Ind. Health* **2011**, *27*, 133–142.

[24] Losi, P.; Briganti, E.; Magera, A.; *et al.* Tissue response to poly(ether)urethane-polydimethylsiloxane-fibrin composite scaffolds for controlled delivery of pro-angiogenic growth factors. *Biomaterials* **2010**, *31*, 5336–5344.

Chapter 9

Conventional and nano-based approaches to prevent bacterial infection

Hilal Yazici[1,2], *Ece Alpaslan*[1], *Garima Bhardwaj*[1] *and Thomas J. Webster*[1,3]

Abstract

Biomedical device-associated infection is still one of the most serious complications in implant surgeries due to the existence of immune depression in the peri-implant area. Despite considerable research and development efforts, the problem of infections related to biomedical devices and implants persists. Many approaches, which understand the pathogenesis of implant-associated infections, have been applied to prevent infections. Among these approaches, conventional treatment and new nano-based approaches are the skeleton of this chapter. Specifically, current strategies such as systemic and local antibiotic treatment, nonfouling surfaces, utilization of antimicrobial agents and antimicrobial peptides (AMPs), and metal (silver) and metal oxide (zinc oxide) nanoparticles have been explained with advantages and disadvantages for treating infection.

9.1 Bacterial infection

Millions of medical devices are used each year, and in spite of many advances in biomaterials, a proportion of each type of device becomes colonized by bacteria [1]. Topical devices (e.g., contact lenses) are colonized by bacteria as soon as they are placed on tissue surfaces; transcutaneous devices (e.g., vascular catheters) are progressively colonized by skin microorganisms; and even surgically implanted devices regularly become the foci of infection [1]. Bacteria may colonize on implanted devices at the time of surgery or via a hematogenous route from a distant source. The most significant factor in the development of device-related infections appears to be the skill of the surgical team; prosthetic hips have been reported to become infected in less than 0.2% of cases, but in as many as 4% of cases in less proficient facilities.

[1]Department of Chemical Engineering, Northeastern University, Boston, MA, USA
[2]TUBITAK-MAM, Genetic Engineering & Biotechnology Research Institute, Gebze, Kocaeli, Turkey
[3]Center of Excellence for Advanced Materials Research, King Abdulaziz University, Jeddah, Saudi Arabia

Generally, large and complex medical devices that require long and complicated surgery for their placement are at high risk of bacterial infection [2]. However, when infection occurs, it can be life threatening.

Device-related infections may occur almost immediately post-surgery or may be very slow to develop, with other symptoms occurring months or even years after the device is implanted. At the cellular level, implant-associated infections include bacterial adhesion to a biomaterial surface and start as a competition for integration of the material into the host tissue or the bacteria into the material. For a successful implant, material integration occurs prior to bacterial adhesion thus preventing colonization of the implant surface. A time period of 6 h post-implantation is usually considered the decisive period, during which all preventative measures are taken to reduce bacterial adhesion. Over this time period, the implant surface is particularly susceptible to colonization and there is a possibility for biofilm formation. Infectious bacteria can be traced to several sources including the ambient atmosphere of the operating room, surgical equipment, clothing worn by medical professionals, resident bacteria on the patient's skin, and bacteria already in the body. Although sterilization and the use of aseptic techniques greatly reduce the levels of bacteria found in hospital settings, pathogenic microorganisms are still found at the site of approximately 90% of all implants. The most common pathogens that cause implant infections include Gram-positive *Staphylococcus aureus* and *Staphylococcus epidermidis*, which are responsible for up to 60% of all prosthetic hip implant infections. *S. aureus* infections proceed rapidly and are generally more severe than *S. epidermidis* infections. However, *S. epidermidis* has more accessibility as an opportunistic pathogen since it is found ubiquitously on the skin. Other bacteria that have been implicated in implant-associated infections include Gram-negative *Escherichia coli*, *Pseudomonas aeruginosa*, and those from the Proteus group (e.g., *Pseudomonas mirabilis* and *Pseudomonas vulgaris*). Most, if not all, of the characteristics of device-related infections can be explained based on characteristics of biofilms [3]. The biofilm concept was developed in environmental microbiology, and it states that bacteria grow preferentially in matrix-enclosed communities attached to surfaces [4]. When implanted medical devices become colonized, the microbial biofilms trigger changes in the surrounding tissues, but symptoms are often slow to develop. Symptoms and other indicators of infection, such as white-blood-cell count, may correlate poorly in an implant infection. The slow development and asymptomatic nature of many device-related infections can be explained by the observation that biofilm bacteria produce few toxins and elicit little inflammatory response. Common bacterial species predominate in device-related infections because they form biofilms very effectively in their natural environments (e.g., skin). The biofilm mode of growth protects the causative agents from both humoral and cell-mediated immunity [5], so the organisms are rarely cleared, even by the most active host–defense mechanisms. Exacerbations of device-related infections are caused by the release of planktonic cells. Antibiotics can kill these floating cells and reverse the symptoms of acute infection, but the infection persists because the causative biofilm is resistant to antibiotics. Good medical management usually dictates removal of colonized or infected devices [6]. The surface attachment of bacteria triggers changes in gene expression that facilitates the growth of a biofilm. This is usually done by the suppression of motility factors and the induction of growth factors. The bacteria interact with each other inducing synthesis of extracellular matrix and generating phenotypic diversity. This helps in the growth and maturation of the biofilm on the medical device or in other words, its bacterial colonization. Due to the fundamental role that bacterial adhesion plays in the development of implant infections, such biofouling has been well documented [7,8]. The adhesion of bacteria is preceded by the surface adsorption of a conditioning film of small organic compounds and macromolecules including proteins.

Subsequently, the physicochemical forces that mediate bacterial adhesion can be divided into two time-dependent phases. Phase I involves reversible cellular association with the surface over the first 1–2 h post-implantation. This nonspecific association is mediated through long (e.g., gravitational, van der Waals, and electrostatic interactions) and short (e.g., hydrogen bonding, dipole–dipole, ionic, and hydrophobic interactions) range forces. Phase II begins approximately 2–3 h later and is characterized by stronger adhesion between the bacteria and the foreign material. Specific chemical reactions between compounds on the cell and substrate surfaces result in irreversible molecular bridging [6–8]. Both polysaccharides and adhesin proteins within the bacterial membrane facilitate attachment to substrate surfaces. Beyond Phase II, certain bacterial strains are capable of forming a biofilm if provided with an appropriate supply of nutrients.

During biofilm formation, bacteria secrete an exopolysaccharide layer that retains nutrients and protects the microorganisms from the immune response. With the protective polysaccharide coating and sequestered nutrients, bacteria in biofilms exhibit extreme resistance to antibiotics [9]. In some cases, it has been found that killing bacteria in a biofilm requires roughly 1,000 times the antibiotic dose necessary to achieve the same results in a suspension of cells. The tremendous resistance of biofilms to conventional antibiotic therapy has prompted a great deal of research on synthetic surfaces and coatings that resist bacterial colonization [7]. Coatings have been developed that reduce bacterial adhesion by altering the physicochemical properties of the substrate so that conditioning films do not form and/or bacteria–substrate interactions are not favorable. These coatings are referred to as "passive" and include surfaces modified with poly(ethylene glycol), poly(ethylene oxide) brushes, and hydrophilic polyurethanes, for example. Unfortunately, the effectiveness of passive coatings for reducing bacterial adhesion is limited and varies greatly depending on bacterial species. The physicochemical properties of the surface (coating) can be masked by an adsorbed conditioning film, thereby diminishing their effectiveness [8]. A recent alternative approach to reducing bacterial adhesion is based on coatings that actively release antibacterial agents [10]. Such "active" coatings have been designed to release high initial fluxes of antibacterial agents during the critical short-term post-implantation period (several hours) to inhibit the initial adhesion of bacteria. Continued release beyond this short-term period is desirable because protective fibrous capsule formation and tissue integration occurs over a longer time period (weeks to months). Active strategies have been shown to prevent biofilm formation both in vitro and in vivo [9].

9.2 Conventional approaches to prevent bacterial infection

9.2.1 Antibiotic treatment

One of the most preferable conventional treatments to prevent implant-associated infection is antibiotic treatment, which are chemical compounds that inhibit the growth of microorganisms, such as bacteria, fungi, or protozoans. The main goal of treating various types of infections should be to reduce the bacterial load in the wound to a level at which the wound healing processes can occur. Conventional systemic delivery of antibiotics involves poor penetration into ischemic and necrotic tissue and can cause systemic toxicity with associated renal and liver complications. This result causes hospitalization for monitoring. Alternative local delivery of antibiotics by either topical administration or by a delivery device may enable the maintenance of a high local antibiotic concentration for an extended duration of release without exceeding systemic toxicity [11,12]. Hence, the controlled release of antibiotics from the implant or medical devices shows many advantages over systemic treatments. This type of treatment increases the local dosage of antibiotics, as well avoiding systemic side effects [13–16].

For nondegradable implants, such as central venous catheters, urinary catheters, and stents, a surface coating is doubtless a simple method, which can improve the antibacterial properties of the surface. Although solvent dipping is the simplest method, the immobilization of drugs on the surface is limited and the release rate cannot be controlled through direct coating. However, the immobilization of antibiotics on the surface can be facilitated by an electrostatic force due to the their structural properties. Antibiotics are synthetically designed as anionic derivatives with carboxylate, phosphate, or sulfate substitutes [17]. Besides immobilization of antibiotics onto implant surfaces, an alternative method is to first mix different antibiotics into the coating materials and then coat the mixture on the surface. Various coating such as biodegradable or nondegradable polymers can be incorporated for implant coatings. For the nondegradable polymer coating, normally polyurethane [18–20] can be an example. The release of the antibiotics is entirely through diffusion in most of the coatings. However, the release profile can be also controlled through the coating strategy. For example, another thin layer can be coated on the top of the layer loaded with antibiotics, thus providing a barrier to restrict the fast initial burst of the release [21]. Controlled release can also be achieved through a concentration gradient of loaded antibiotics. Antibiotic can be loaded with a higher concentration in the inner layer and a lower concentration in the outer layer. Biodegradable polymer coatings is one of the applicable materials for antibiotic loading due to their potential releasing control mechanisms [22–24]. The release of antibiotics is not only through diffusion, but also influenced by the degradation rate and mechanism of the coating. Some of FDA-approved biodegradable polymers include PLA, PGA, PLGA, and polyanhydride. All of these show great potential as matrices for the controlled release of antibiotics [25,26].

Antibiotics have been also merged into nondegradable PMMA bone cement and beads for prophylaxis and the treatment of total joint arthroplasty infection, treatment of chronic osteomyelitis, and prophylaxis of infections in open fracture repair [27]. Although these have been put into clinic applications, the release of most antibiotics from the PMMA cements or beads is far from satisfaction and success [28,29]. For example, less than 50% was observed to release from implants after 4 weeks for gentamicin and no continuous release was found after initial release. The failure in continuous release may be due to the solid, glassy, nonswelling properties of the PMMA matrix [30–33]. Besides PMMA, biodegradable implants degrade over time, eliminating the need for a second surgery to take them out, and thereby preventing the possibility of re-infection during a second surgery. Moreover, the loaded antibiotics can be totally released out of the implants once they degrade. The degradation rate can be regulated through the selection of the biodegradable chemical structure and through building up a concentration gradient in the implants.

Bioactive ceramics as a coating for metal implants is a preferable material as an antibiotic carrier. The utilization of a bioactive ceramic coating containing hydroxyapatite (HA), calcium phosphate, and other osteo-conductive materials offers the physiochemical environment and structural scaffold required for bone-implant integration. Thus some antibiotic loading studies were reported in vitro from HA coated implants for chlorhexidine, vancomycin, gentamicin, tobramycin, and several other antibiotics [12,34].

In all application of antibiotic treatment, the main concern for all antibiotics is the development of resistant bacteria. Many important pathogens, especially *S. aureus*, have long been recognized to exhibit always more alarming levels of antibiotic resistance [35,36]. Moreover, bacteria forming biofilms on prosthetic surfaces are particularly resistant to antimicrobials and tend to survive to aggressive chemotherapy even in the absence of specific antibiotic resistance factors. Depending on this, it is clear how important it is to survey the presence of antibiotic resistant strains in clinical settings, not uniquely with the scope to decide the patient treatment regimen [37].

The mechanisms of resistance to antibiotics in bacterial biofilms are beginning to be elucidated [38]. Figure 9.1 shows three main hypotheses. The first hypothesis is the possibility of slow penetration of the antibiotic into the biofilm. Measurements of antibiotic penetration into biofilms in vitro have shown that some antibiotics readily pervade bacterial biofilms [39]. There is no generic barrier to the diffusion of solutes for the size of the antibiotics through the biofilm matrix, which is mostly water [40].

The second hypothesis depends on an altered chemical microenvironment within the biofilm. Gradients of the nutrient concentrations are a well-known feature of biofilms. The bottom parts of the biofilms are very poor in terms of oxygen concentrations. Studies in the literature have shown that oxygen can be completely consumed in the surface layers of a biofilm. This leads to anaerobic zones in the deep layers of the biofilm [42]. Concentration gradients in metabolic products mirror those of the substrates. Local accumulation of acidic waste products might lead to pH differences greater than between the bulk fluid and the biofilm interior [43], which could directly antagonize the action of an antibiotic.

A third and still speculative mechanism of antibiotic resistance is that a subpopulation of microorganisms in a biofilm forms a unique, and highly protected,

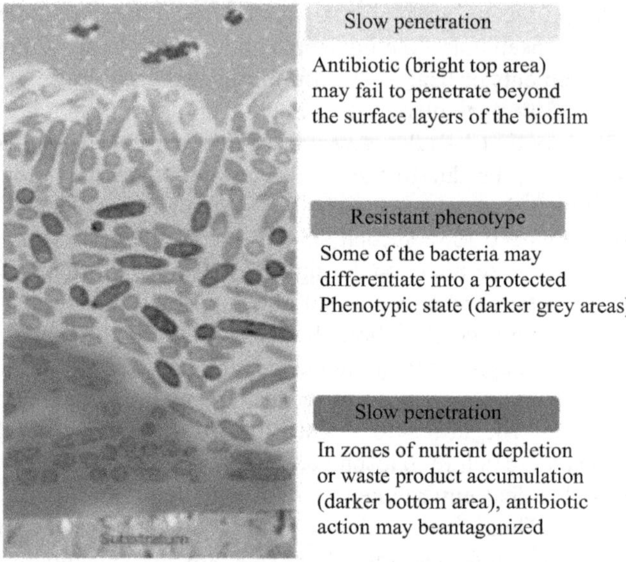

Figure 9.1 Three hypotheses for mechanisms of antibiotic resistance in biofilms [41]

phenotypic state – a cell differentiation similar to spore formation. This hypothesis indicates a barrier to the penetration of either an antimicrobial agent or metabolic substrates. Additionally, most bacteria in the biofilm, but not all, are rapidly killed by antibiotics [44]. Survivors, which might consist of 1% or less of the original population, resist exposure of the antibiotic. The hypothesis of a spore-like state entered into by some of the bacteria in a biofilm provides a powerful, and generic, explanation for the reduced susceptibility of biofilms to antibiotics and disinfectants of widely different chemistries [41].

Besides these mechanism steps, a number of studies have reported the existence of a strong association between genomic DNA mobile elements termed transposons and important virulence factors as well as most of the genes for the resistance to antimicrobials. Movement of transposons across the microbial population causes transferring clusters of genes. Integration of new transposons into bacterial DNA triggers bacteria species to do harm and survive to medical treatments [37,45].

9.2.2 Antimicrobial agents

Another alternative way to using antibiotics that involve an antiseptic that tends to have broad-spectrum activity against Gram-positive and Gram-negative bacteria, such as chlorhexidine or quaternary ammonia compounds. Chlorhexidine is commonly used in surgical scrub, hand washes, and to clean wounds, but has been used in association with external fixator pins [46], polymeric coatings on titanium [47], and other indwelling implants [48,49]. There are limitations reported for using chlorhexidine in association with implants, including the toxicity of the antiseptic to the host cells. In vitro, chlorhexidine diacetate has been reported to be lethal to fibroblasts [50], probably because chlorhexidine diacetate lysis the host cell membranes in the same way as it does bacterial cell membranes. Hypersensitive reactions are also known to occur when patients are exposed to chlorhexidine diacetate [51,52]. A second disadvantage is that chlorhexidine diacetate resistance has been observed in various staphylococci and Gram-negative bacteria, particularly with antibiotic-resistant bacteria [53]. Quaternary ammonia compounds (QAC) have been applied through covalent binding of the antimicrobial directly onto the implant surface [54]. QAC work by disrupting the cell membrane of both Gram-positive and Gram-negative bacteria [55].

Another alternative method is silver ion- and silver nitrate-based coatings on implants, since silver is known to have broad-spectrum antibacterial properties. Silver-based coatings have been studied in association with polymer and metal implants, particularly on external fixation pins [56,57]. Coating stainless steel pins with silver has been found to decrease the adhesion and colonization of *S. aureus*, *S. epidermidis*, *E. coli*, and *P. aeruginosa* around the coated implant [58,59]. However, there are reports that silver is toxic to host cells, causing severe inflammatory responses [60]. Due to the toxic properties of silver, "NanoSilver," which is a new form of silver consisting of silver particles (5–50 nm), has been tested and in vitro results have shown that it was effective against several type of bacteria and has indicated that it is not cytotoxic due to its high porosity and active surface [59].

Another possible solution is using lysostaphin instead of conventional antibiotics. Lysostaphin is an antibacterial enzyme that cleaves pentaglycine bridges in the cell wall of staphylococci. Wu *et al.* [61] indicated that applying lysostaphin to *S. aureus* and *S. epidermidis* biofilms not only killed the bacteria, but also disrupted the extracellular polysaccharide matrix surrounding the bacteria [25].

9.2.3 Nonfouling surfaces

An essential step in preventing infections is to inhibit the adhesion of proteins, biomolecules, and bacteria onto the implant surfaces. PEO or polyethyleneglycol (PEG) and its derivatives are examples of a class of hydrophilic polymers that can be used to create anti-adhesive surfaces [13,25,62–64]. These hydrophilic polymers have highly dynamic chain structures, when locating on surfaces. They present large exclusion volumes in an aqueous environment (such as in circulation) by retaining a surrounding hydrous layer, thereby repelling the adhesion of molecular species [65], cells [66], and bacteria from the surfaces.

PEO or PEG type of polymers can be introduced onto surfaces via physical or chemical immobilization approaches. Using physical approaches, PEG is coated onto material surfaces by direct deposition such as dipping. However, the PEG layer formed by physical approaches is unstable in an aqueous environment. Because PEG has good solubility in water and poor wettability to the surface. Alternatively, PEG may be chemically linked to surfaces by crosslinking or grafting. PEG can be cross-linked by argon radio frequency (RF)-plasma after spin coating on a stainless steel (SS) surface [67] or by using dicumyl peroxide (DCP) as a crosslinking agent on the surface of a polyurethane-containing PEO segment [68]. The cross-linked PEG surface was shown to have decreased bacterial adhesion. PEG with a functional end group can be chemically grafted onto the substrate surface, such as polyurethane, glass, silicon, and poly(ethylene terephthalate). The coating density and PEO chain length greatly influence the antibacterial ability. When the density of PEO is high enough, the molecules are forced to stretch out and form a layer called a "molecular brush." In general, better antimicrobial activity is strongly related to a higher coating density and longer PEO chain length [13].

Various types of PEG have been testing their effectiveness against different bacteria type as well as osteoblast, and fibroblast function. For example, osteoblast and fibroblast cell adhesion and spreading on metal oxide surfaces coated with PLL-*g*-PEG are strongly reduced in comparison to uncoated oxide surfaces [69]. In vitro PLL-*g*-PEG coatings are also effective at inhibiting *S. aureus*, *S. epidermidis*, *P. aeruginosa*, and *S. mutans* colonization [63,69–71]. A biofunctionalized version of PLL-*g*-PEG copolymer with the cell-adhesive peptide Arg-Gly-Asp (RGD) motif has been synthesized and used to induce specific attachment of fibroblasts and osteoblasts to surfaces [69,72]. The RGD motif is present in proteins such as fibronectin, vitronectin, and fibrinogen, and is known to interact specifically with a number of integrin cell receptors, while being resistant to the adsorption of non specific biomolecules. The advantage of this coating is that it enhances fibroblasts and osteoblasts, but remains inhibitory to *S. aureus*, *S. epidermidis*, *P. aeruginosa*, and *Streptococcus mutans* adhesion. Such surfaces (PLL-*g*-PEG and PLL-*g*-PEG/RGD) are believed to have potential as implant coatings, but they currently need improvement in the coating adhesion strength to the implant surface [25].

9.3 New approaches to prevent bacterial infection

9.3.1 Antimicrobial peptides (AMPs)

A novel class of peptides, the AMPs, has received attention because of their potential use as therapeutic agents. AMPs are an evolutinary conserved constituent of the innate immune response and have been found in all walks of life including insects, plants, and animals [73–76].

AMPs can be called as natural antibiotics because they are effective against a broad range of microorganisms [77–80], including multidrug resistant bacteria. One of the advantage of AMPs is their ability to work synergestically with classical antibiotics, possibly facilitating access of antibiotics to the infection site, and thus providing an avenue for a more aggressive treatment approach against biofilms [73]. The ability of AMPs to inhibit and/or kill multidrug resistant bacteria increases their importance as a therapeutic agent.

Due to the wide distribution and characteristics of AMPs, we can categorize them broadly on the basis of their secondary stucture and aminoacid composition. AMP characteristics and their AMP activity and specificity against a wide specturum of organisms are influenced by the size, sequence, charge, conformation and structure, hydrophobicity, and ampipathicity of the peptide [74].

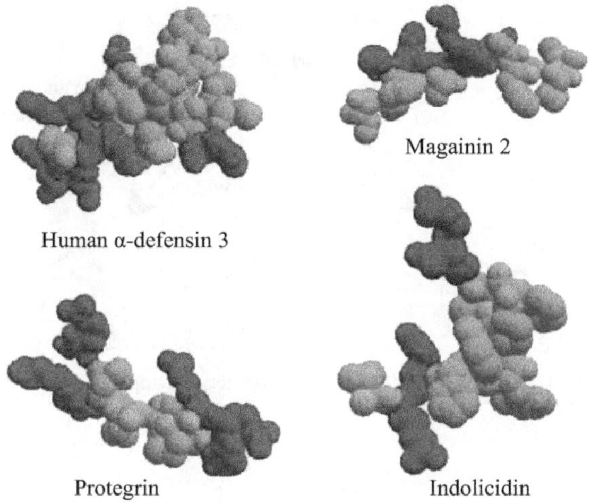

Figure 9.2 Clustering of cationic and hydrophobic amino acids into distinct domains in several AMPs of different structural classes. This "amphipathic" design is evident in many, but not all, AMPs. Darker grey, basic (positively charged) amino acids; lighter grey, hydrophobic ("oily") amino acids. Other amino acids are not shown. Magainin is depicted in its α-helical configuration [75]

The fundamental structural principle underlying all classes is the ability of the molecule to adapt a shape based on hydrophobic and cationic amino acid spatial organization (Figure 9.2) [75]. Based on structural properties, they have been classified into four major groups: β-sheet, α-helical, loop, and extended peptides. Among these groups, the β-sheet and α-helical are the most common in nature [73]. Several idealized peptides have been synthesized since the explosive interest in the group of peptides as a therapeutic target [81,82].

The exact mechanism of action for AMPs has not been known yet. Nevertheless, most AMPs are believed to target the negatively charged cell membrane of bacteria (and possibly the embedded lipids bearing phospholipid head-groups) and mediate killing by membrane disruption or pore formation [73–75]. For some AMPs an intracellular target has been proposed that may be the sole mechanism of action of the peptide, or may work synergistically with membrane disruption [73].

According to the described variety of mechanisms on AMP activity, these peptides have targeted clearly the difference in design of membranes of microbes and multicellular animals. Bacterial membranes are organized in leaflets of lipid bilayers. The surface exposed to the outer world is heavily populated by lipids with negatively charged phospholipid headgroups. In contrast, the outer leaflet of the membranes of plants and animals is composed principally of lipids with no net charge; most of the lipids with negatively charged headgroups are segregated into the inner leaflet, facing the cytoplasm (Figure 9.3) [75].

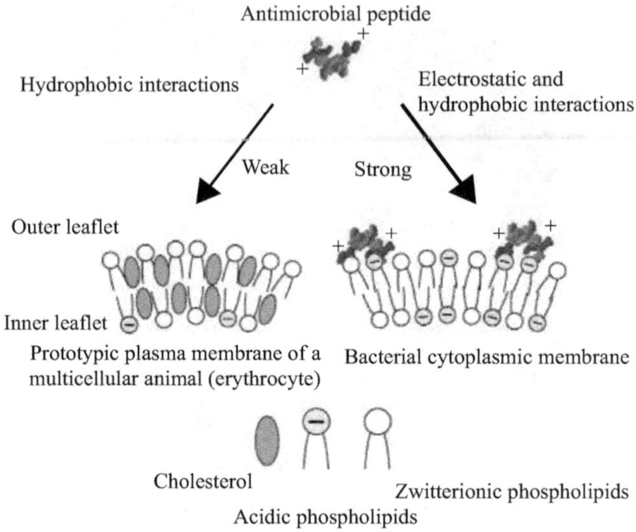

Figure 9.3 The membrane target of AMPs of multicellular organisms and the basis of specificity [75]

The Shai-Matsuzaki-Huang (SMH) model (Figure 9.4) is one of the known models that explain the activity of most AMPs. The model proposes several steps that indicate mechanism of AMPs. The first step is interaction of the peptide with the membrane, followed by displacement of lipids, alteration of membrane structure, and in certain cases entry of the peptide into the interior of the target cell. Cholesterol is one of the key molecules in the target membrane in general reduces the activity of AMPs, due either to stabilization of the lipid bilayer or to interactions between cholesterol and the peptide. On the other hand, it is believed that increasing ionic strength, which in general reduces the activity of most AMPs, does so in part by weakening the electrostatic charge interactions required for the initial interaction [75].

Many hypotheses have been presented on the actual mechanism of AMP to kill the microbes which include: fatal depolarization of the normally energized bacterial membrane; the creation of physical holes that cause cellular contents to leak out; the activation of deadly processes such as induction of hydrolases that degrade the cell wall; the scrambling of the usual distribution of lipids between the leaflets of the bilayer, resulting in disturbance of membrane functions; and the damaging of critical intracellular targets after internalization of the peptide [83,84].

The use of AMPs in therapeutical approaches is vastly underway. Appendini *et al.* covalently linked an AMP onto the surface of polystyrene resin beads that had PEG grafted onto them and found it to be microcidal against several microorganisms [79]. Etienne *et al.* developed a way to insert an AMP into polyelectrolyte multilayer films that was reported to effectively inhibit *Micrococcus luteus* and *E. coli* D22 when 10 AMP layers were applied [80]. Kwakman *et al.* locally administered an AMP along the site of insertion of a catheter in a murine model and

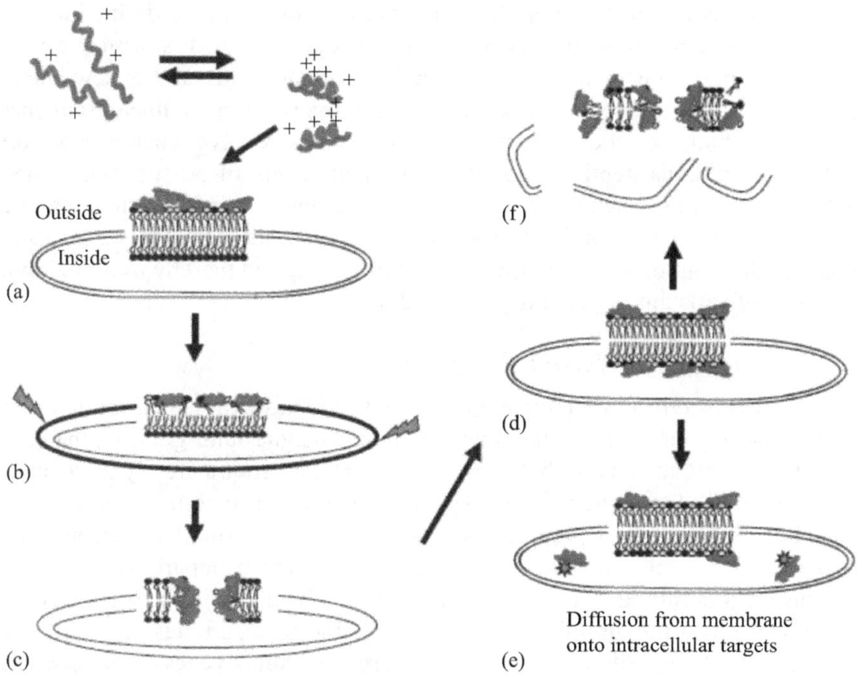

Figure 9.4 The SMH model of the mechanism of action of an AMP. An α-helical peptide is depicted. (a) Carpeting of the outer leaflet with peptides. (b) Integration of the peptide into the membrane and thinning of the outer leaflet. The surface area of the outer leaflet expands relative to the inner leaflet, resulting in strain within the bilayer (jagged arrows). (c) Phase transition and "wormhole" formation. Transient pores form at this stage. (d) Transport of lipids and peptides into the inner leaflet. (e) Diffusion of peptides onto intracellular targets (in some cases). (f) Collapse of the membrane into fragments and physical disruption of the target cell's membrane. Lipids with light color headgroups are acidic, or negatively charged. Lipids with black headgroups have no net charge [75]

found the peptide to have a broad in vivo potentency against microorganisms and reduced inflammed tissue [85].

At industrial level, several companies are focused on the development of AMPs with several molecules both at the preclinical and clinical stage. This demonstrates that despite the first clinical trials failing (Pexiganan and Iseganan), there is still a lot of general optimism for their use in future clinical practice. Some of the challenges facing the development of peptidic drugs have already been overcome, starting from the industrial production of T-20 peptide (Fuzeon) at the multi-ton scale production, incredibly boosting the production of new peptides on a large scale with beneficial effects on the cost of all starting materials. Other challenges are common to other class of molecules that may be defined as innovative. Several strategies have been created to optimize AMPs with promising activity, ranging from inclusion of non-natural amino acids and a new method applicable to high-throughput screening to multimerization of linear sequences. Even if we cannot exclude the fact that resistance may evolve whenever bacterial populations are consistently exposed to elevated levels of AMPs, this concern should not discourage their further study and development; instead, they may help us to rationalize their use in future, for example preventing the big mistake made in the past of distribution of large amounts of antibiotics, and thereby minimizing the emergence of resistant organisms [73–76,82,83].

9.3.2 Nanoparticle-based therapies

To date different types of nanomaterials have been studied for their antimicrobial activity such as zinc, copper, titanium, selenium, magnesium, gold, alginate, and silver. It is important to note that some of these metals are inherently antibacterial even in their bulk form, such as zinc, silver, copper, other materials like iron oxide only exhibit antimicrobial properties in its nanoparticle form. The mechanism of action of nanoparticles is different for different kinds nanoparticles. Thus, the mechanism is not fully understood for all varieties of nanoparticles. Some proposed mechanisms relate the antimicrobial efficacy of the nanoparticles to the physical structure of the nanoparticles as, membrane damaging abrasiveness of the particles, and others relate to the release of metal ions from the surface of the nanoparticles. Greater surface area of the nanoparticles compared to their bulk counterparts makes inherently antibacterial nanoparticles more toxic to bacteria at the nanoscale [86].

This chapter will try to elucidate on the mechanism of action of some of the major antibacterial nanoparticles as silver, and some metal oxide nanoparticles as zinc oxide in detail.

9.3.2.1 Silver nanoparticles

Nanoparticles in general have proven to be more efficient compared to salts of the same kind for several reasons. The first reason is the extremely large surface area of nanoparticles, which provides a better contact with microorganism [87]. Second is interfering effects of salts and antimicrobial mechanism. These limitations of silver ions and silver salts can be overcome with the use of silver nanoparticles [88]. Nowadays highly concentrated and non hazardous silver nanoparticles can be prepared in a cost efficient manner [89,90]. For the scope of this chapter, the antimicrobial properties of nanoparticles will be discussed.

The bactericidal efficacy of silver nanoparticles was found to be dependent on several factors including size and shape of the particles and the type of the bacteria [91–94]. The particle size is one of the major factors that affect the toxicity of silver nanoparticles. Smaller particles have been found to be more toxic compared to their larger counterparts since small particles have larger surface area to come in contact with the bacterial membrane. The usual interaction pathway of silver nanoparticles with bacteria is through the cell membrane. Silver nanoparticles can attach to the membrane of bacteria and penetrate inside the cell.

Kim *et al.* investigated the antibacterial efficacy of sub 15 nm silver nanoparticles against yeast, *E. coli* and *S. aureus* [88]. Their findings suggest that silver nanoparticles are more effective in killing *E. coli* and yeast cells, whereas particles showed mild inhibitory results against *S. aureus*. Their results indicate that the antibacterial efficacy of silver nanoparticles may be associated with certain types of bacterial species. Gram-negative (*E. coli*) and Gram-positive (*S. aureus*) bacteria have very distinct cell membrane structure, especially in terms of the peptidoglycan layer thickness [91].

Despite the vast number of research suggesting highly effective antibacterial properties of silver nanoparticles, only a few of them try to elucidate the mechanism by which silver nanoparticles exert toxicity to bacteria. As a result mechanisms are not vastly understood or agreed upon. The common proposed mechanism is release of silver ions followed by generation of reactive oxygen species (ROS) and cell membrane damage (Figure 9.5) [95].

(i) Release of silver ions

Silver nanoparticles have been reported to dissociate into silver ions (Ag^+) when interacting with H_2O_2 in vivo. Asharani *et al.* proposed that the following reaction could take place for silver nanoparticle dissolution [95,96]. Asharani proposes that in eukaryotic cells this reaction may take place in mitochondria where H^+ ions are abundant.

$$2Ag + H_2O_2 \rightarrow 2H^+ \rightarrow 2Ag^+ + 2H_2O \qquad E^0 = 0.17\ V$$

Choi *et al.* have proposed another mechanism for the release of silver ions where silver ions are reduced in the presence of oxygen [94,97].

$$4Ag + O_2 + 2H_2O \rightarrow 4Ag^+ + 4OH^-$$

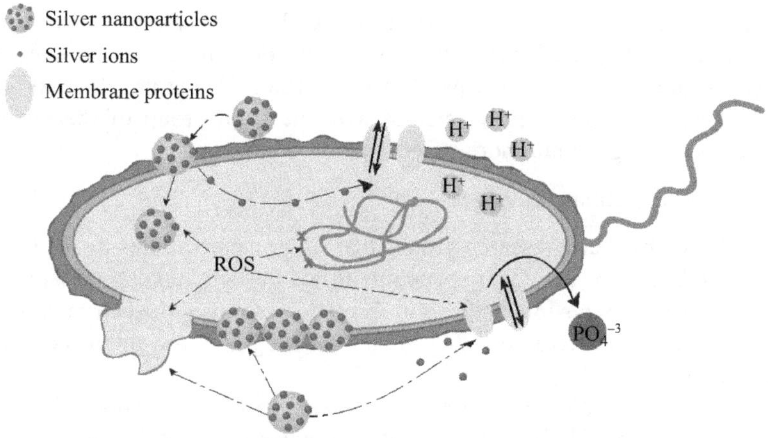

Figure 9.5 Schematic of mechanism of interaction of silver nanoparticles with bacteria

The resultant silver ions could harm the bacteria in several ways, in which enzyme inactivation occurs through its strong interaction with thiol, carboxylates, phosphates and hydroxyl groups [98]. Even though some studies claim that silver ions do not harm the DNA, there is also evidence that with the introduction of silver ions DNA molecules become condensed and lose its replication capability [91,99]. In each study the amount of silver ions in the environment was measured differently. Choi *et al.* measured it to be 2.2% of the total silver concentration. Regardless ionic silver has been known to be a good antibacterial agent for so long; thus generated silver ions are responsible at least in part of the efficient antibacterial properties of silver nanoparticles [87,91,99].

According to Schreurs and Rosenberg *et al.* silver ions also prohibit the uptake and exchange of phosphate (PO_4^{-3}) in *E. coli*, leading to efflux of accumulated phosphate, mannitol, succinate, glutamine, and proline [98,100]. Additionally, Ag^+ also collapses the proton motive force in *Vibrio cholera*. Elimination of the proton motive force leads to cell death since it has crucial importance in overall microbial metabolism [98,101].

(ii) ROS generation

ROS are generated as natural by products of living organisms [94,102]. While low levels of ROS are useful for cells and can be controlled by antioxidant defenses as glutathione/glutathione disulfide (GSH/GSSG) within the cells, at higher concentrations they are highly toxic to the cells due to the generation of oxidative stress. An excess amount of ROS can attack lipids, and cause membrane damage as well as detrimental effects on DNA by inducing mutations.

In the presence of dissolved oxygen, metals can act as a catalyst and produce ROS [103]. Thus, in the presence of oxygen silver nanoparticles may also lead to the generation of ROS. Moreover, studies have shown that in eukaryotic cells silver nanoparticles inhibit antioxidant defenses by interacting directly with GSH, GSSG, and other enzymes. This could lead to a reduction in the GSH/GSSG ratio, and a subsequent increase in ROS levels within the cell. Another mechanism for ROS generation may be caused by elution of silver ions from the particle surface and chemisorption of those ions back to the surface. According to Park *et al.*'s findings, these ions can directly interact with the thiol group of the respiratory chain enzymes and generate more ROS [104].

(iii) Direct cell membrane damage

It is now proven by many research groups that silver nanoparticles interact with the bacterial membrane and are able to penetrate inside the cell. Several research groups were able to visualize this interaction and formation of pits in the cell membrane via transmission electron microscopy (TEM) or scanning electron microscopy (SEM) (see Figure 9.6).

So far several studies have linked the toxicity of silver nanoparticles to the release of silver ions. Hwang *et al.* tried to elucidate the mechanism by treating stress-specific bioluminescent bacteria with silver nanoparticles. Their study demonstrated that silver nanoparticles led to the generation of silver ions and

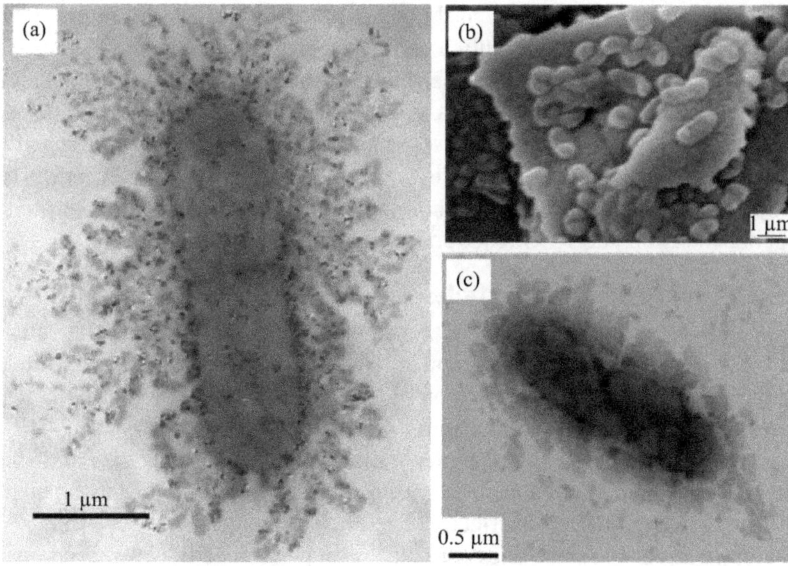

Figure 9.6 Electron microscopy images denoting interaction of silver nanoparticles with bacterial cell membranes. (a) TEM micrograph of E. coli *cell treated with 50 μg/cm³ of silver nanoparticles in liquid LB medium for 1 h [105]. (b) SEM micrograph of* E. coli *treated with silver nanoparticles [106]. (c) cell structure of* E. coli *cells treated with 50 μg/mL silver nanoparticles [98]*

subsequently superoxide radicals. It was found that membrane and protein structure was damaged but not DNA. Moreover, this damage was found to be associated with particle size. Larger sized particles (micron sized) did not exert any toxicity compared to nanosized silver particles [107]. Their findings on silver nanoparticles not damaging DNA were supported by Rai *et al.* as well. They suggest that once silver nanoparticle gets in the cell it forms a low molecular weight region in the center of bacteria to which bacteria conglomerates. This may mean bacteria protect its DNA from interacting with silver ions [87].

The third mechanism that plays a role in killing bacteria with silver nanoparticle is the membrane damage. Smetana *et al.* found that 5–10 nm silver nanocrystals show the best antibacterial activity by destroying both Gram-positive and Gram-negative bacteria membrane [108]. According to Raffi *et al.*'s results, 16 nm silver nanoparticles inhibit *E. coli* growth and multiplication by penetrating through the cell membrane at a concentration as low as 60 μg/mL [106]. According to their study, during this penetration several morphological changes took place in the cell membrane morphology resulting in significant increases in its permeability, thus transport through the plasma membrane is affected.

Silver and its derivatives have been utilized as antimicrobial agents for so long. Silver in nanoparticle form is more effective since it has much greater area in contact with bacteria cell membranes. Even though it has been utilized as an antimicrobial agent, the mechanism behind its antibacterial property was elucidated to some extent. So far there are three mechanisms that play a role in the activity of silver against bacteria including release of silver ions, generation of ROS and damaging the bacteria membrane. There is still a lot to understand to have a better control on the activity of silver nanoparticles. Park *et al.* suggests that bacterial activity of silver ions is related to the presence of O_2, and to develop more efficient silver based disinfection systems, the O_2 concentration and ROS reaction should be considered [104].

9.3.2.2 Zinc oxide nanoparticles

Inorganic metal oxide nanoparticles are another class of materials that are being developed for use in health-related applications. Metal oxide nanoparticles are highly ionizable, and not only attractive for their physical and chemical properties as stability, robustness, and long-shelf life but also for their antibacterial activity [109–112]. And with the recent advancements in nanotechnology, the control over size and shape of the particles have greatly enhanced. Several groups have talked about using elemental or metal oxide nanoparticles as effective bactericidal materials [109,110,113].

A comparative study by Jones *et al.* has been performed in order to compare the antibacterial efficacy of the six different metal oxide nanoparticles [113]. The metal oxide nanoparticles that were used in this study are MgO, TiO_2, Al_2O_3, CuO, CeO_2, and ZnO which were 50–70 nm in diameter. These particles were tested primarily against *S. aureus*. Initial results indicated that MgO, TiO_2, CuO, and CeO_2 do not have any effect in the inhibition of bacteria growth up to a 10 mM concentration [113]. Previously other metal oxide nanoparticles including MgO and CuO were reported to be toxic against *E. coli.* According to Jones *et al.* these particles did not show any toxicity against *S. aureus*. In the case of TiO_2 it is known to be toxic for many strains of bacteria including *S. aureus*, but in order for TiO_2 to show its enhanced activity it needs to be excited with UV-lighting. It is reported that under regular laboratory lighting only 20% inhibition ingrowth was observed which is under its antibacterial activity. On the other hand when treated with 50–70 nm in diameter Al_2O_3, and ZnO nanoparticles *S. aureus* growth was inhibited even at low concentrations. Studies with Al_2O_3 were not pursued since it is found to be highly toxic to the most living cells.

Even though some research has been done about the antibacterial efficiency of regular ZnO against several strains of bacteria, a little is known about its antibacterial efficacy when it is in nanoparticle form. So far it has been tested against pathogenic and nonpathogenic bacteria strains as *S. aureus* and *E. coli*. Moreover, ZnO nanoparticles are known to be non toxic, biosafe, and biocompatible. These particles have also been used as drug carriers, in cosmetic products, bioimaging, and as fillings in medical devices [112,114,115]. There are some reports addressing to the toxicity of ZnO nanoparticles against living cells, but comparatively low concentrations of these particles are non toxic to living cells [112,116]. Thus, ZnO nanoparticles are good candidates for further investigation on their antibacterial activity.

It has been reported that their antibacterial activity changes as a function of strain of bacteria, particle morphology (i.e. size and shape), and presence of visible light [109,111–113]. Some metal oxides such as ZnO, CuO, and Fe_2O_3 were found to be more effective against Gram-positive bacteria [111]. In Jones *et al.*'s study a comparison between the antibacterial properties of 8 and 50 nm ZnO nanoparticles has been performed. Results suggested that 8 nm ZnO nanoparticles are much more effective in killing bacteria compared to their 50 nm counterparts. They also performed another comparison with the same sized particles, by just changing the lightening of the environment. Their experiments were performed under normal ambient laboratory light conditions, and in the dark. Under the normal laboratory light conditions, no bacteria growth was observed, and a delay in the growth curve of *S. aureus* was recorded when treated with 8 nm and 50 nm ZnO nanoparticles respectively [113].

In Xie *et al.*'s study on antibacterial activity and mechanism of action of ZnO nanoparticles against *Campylobacter jejuni*, the minimum inhibitory concentration (MIC) was found somewhere in between 0.05 and 0.025 mg/mL which denotes extreme sensitivity of *C. jejuni* to ZnO nanoparticles [109]. According to Xie *et al.*'s mechanistic studies the occurrence of cell death was due to several reasons: (i) significant morphological changes, (ii) leakage in the cell membrane, and (iii) increases in oxidative stress gene expression. And a possible mechanism that plays a role in this scenario could be explained by direct interaction of ZnO nanoparticles with cell surfaces. Membrane permeability was affected through the nanoparticle pathway, and an oxidative stress was induced, subsequently resulting in the inhibition of cell growth and eventually leading to cell death. Raghupathi *et al.* also reported on commercially available 8 nm ZnO particle toxicity toward different microorganisms as *S. aureus*, *E. coli*, and *B. subtilis* [116].

To elucidate the effect of ZnO particle size on its antibacterial activity, another detailed study was performed by Raghupathi *et al.* A MIC of 4–7 mM depending on the bacteria strain has been proposed. For *S. aureus* 6 mM ZnO nanoparticles between the sizes of 30 and 307 nm were tested. Results suggested that 30 nm particles are much more effective in killing *S. aureus*. These results are consistent with earlier studies that were performed. Several mechanisms have been proposed for antibacterial activity of ZnO nanoparticles including ionization of ZnO when suspended in water or the pH change within the solution. Moreover, the solubility of metal oxide nanoparticles increases with decreasing particle size, and the improved activity of smaller size can be attributed to presence of more Zn^{+2}.

The photocatalytic activity of ROS has been a major contributor to antibacterial activities of various metal oxides [112]. Several studies have linked the antibacterial activity of ZnO nanoparticles with the generation of different types of ROS, including superoxide anion (O_2), hydrogen peroxide (H_2O_2), and hydroxide (OH^-). The resultant highly reactive molecules are responsible from destruction of cellular components as lipids, DNA and proteins. However, the generation of ROS creates controversies between different groups. Raghupathi *et al.* showed advanced ZnO antibacterial activity due to elevated ROS generation by ZnO under UV-exposure [116], whereas alternative studies by Hirota *et al.* observed ROS generation by ZnO in the dark by testing ZnO against *E. coli* [117].

Padmavathy and Vijayaraghavan performed a detailed size comparative study of ZnO nanoparticles at 12 nm, 45 nm, and 2 μm [118]; 12 nm ZnO particle was found to be the most effective size among others. Better antibacterial properties of small sized particles were attributed to ROS generation at the surface of the particles under both the UV and the visible light. According to their study generation of ROS is happening through the following mechanism. Both UV and visible light can create electron and hole pairs (e^- and h^+) on the surface of ZnO [1]. Interaction of h^+ with water molecules dissociates water into $^\bullet OH$ and H^+ [2]. Dissolved oxygen molecules are then transformed into superoxide radical ($^\bullet O_2{}^-$) [3] which then react with H^+ and generate $H_2O^\bullet$ radicals [4] which turn into hydrogen peroxide anion (HO_2-) with subsequent collisions with electrons. Then they react with H^+ again to generate H_2O_2 [5]. The larger surface area of nanoparticles induces the generation of ROS. Thus, enhanced toxicity of the smaller sized ZnO could be explained by larger surface area at a smaller size.

$$ZnO + h\nu \rightarrow e^- + h^+ \tag{9.1}$$

$$h^+ + H_2O \rightarrow {}^\bullet OH + H^+ \tag{9.2}$$

$$e^- + O_2 \rightarrow {}^\bullet O_2{}^- \tag{9.3}$$

$${}^\bullet O_2 + H^+ \rightarrow H_2O^\bullet \tag{9.4}$$

$$H_2O^\bullet + H^+ + e^- \rightarrow H_2O_2 \tag{9.5}$$

Not every ROS is able to pass through the bacteria cell membrane. Specifically, $^\bullet OH$ and $^\bullet O_2{}^-$ cannot penetrate inside the cells since they are both negatively charged. By contrast, H_2O_2 is able to pass through the cell membrane and cause damage inside the bacterial cell and subsequently lead to cell death, while other types of ROS can be found on the outer surface of the cell wall.

Another mechanism that is being proposed for ZnO nanoparticle antibacterial activity is the release of zinc ions (Zn^{+2}) from the nanoparticle surface. The released ions have inhibitory effects on the active transport and disruption in the amino acid metabolism and enzyme system. The release of Zn^{+2} is also size dependent. The smaller the size of the particle, the larger the surface area, and the larger the surface area, the more atoms present at the surface of the particles. According to Parquet *et al.* in addition to particle size, there are other parameters that play a role in Zn^{+2} release (i) the physiochemical parameters as porosity, concentration, and morphology and (ii) the chemistry of the media: the pH, UV illumination, exposure time, and existence of other elements. However, the way that these parameters contribute has not been fully understood. Regardless, it was elucidated that Zn^{+2} released from ZnO nanoparticles is capable of interacting with proteins.

The promising properties of ZnO nanoparticles have developed a global interest to study their antibacterial activity. The noble physical and chemical characteristics of ZnO nanoparticles which show significant toxicity to bacteria, while not showing as much to mammalian cells have made ZnO a successful candidate among the other metal oxides. The extent of antibacterial activity can be enhanced by adjusting the parameters like UV-illumination, nanoparticle size, concentration, and morphology, and surface morphology. Induction of intracellular ROS, release of Zn^{+2} ions, and adhesion on the bacterial cell membrane have been shown to be the main mechanism that leads to cell death.

References

[1] Costerton, B. Microbial ecology comes of age and joins the general ecology community. *Proceedings of the National Academy of Sciences of the United States of America.* **2004**, *101*, 16983–16984.

[2] Bazaka, K.; Jacob, M. V.; Chrzanowski, W.; Ostrikov, K. Anti-bacterial surfaces: natural agents, mechanisms of action, and plasma surface modification. *RSC Advances*. **2015**, *5*, 48739–48759.

[3] Campoccia, D.; Montanaro, L.; Arciola, C. R. A review of the biomaterials technologies for infection-resistant surfaces. *Biomaterials.* **2013**, *34*, 8533–8554.

[4] Roos-Jansaker, A. M.; Renvert, S.; Egelberg, J. Treatment of peri-implant infections: a literature review. *Journal of Clinical Periodontology.* **2003**, *30*, 467–485.

[5] Francolini, I.; Donelli, G. Prevention and control of biofilm-based medical-device-related infections. *FEMS Immunology & Medical Microbiology.* **2010**, *59*, 227–238.

[6] Hall-Stoodley, L.; Costerton, J. W.; Stoodley, P. Bacterial biofilms: from the natural environment to infectious diseases. *Nature Reviews Microbiology.* **2004**, *2*, 95–108.

[7] Zhao, L. Z.; Chu, P. K.; Zhang, Y. M.; Wu, Z. F. Antibacterial coatings on titanium implants. *Journal of Biomedicine Materials Research Part B.* **2009**, *91B*, 470–480.

[8] Surmenev, R. A.; Surmeneva, M. A.; Ivanova, A. A. Significance of calcium phosphate coatings for the enhancement of new bone osteogenesis – a review. *Acta Biomaterialias*. **2014**, *10*, 557–579.

[9] Alt, V.; Bitschnau, A.; Osterling, J.; *et al.* The effects of combined gentamicin-hydroxyapatite coating for cementless joint prostheses on the reduction of infection rates in a rabbit infection prophylaxis model. *Biomaterials.* **2006**, *27*, 4627–4634.

[10] Carter, D. R.; Beaupre, G. S.; Giori, N. J.; Helms, J. A. Mechanobiology of skeletal regeneration. *Clinical Orthopaedics and Related Research.* **1998**, S41–S55.

[11] Wu, P.; Grainger, D. W. Drug/device combinations for local drug therapies and infection prophylaxis. *Biomaterials.* **2006**, *27*, 2450–2467.

[12] Zilberman, M.; Elsner, J. J. Antibiotic-eluting medical devices for various applications. *Journal of Controlled Release.* **2008**, *130*, 202–215.

[13] Qiu, Y.; Zhang, N.; An, Y. H.; Wen, X. Biomaterial strategies to reduce implant-associated infections. *International Journal of Artificial Organs.* **2007**, *30*, 828–841.

[14] Trampuz, A.; Widmer, A. F. Infections associated with orthopedic implants. *Current Opinion in Infectious Diseases.* **2006**, *19*, 349–356.

[15] Darouiche, R. O. Antimicrobial approaches for preventing infections associated with surgical implants. *Clinical Infectious Diseases.* **2003**, *36*, 1284–1289.

[16] Trampuz, A.; Zimmerli, W. Antimicrobial agents in orthopaedic surgery – prophylaxis and treatment. *Drugs*. **2006**, *66*, 1089–1105.

[17] Jagpal, R.; Greco, R. S. Studies of a graphite-benzalkonium-oxacillin surface. *American Surgeon*. **1979**, *45*, 774–779.

[18] Piozzi, A.; Francolini, I.; Occhiaperti, L.; Di Rosa, R.; Ruggeri, V.; Donelli, G. Polyurethanes loaded with antibiotics: influence of polymer-antibiotic interactions on in vitro activity against *Staphylococcus epidermidis*. *Journal of Chemotherapy*. **2004**, *16*, 446–452.

[19] Piozzi, A.; Francolini, I.; Occhiaperti, L.; Venditti, M.; Marconi, W. Antimicrobial activity of polyurethanes coated with antibiotics: a new approach to the realization of medical devices exempt from microbial colonization. *International Journal of Pharmacy*. **2004**, *280*, 173–183.

[20] Kwok, C. S.; Wan, C. X.; Hendricks, S.; Bryers, J. D.; Horbett, T. A.; Ratner, B. D. Design of infection-resistant antibiotic-releasing polymers: I. Fabrication and formulation. *J. Controlled Release*. **1999**, *62*, 289–299.

[21] Kwok, C. S.; Horbett, T. A.; Ratner, B. D. Design of infection-resistant antibiotic-releasing polymers – II. Controlled release of antibiotics through a plasma-deposited thin film barrier. *Journal of Controlled Release*. **1999**, *62*, 301–311.

[22] Liu, S. J.; Chi, P. S.; Lin, S. S.; Ueng, S. W. N.; Chan, E. C.; Chen, J. K. Novel solvent-free fabrication of biodegradable poly-lactic-glycolic acid [PLGA] capsules for antibiotics and Rhbmp-2 delivery. *International Journal of Pharmacy*. **2007**, *330*, 45–53.

[23] Suzuki, A.; Terai, H.; Toyoda, H.; *et al.* A biodegradable delivery system for antibiotics and recombinant human bone morphogenetic protein-2: a potential treatment for infected bone defects. *Journal of Orthopaedic Research*. **2006**, *24*, 327–332.

[24] Price, J. S.; Tencer, A. F.; Arm, D. M.; Bohach, G. A. Controlled release of antibiotics from coated orthopedic implants. *Journal of Biomedical Materials Research*. **1996**, *30*, 281–286.

[25] Harris, L. G.; Richards, R. G. Staphylococci and implant surfaces: a review. *Injury-International Journal of the Care of the Injured*. **2006**, *37*, 3–14.

[26] Hendriks, J. G. E.; van Horn, J. R.; van der Mei, H. C.; Busscher, H. J. Backgrounds of antibiotic-loaded bone cement and prosthesis-related infection. *Biomaterials*. **2004**, *25*, 545–556.

[27] Wininger, D. A.; Fass, R. J. Antibiotic-impregnated cement and beads for orthopedic infections. *Antimicrobial and Agents Chemotherapy*. **1996**, *40*, 2675–2679.

[28] Neut, D.; van de Belt, H.; van Horn, J. R.; van der Mei, H. C.; Busscher, H. J. The effect of mixing on gentamicin release from polymethylmethacrylate bone cements. *Acta Orthopaedica Scandinavica*. **2003**, *74*, 670–676.

[29] Neut, D.; van de Belt, H.; van Horn, J. R.; van der Mei, H. C.; Busscher, H. J. Residual gentamicin-release from antibiotic-loaded polymethylmethacrylate beads after 5 years of implantation. *Biomaterials*. **2003**, *24*, 1829–1831.

[30] Garvin, K.; Feschuk, C. Polylactide-polyglycolide antibiotic implants. *Clinical Orthopaedics and Related Research.* **2005**, 105–110.

[31] Gerhart, T. N.; Roux, R. D.; Hanff, P. A.; Horowitz, G. L.; Renshaw, A. A.; Hayes, W. C. Antibiotic-loaded biodegradable bone-cement for prophylaxis and treatment of experimental osteomyelitis in rats. *Journal of Orthopaedic Research.* **1993**, *11*, 250–255.

[32] McLaren, A. C. Alternative materials to acrylic bone cement for delivery of depot antibiotics in orthopaedic infections. *Clinical Orthopaedics and Related Research.* **2004**, 101–106.

[33] Schnieders, J.; Gbureck, U.; Thull, R.; Kissel, T. Controlled release of gentamicin from calcium phosphate – poly[lactic acid-co-glycolic acid] composite bone cement. *Biomaterials.* **2006**, *27*, 4239–4249.

[34] Stigter, M.; Bezemer, J.; de Groot, K.; Layrolle, P. Incorporation of different antibiotics into carbonated hydroxyapatite coatings on titanium implants, release and antibiotic efficacy. *Journal of Controlled Release.* **2004**, *99*, 127–137.

[35] Witte, W. Antibiotic resistance in gram-positive bacteria: epidemiological aspects. *Journals of Antimicrobial Chemotherapy.* **1999**, *44*, 1–9.

[36] Gold, H. S.; Moellering, R. C. Drug therapy – antimicrobial-drug resistance. *New England Journal of Medicine.* **1996**, *335*, 1445–1453.

[37] Campoccia, D.; Montanaro, L.; Arciola, C. R. The significance of infection related to orthopedic devices and issues of antibiotic resistance. *Biomaterials.* **2006**, *27*, 2331–2339.

[38] Mah, T. F. C.; O'Toole, G. A. Mechanisms of biofilm resistance to antimicrobial agents. *Trends in Microbiology.* **2001**, *9*, 34–39.

[39] Stewart, P. S. Theoretical aspects of antibiotic diffusion into microbial biofilms. *Antimicrobial and Agents Chemotherapy.* **1996**, *40*, 2517–2522.

[40] Stewart, P. S. A review of experimental measurements of effective diffusive permeabilities and effective diffusion coefficients in biofilms. *Biotechnology Bioengineering*. **1998**, *59*, 261–272.

[41] Stewart, P. S.; Costerton, J. W. Antibiotic resistance of bacteria in biofilms. *Lancet.* **2001**, *358*, 135–138.

[42] Debeer, D.; Stoodley, P.; Roe, F.; Lewandowski, Z. Effects of biofilm structures on oxygen distribution and mass-transport. *Biotechnology Bioengineering*. **1994**, *43*, 1131–1138.

[43] Zhang, T. C.; Bishop, P. L. Evaluation of substrate and pH effects in a nitrifying biofilm. *Water Environmental Research.* **1996**, *68*, 1107–1115.

[44] Brooun, A.; Liu, S. H.; Lewis, K. A Dose–response study of antibiotic resistance in *Pseudomonas aeruginosa* biofilms. *Antimicrobial and Agents Chemotherapy.* **2000**, *44*, 640–646.

[45] Hanssen, A. M.; Kjeldsen, G.; Sollid, J. U. E. Local variants of staphylococcal cassette chromosome mec in sporadic methicillin-resistant *Staphylococcus aureus* and methicillin-resistant coagulase-negative staphylococci: evidence of horizontal gene transfer? *Antimicrobial and Agents Chemotherapy.* **2004**, *48*, 285–296.

[46] DeJong, E. S.; DeBerardino, T. M.; Brooks, D. E.; *et al.* Antimicrobial efficacy of external fixator pins coated with a lipid stabilized hydroxyapatite/chlorhexidine complex to prevent pin tract infection in a goat model. *Journal of Trauma-Injury Infection and Critical Care.* **2001**, *50*, 1008–1013.

[47] Harris, L. G.; Patterson, L. M.; Bacon, C.; ap Gwynn, I.; Richards, R. G. Assessment of the cytocompatibility of different coated titanium surfaces to fibroblasts and osteoblasts. *Journal of Biomedicine Materials Research, Part A.* **2005**, *73A*, 12–20.

[48] Sampath, L. A.; Tambe, S. M.; Modak, S. M. In vitro and in vivo efficacy of catheters impregnated with antiseptics or antibiotics: evaluation of the risk of bacterial resistance to the antimicrobials in the catheters. *Infection Control and Hospital Epidemiology.* **2001**, *22*, 640–646.

[49] Tambe, S. M.; Sampath, L.; Modak, S. M. In vitro evaluation of the risk of developing bacterial resistance to antiseptics and antibiotics used in medical devices. *Journal of Antimicrobial Chemotherapy.* **2001**, *47*, 589–598.

[50] Goldschmidt, P.; Cogen, R.; Taubman, S. Cytopathologic effects of chlorhexidine on human cells. *Journal of Periodontology.* **1977**, *48*, 212–215.

[51] Cheung, J.; Oleary, J. J. Allergic reaction to chlorhexidine in an anesthetized patient. *Anaesthesia and Intensive Care.* **1985**, *13*, 429–439.

[52] Ohtoshi, T.; Yamauchi, N.; Tadokoro, K.; *et al.* IGE antibody-mediated shock reaction caused by topical application of chlorhexidine. *Clinical Allergy.* **1986**, *16*, 155–161.

[53] Thomas, L.; Russell, A. D.; Maillard, J. Y. Antimicrobial activity of chlorhexidine diacetate and benzalkonium chloride against *Pseudomonas aeruginosa* and its response to biocide residues. *Journal of Applied Microbiology.* **2005**, *98*, 533–543.

[54] Gottenbos, B.; van der Mei, H. C.; Klatter, F.; Nieuwenhuis, P.; Busscher, H. J. In vitro and in vivo antimicrobial activity of covalently coupled quaternary ammonium silane coatings on silicone rubber. *Biomaterials.* **2002**, *23*, 1417–1423.

[55] Gilbert, P.; Altaae, A. Antimicrobial activity of some alkyltrimethylammonium bromides. *Letters Applications Microbiology.* **1985**, *1*, 101–104.

[56] Simonetti, N.; Simonetti, G.; Bougnol, F.; Scalzo, M. Electrochemical Ag^+ for preservative use. *Applied and Environmental Microbiology.* **1992**, *58*, 3834–3836.

[57] Bosetti, M.; Masse, A.; Tobin, E.; Cannas, M. Silver coated materials for external fixation devices: in vitro biocompatibility and genotoxicity. *Biomaterials.* **2002**, *23*, 887–892.

[58] Wassall, M. A.; Santin, M.; Isalberti, C.; Cannas, M.; Denyer, S. P. Adhesion of bacteria to stainless steel and silver-coated orthopedic external fixation pins. *Journal of Biomedical Materials Research.* **1997**, *36*, 325–330.

[59] Alt, V.; Bechert, T.; Steinrucke, P.; *et al.* An in vitro assessment of the antibacterial properties and cytotoxicity of nanoparticulate silver bone cement. *Biomaterials.* **2004**, *25*, 4383–4391.

[60] Kraft, C. N.; Hansis, M.; Arens, S.; Menger, M. D.; Vollmar, B. Striated muscle microvascular response to silver implants: a comparative in vivo study with titanium and stainless steel. *Journal of Biomedical Materials Research.* **2000**, *49*, 192–199.

[61] Wu, J. A.; Kusuma, C.; Mond, J. J.; Kokai-Kun, J. F. Lysostaphin disrupts *Staphylococcus aureus* and *Staphylococcus epidermidis* biofilms on artificial surfaces. *Antimicrobial and Agents Chemotherapy.* **2003**, *47*, 3407–3414.

[62] Ratner, B. D.; Bryant, S. J. Biomaterials: where we have been and where we are going. *Annual Review of Biomedical Engineering.* **2004**, *6*, 41–75.

[63] Harris, L. G.; Tosatti, S.; Wieland, M.; Textor, M.; Richards, R. G. *Staphylococcus aureus* adhesion to titanium oxide surfaces coated with non-functionalized and peptide-functionalized poly[L-lysine]-grafted-poly[ethylene glycol] copolymers. *Biomaterials.* **2004**, *25*, 4135–4148.

[64] Harbers, G. M.; Emoto, K.; Greef, C.; *et al.* Functionalized poly[ethylene glycol]-based bioassay surface chemistry that facilitates bio-immobilization and inhibits nonspecific protein, bacterial, and mammalian cell adhesion. *Chemical Materials.* **2007**, *19*, 4405–4414.

[65] Chen, H.; Brook, M. A.; Chen, Y.; Sheardown, H. Surface properties of peo-silicone composites: reducing protein adsorption. *Journal of Biomaterials Science-Polymer Edition.* **2005**, *16*, 531–548.

[66] Groll, J.; Fiedler, J.; Engelhard, E.; *et al.* A novel star peg-derived surface coating for specific cell adhesion. *Journal of Biomedicine Materials Research, Part A.* **2005**, *74A*, 607–617.

[67] Dong, B. Y.; Manolache, S.; Somers, E. B.; Wong, A. C. L.; Denes, F. S. Generation of antifouling layers on stainless steel surfaces by plasma-enhanced crosslinking of polyethylene glycol. *Journal of Applied Polymer Science.* **2005**, *97*, 485–497.

[68] Yoo, H. J., Kim, H. D. Properties of crosslinked blends of pellethene and multiblock polyurethane containing poly[ethylene oxide] for biomaterials. *Journal of Applied Polymer Science.* **2004**, *91*, 2348–2357.

[69] VandeVondele, S.; Voros, J.; Hubbell, J. A. RGD-grafted poly-L-lysine-graft-[polyethylene glycol] copolymers block non-specific protein adsorption while promoting cell adhesion. *Biotechnology Bioengineering.* **2003**, *82*, 784–790.

[70] Park, K. D.; Kim, Y. S.; Han, D. K.; *et al.* Bacterial adhesion on PEG modified polyurethane surfaces. *Biomaterials.* **1998**, *19*, 851–859.

[71] Desai, N. P.; Hossainy, S. F. A.; Hubbell, J. A. Surface-immobilized polyethylene oxide for bacterial repellence. *Biomaterials.* **1992**, *13*, 417–420.

[72] Tosatti, S.; De Paul, S. M.; Askendal, A.; *et al.* Peptide functionalized poly [L-lysine]-G-poly[ethylene glycol] on titanium: resistance to protein adsorption in full heparinized human blood plasma. *Biomaterials.* **2003**, *24*, 4949–4958.

[73] Giuliani, A.; Pirri, G.; Nicoletto, S. F. Antimicrobial peptides: an overview of a promising class of therapeutics. *Central European Journal of Biology.* **2007**, *2*, 1–33.

[74] Brogden, K. A. Antimicrobial peptides: pore formers or metabolic inhibitors in bacteria? *Nature Reviews Microbiology.* **2005**, *3*, 238–250.

[75] Zasloff, M. Antimicrobial peptides of multicellular organisms. *Nature.* **2002**, *415*, 389–395.

[76] Marshall, S. H.; Arenas, G. Antimicrobial peptides: a natural alternative to chemical antibiotics and a potential for applied biotechnology. *Electronic Journal of Biotechnology.* **2003**, *6*, 271–284.

[77] Haynie, S. L.; Crum, G. A.; Doele, B. A. Antimicrobial activities of amphiphilic peptides covalently bonded to a water-insoluble resin. *Antimicrobial and Agents Chemotherapy.* **1995**, *39*, 301–307.

[78] Appendini, P.; Hotchkiss, J. H. Antimicrobial activity of a 14-residue peptide against *Escherichia coli* O157:H7. *Journal of Applied Microbiology.* **1999**, *87*, 750–756.

[79] Appendini, P.; Hotchkiss, J. H. Surface modification of poly[styrene] by the attachment of an antimicrobial peptide. *Journal of Applied Polymer Science.* **2001**, *81*, 609–616.

[80] Etienne, O.; Picart, C.; Taddei, C.; *et al.* Multilayer polyelectrolyte films functionalized by insertion of defensin: a new approach to protection of implants from bacterial colonization. *Antimicrobial and Agents Chemotherapy.* **2004**, *48*, 3662–3669.

[81] Blondelle, S. E.; Houghten, R. A. Design of model amphipathic peptides having potent antimicrobial activities. *Biochemistry.* **1992**, *31*, 12688–12694.

[82] Hancock, R. E. W.; Lehrer, R. Cationic peptides: a new source of antibiotics. *Trends Biotechnol.* **1998**, *16*, 82–88.

[83] Powers, J. P. S.; Hancock, R. E. W. The relationship between peptide structure and antibacterial activity. *Peptides.* **2003**, *24*, 1681–1691.

[84] Yeaman, M. R.; Yount, N. Y. Mechanisms of antimicrobial peptide action and resistance. *Pharmacological Reviews.* **2003**, *55*, 27–55.

[85] Kwakman, P. H. S.; Velde, A. A. T.; Vandenbroucke-Grauls, C. M. J. E.; Van Deventer, S. J. H.; Zaat, S. A. J. Treatment and prevention of *Staphylococcus epidermidis* experimental biomaterial-associated infection by bactericidal peptide 2. *Antimicrobial and Agents Chemotherapy.* **2006**, *50*, 3977–3983.

[86] Seil, J. T.; Webster, T. J. Antimicrobial applications of nanotechnology: methods and literature. *International Journal of Nanomedicine.* **2012**, *7*, 2767.

[87] Rai, M.; Yadav, A.; Gade, A. Silver nanoparticles as a new generation of antimicrobials. *Biotechnology Advances.* **2009**, *27*, 76–83.

[88] Kim, J. S.; Kuk, E.; Yu, K. N.; *et al.* Antimicrobial effects of silver nanoparticles. *Nanomedicine: Nanotechnology, Biology and Medicine*. **2007**, *3*, 95–101.

[89] Jain, D.; Daima, H. K.; Kachhwaha, S.; Kothari, S. Synthesis of plant-mediated silver nanoparticles using papaya fruit extract and evaluation of their anti microbial activities. *Digest Journal of Nanomaterials and Biostructures.* **2009**, *4*, 557–563.

[90] Sharma, V. K.; Yngard, R. A.; Lin, Y. Silver nanoparticles: green synthesis and their antimicrobial activities. *Advances in Colloid and Interface Science.* **2009**, *145*, 83–96.

[91] Morones, J. R.; Elechiguerra, J. L.; Camacho, A.; *et al.* The bactericidal effect of silver nanoparticles. *Nanotechnology.* **2005**, *16*, 2346.

[92] Pal, S.; Tak, Y. K.; Song, J. M. Does the antibacterial activity of silver nanoparticles depend on the shape of the nanoparticle? A study of the Gram-negative bacterium *Escherichia coli*. *Applied and Environmental Microbiology.* **2007**, *73*, 1712–1720.

[93] Das, B.; Dash, S. K.; Mandal, D.; *et al.* Green synthesized silver nanoparticles destroy multidrug resistant bacteria via reactive oxygen species mediated membrane damage. *Arabian Journal of Chemistry*. **2015**.

[94] Choi, O.; Hu, Z. Size dependent and reactive oxygen species related nanosilver toxicity to nitrifying bacteria. *Environmental Science & Technology.* **2008**, *42*, 4583–4588.

[95] Marambio-Jones, C.; Hoek, E. M. A Review of the antibacterial effects of silver nanomaterials and potential implications for human health and the environment. *Journal of Nanoparticle Research.* **2010**, *12*, 1531–1551.

[96] AshaRani, P.; Low Kah Mun, G.; Hande, M. P.; Valiyaveettil, S. Cytotoxicity and genotoxicity of silver nanoparticles in human cells. *ACS Nano.* **2008**, *3*, 279–290.

[97] Choi, O.; Deng, K. K.; Kim, N.-J.; Ross, L.; Surampalli, R. Y.; Hu, Z. The inhibitory effects of silver nanoparticles, silver ions, and silver chloride colloids on microbial growth. *Water Research*. **2008**, *42*, 3066–3074.

[98] Li, W.-R.; Xie, X.-B.; Shi, Q.-S.; Zeng, H.-Y.; You-Sheng, O.-Y.; Chen, Y.-B. Antibacterial activity and mechanism of silver nanoparticles on *Escherichia coli*. *Applied Microbiology and Biotechnology.* **2010**, *85*, 1115–1122.

[99] Feng, Q.; Wu, J.; Chen, G.; Cui, F.; Kim, T.; Kim, J. A mechanistic study of the antibacterial effect of silver ions on *Escherichia coli* and *Staphylococcus aureus*. *Journal of Biomedical Materials Research.* **2000**, *52*, 662–668.

[100] Schreurs, W.; Rosenberg, H. Effect of silver ions on transport and retention of phosphate by *Escherichia coli*. *Journal of Bacteriology.* **1982**, *152*, 7–13.

[101] Dibrov, P.; Dzioba, J.; Gosink, K. K.; Häse, C. C. Chemiosmotic mechanism of antimicrobial activity of Ag^+ in *Vibrio cholerae*. *Antimicrobial Agents and Chemotherapy.* **2002**, *46*, 2668–2670.

[102] Giorgio, M.; Trinei, M.; Migliaccio, E.; Pelicci, P. G. Hydrogen peroxide: a metabolic by-product or a common mediator of ageing signals? *Nature Reviews Molecular Cell Biology.* **2007**, *8*, 722–728.

[103] Wu, H.; Yin, J.-J.; Wamer, W. G.; Zeng, M.; Lo, Y. M. Reactive oxygen species-related activities of nano-iron metal and nano-iron oxides. *Journal of Food and Drug Analysis.* **2014**, *22*, 86–94.

[104] Park, H.-J.; Kim, J. Y.; Kim, J.; *et al.* Silver-ion-mediated reactive oxygen species generation affecting bactericidal activity. *Water Research*. **2009**, *43*, 1027–1032.

[105] Sondi, I.; Salopek-Sondi, B. Silver nanoparticles as antimicrobial agent: a case study on *E. coli* as a model for Gram-negative bacteria. *Journal of Colloid and Interface Science.* **2004**, *275*, 177–182.
[106] Raffi, M.; Hussain, F.; Bhatti, T.; Akhter, J.; Hameed, A.; Hasan, M. Antibacterial characterization of silver nanoparticles against *E. coli* ATCC-15224. *Journal of Materials Science and Technology.* **2008**, *24*, 192–196.
[107] Hwang, E. T.; Lee, J. H.; Chae, Y. J.; *et al.* Analysis of the toxic mode of action of silver nanoparticles using stress-specific bioluminescent bacteria. *Small.* **2008**, *4*, 746–750.
[108] Smetana, A. B.; Klabunde, K. J.; Marchin, G. R.; Sorensen, C. M. Biocidal activity of nanocrystalline silver powders and particles. *Langmuir.* **2008**, *24*, 7457–7464.
[109] Xie, Y.; He, Y.; Irwin, P. L.; Jin, T.; Shi, X. Antibacterial activity and mechanism of action of zinc oxide nanoparticles against *Campylobacter Jejuni*. *Applied and Environmental Microbiology.* **2011**, *77*, 2325–2331.
[110] Stoimenov, P. K.; Klinger, R. L.; Marchin, G. L.; Klabunde, K. J. Metal oxide nanoparticles as bactericidal agents. *Langmuir.* **2002**, *18*, 6679–6686.
[111] Azam, A.; Ahmed, A. S.; Oves, M.; Khan, M. S.; Habib, S. S.; Memic, A. Antimicrobial activity of metal oxide nanoparticles against Gram-positive and Gram-negative bacteria: a comparative study. *International Journal of Nanomedicine.* **2012**, *7*, 6003.
[112] Sirelkhatim, A.; Mahmud, S.; Seeni, A.; *et al.* Review on zinc oxide nanoparticles: antibacterial activity and toxicity mechanism. *Nano-Micro Letters*. 1–24.
[113] Jones, N.; Ray, B.; Ranjit, K. T.; Manna, A. C. Antibacterial activity of ZnO nanoparticle suspensions on a broad spectrum of microorganisms. *FEMS Microbiology Letters.* **2008**, *279*, 71–76.
[114] Xiong, H. M. Zno nanoparticles applied to bioimaging and drug delivery. *Advanced Materials.* **2013**, *25*, 5329–5335.
[115] Zhang, Y.; Nayak, T. R.; Hong, H.; Cai, W. Biomedical applications of zinc oxide nanomaterials. *Current Molecular Medicine.* **2013**, *13*, 1633.
[116] Raghupathi, K. R.; Koodali, R. T.; Manna, A. C. Size-dependent bacterial growth inhibition and mechanism of antibacterial activity of zinc oxide nanoparticles. *Langmuir.* **2011**, *27*, 4020–4028.
[117] Hirota, K.; Sugimoto, M.; Kato, M.; Tsukagoshi, K.; Tanigawa, T.; Sugimoto, H. Preparation of zinc oxide ceramics with a sustainable antibacterial activity under dark conditions. *Ceramics International.* **2010**, *36*, 497–506.
[118] Padmavathy, N.; Vijayaraghavan, R. Enhanced bioactivity of Zno nanoparticles—an antimicrobial study. *Science and Technology of Advanced Materials.* **2008**, *9*, 035004.

Chapter 10

Nanomaterials used in implant technology and their toxicity

Mine Altunbek[1], *Gamze Kuku*[1] *and Mustafa Culha*[1]

Abstract

The use of nanomaterials in materials science and engineering has provided the modern world an enhanced inventory of products due to the nanomaterials' unique physicochemical properties. The similar size scale between nanomaterials and biological molecules has opened up many application areas in life sciences such as drug delivery, gene therapy, and tissue engineering. Moreover, the surface and mechanical properties of these novel materials have allowed enhanced biocompatibility and strength. Therefore, they have been extensively investigated for their applications in implants and tissue engineering. The most widely used nanomaterials in these fields are titanium, silver and polymeric nanoparticles, and carbon nanotubes (CNTs). Although these nanomaterials are widely used in the mentioned areas, there is no consensus about their toxicity yet. Therefore, their toxicity should be carefully considered before using these nanomaterials. This chapter aims to summarize the toxicity of nanomaterials used in tissue engineering and implant technology.

10.1 Introduction

Nanomaterials (NMs) are composed of particles with at least one dimension of up to 100 nm that possess unique physicochemical properties compared to their bulk forms. The properties including improved conductivity, strength, and charge paved the way for their use in implant technologies. A good implant is described as to have increased integration to its surrounding tissue, optimal healing time, anti-bacterial properties, and a minimized induction of anti-inflammatory response. These properties require biocompatible and corrosion resistant implants. The failure from possessing these characteristics may lead to complications in the patients. For instance, it may result with implant loss due to osteolysis and toxic accumulation of

[1]Yeditepe University, Genetics and Bioengineering Department, Atasehir, Istanbul, Turkey

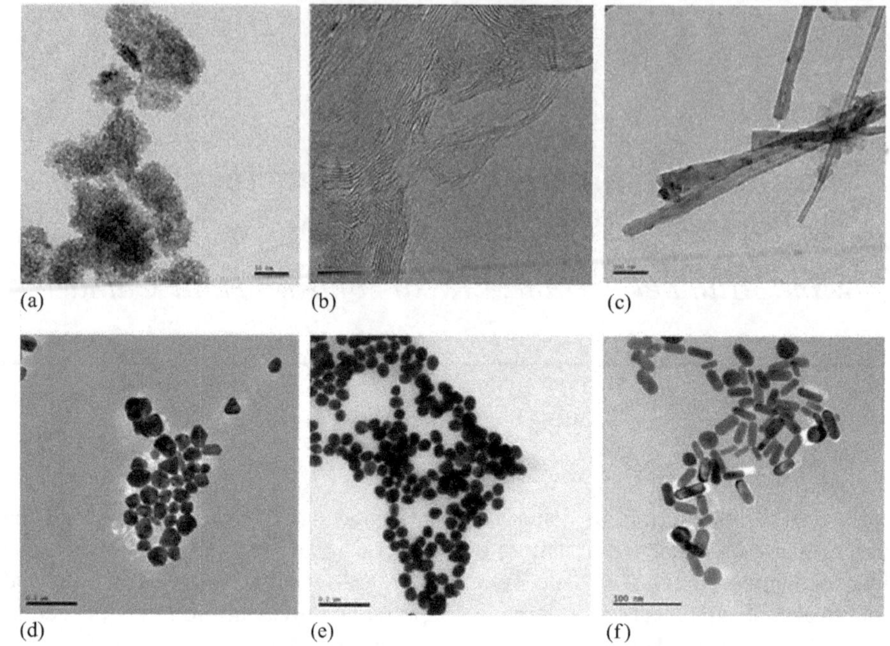

Figure 10.1 TEM image of (a) TiO_2 NPs; (b) aggregated SWCNTs; (c) BNNT; (d) AgNPs; (e) AuNPs; and (f) Au nanorods. The scale bars are 50 nm, 5 nm, 100 nm, 200 nm, 200 nm, and 100 nm, respectively

the wear debris in body organs. The toxicity of the nanomaterials may depend on various physicochemical factors such as size, shape, surface chemistry, concentration, and chemical nature. Their possible adverse effects can be evaluated in vitro through determining cell membrane integrity, mitochondrial activity, lipid peroxidation, and cell proliferation or by in vivo methods through the evaluation of blood contact response, biodistribution, toxicokinetics, and immune system response. In this chapter, commonly used NMs, titanium dioxide (TiO_2), CNTs, boron nitride nanotubes (BNNTs), silver nanoparticles (AgNPs), and gold nanoparticles (AuNPs) in body implants, their toxicity, and biodistribution will be discussed in the light of present knowledge. Figure 10.1 shows the representative TEM images of these NMs as examples.

10.2 Nanomaterials in implants

10.2.1 TiO_2 nanoparticles

Titanium is a widely used material in many industries ranging from food and biomedical to cosmetics and plastics. According to the U.S. Geological Survey Mineral Resources Program, worldwide TiO_2 production capacity was 6.5×10^6

metric tons in 2014 and there is a growing demand in the industry [1]. What makes TiO_2 attractive for many fields are mostly its light scattering properties; it possesses excellent opacity allowing it to be used as a coloring additive, as well as its high corrosion resistance and being lighter than stainless steel, both of which are advantages to be used as a body implant [2].

Titanium and its alloys have been used for decades in body implants specifically in bone and dental implants. The earliest report on titanium bone implants is by Bothe *et al.* in 1940, where they tested seven types of metals as implants on cat femurs and found that the titanium is the most corrosion resistant and strongest to provide enough support to its surrounding environment [3]. Following this report, in 1951, Leventhal showed on rat femurs that the implanted screws were almost impossible to remove from the bones even after 16 weeks [2]. This phenomenon was named as "osseointegration" by Brånemark, who successfully used titanium dental implants on human subjects for the first time and proved their efficiency by a 24-year follow-up study [4–6].

The success of Ti-based bone implants is mostly attributed to the thin oxide layer that naturally occurs on the implant surface due to its high affinity for oxygen. The TiO_2 layer is about 5 nm thick and possesses TiO_2 nanoparticle (TiO_2 NP) characteristics because it has at least one dimension smaller than 100 nm. The TiO_2 NP layer protects the metallic implant to release wear debris or ions and protects it from corrosion. However, the former is not guaranteed for implants such as hip or knee joints, where constant mechanical stress is applied that result in wear debris. This brings out an immediate question about TiO_2 NP implant toxicity both to the surrounding tissue and to the patients themselves.

10.2.1.1 Toxicity and biodistribution of TiO_2 NPs

In vitro cytotoxicity studies reveal that TiO_2 NPs might induce endoplasmic reticulum (ER) stress by elevating reactive oxygen species (ROS) levels, which in turn result in disruption of mitochondria-associated ER membranes (MAM) [7]. Moreover, TiO_2 NPs might induce genotoxicity by DNA adduct and micronuclei formations [8,9].

At a broader level, in vivo, TiO_2 NP-related toxicity studies show varying complications in rodents. For instance, a portion of intranasal instilled TiO_2 NPs (80 nm rutile) were shown to end up in the brain hippocampus regions of mice [10]. In another mice study, 5 nm-sized TiO_2 NPs distributed into the liver, kidney, spleen, lung, brain, and heart in the order of decreasing amounts, causing severe damage to the liver, kidney, and myocardium [11]. Blood sugar and lipid levels were also disturbed. Intraperitoneal injection of 80–100 nm TiO_2 NPs accumulated in 2 weeks, in the spleen, liver, kidneys, and lungs of mice in the order of decreasing amounts [12].

Although there is an increasing number of reports on TiO_2 NP toxicity studies that were carried out on rodents, the knowledge based on human subjects is still premature and is mostly about occupational exposure to TiO_2 NPs [13–22]. In the light of the information obtained from these reports, TiO_2 NPs are considered as "possibly carcinogenic for humans" [23].

Although there is ambiguity on the toxicity of TiO_2 NPs, the existing knowledge gives clues about the possible toxic response of titanium containing body implants. First of all, the major source of translocated TiO_2 NPs is the wear debris from the implant surface. The lack of fluid exchange around the implant area also plays an important role in the TiO_2 NP implant toxicity, especially when combined with fretting. Distribution of the naturally occurring TiO_2 NP layer (depassivation) leads to the release of metal ions as well as TiO_2 NPs from the implant surface. In a study by Urban *et al.*, both micro- and nanosized TiO_2 particles were found in the liver and spleen of patients that carry titanium body implants [24]. The depassivation forces the implant surface to a repassivation process. This process yields negatively charged chloride ions and, in the lack of oxygen, which happens in parallel with the lack of fluid exchange, these ions form titanium chloride that hydrolyzes and contributes to a pH decrease in the proximity of the implant surface and causes severe corrosion, which may even end up with implant loss [25]. To overcome this issue, coating the implant surface with a thicker TiO_2 NP layer by several methods including anodic oxidation and hydrothermal treatment were proposed [26,27]. It was possible to improve corrosion resistance and biocompatibility this way.

Even though most of the follow-up studies that were carried out on patients with hip replacement operations did not reveal a direct relation with cancer formation, in vitro studies on human cell lines indicate possible risk to human health, as mentioned earlier. This forces most of the follow-up studies to conclude with a warning sentence highlighting that the possible toxicity should still be considered [9,28–36].

10.2.2 Carbon nanotubes (CNTs)

CNTs are hollow cylinders formed from coiled carbon atoms that are arranged in condensed aromatic rings [37]. They can be formed into a single layer (single-walled CNTs – SWCNTs) or multiple layers (multi-walled CNTs – MWCNTs) coiled structures [37,39]. The diameter of SWCNTs is in the range of 0.4–2 nm whereas the diameter of MWCNTs is much larger, varying from 20 to 100 nm. The unique mechanical and electrical properties of CNTs depend on their structure [38]. Due to the multiple layers with a larger diameter, MWCNTs have a more rigid structure with a lower mechanical tensile strength and electrical property compared to SWCNTs [40].

CNTs are known as relatively biocompatible nanomaterials. Due to the mechanical and electrical properties compared to the biocompatible metal oxide nanoparticles such as TiO_2, they are considered to be used in implants to enhance their strength, flexibility, or conductivity. The CNTs have been used in bone (MWCNTs), heart, and neural (SWCNTs) tissue engineering applications [41–46]. The properties of CNTs such as length, diameter, roughness, surface energy, functional groups, and electrostatic charge are important factors for implant-quality and functionality, influencing the adhesion, morphology, proliferation, and differentiation of the cells. The smaller size and thinner diameter of CNT films are found to be more effective to improve the mechanical properties of the implants, which provide sufficient interaction with the tissue, enhancing cell adhesion and proliferation. For instance, the use of SWCNT films was found to be more effective for the proliferation of osteoblast-like Saos-2 cells by keeping the morphological characteristics compared to MWCNT films [47,48]. The concentration of MWCNTs was found directly proportional with porosity and water uptake capacity of the scaffold, enhancing the adhesion of osteoblast cells to the surface and influencing the calcium-phosphate mineral formation [50]. In addition, the stimulation of the cells on CNT composites with an electrical current also accelerated osteoblast cell proliferation [51]. The electrical property of CNT can also be used for the stimulation of neuronal cell growth, differentiation, or regeneration [44].

The surface change and morphology of the CNT used in thin films also play an important role for the initial attachment and proliferation of the cells, respectively. For instance, the CNT thin films with a rougher surface increased the protein adsorption, which improved the adhesion, proliferation, and differentiation of human mesenchymal stem cells (hMSC) into a bone lineage [49,52]. The importance of the surface electrostatic property was observed with the increase in the osteoblast cell adhesion and proliferation on the thin films prepared with neutral SWCNT compared with the thin films prepared with positively and negatively charged SWCNT [53].

The modification of CNT surfaces with molecules carrying different functional groups also influences the scaffold properties, which, in turn, may affect the desired action on the cells. For instance, the scaffold prepared with MWCNTs functionalized with a carboxyl group (–COOH), hydroxyl group (–OH) and poly (ethylene glycol) (PEG) showed different effects on MSCs differentiation. The PEG functionalized MWCNT films were found to be more effective at differentiating the MSCs through the rapid induction of a bone matrix mineralization and proliferation with a correct morphology compared to the –COOH modified MWCNTs [54]. In another study, –COOH modified MWCNTs films inhibited the proliferation and differentiation into osteoblast cells [55].

The polymer composites prepared with CNTs showed selective enhancement for the adhesion and proliferation of the osteoblast cells [56,57]. For example, the osteoblast cells selectively adhered and proliferated on polycarbonate urethane (PU)-CNT composites while the fibroblast cell number decreased [56]. Another study showed that compared to their pristine forms, CNT-blended polymers such as polycaprolactone (PCL) or poly(3-hydroxybutyrate) (P3HB) enhanced the MSC differentiation and proliferation [58,59].

10.2.2.1 Toxicity and biodistribution of CNTs

There are conflicting reports about the safety or toxicity of CNTs due to the variation in their diameter, size, large surface area, and surface properties. The exposure route such as injection, inhalation, or release from implants also influences their side effects. The injection of MWCNT induced mesothelioma, like asbestos, whereas the inhalation performed at a safety level, a level that is specified by the National Institute for Occupational Safety and Health (NIOSH), that did not induce any tumorigenesis in the in vivo studies [60–62]. CNT-toxicity can be induced due to a different combination of physicochemical properties. One of the reasons for the CNT toxicity is attributed to the remaining impurities from the various metals, used as catalysts in the synthesis of CNTs such as iron (Fe), nickel oxide (NiO), molybdenum (Mo), and cobalt (Co) [63,64]. For instance, Fe impurities cause hydroxyl radical generation whereas NiO affects the redox properties [65]. When the source of CNT toxicity was investigated by comparing the non-purified, purified, and modified SWCNTs, the removal of the impurities showed a decrease in their toxicity [66,67]. However, structural defects of CNTs were formed during synthesis and modification or removal of impurities caused less-stable particles with reactive groups on the surface. The unstable CNTs were found to induce toxicity through inducing ROS production [68].

The fate and toxicity of CNTs are related with their adsorption, distribution, and clearance capacities. MWCNTs did not induce any significant inflammatory response themselves. However, the accumulation of the CNTs in the intracellular environment at a certain level is the pronounced reason for their cytotoxicity since the aggregation influences the shape and surface area of CNTs as well as rigidity, whereas dispersed CNTs were more flexible. An in vivo follow-up study on mice showed that the accumulation of MWCNTs inside cells at long-term exposure caused the induction of apoptosis and activation of the tumor suppressor protein [72]. The surface chemistry of CNTs plays an important role in their dispersibility, interaction with the cells, and the clearance of CNTs from the body [69–71]. For example, the hydroxylated SWCNTs (SWCNTs-OH) rapidly distributed to the entire body and their accumulation in bone took a longer time since their hydrophilic property was enhanced after hydroxylation [73]. The carboxylated CNTs were degraded and more easily eliminated by the body compared to non-carboxylated CNTs in the other study [75,76].

The shape of the CNTs is another reason for the toxicity. The larger, longer, and needle-like structures of CNTs can pierce the cell membrane and induce inflammation through the release of IL-1α and IL-1β [77]. It was reported that an 825 nm long-CNT in length induced more inflammation compared to the 220 nm long ones [78]. The diameter of the CNTs also has an impact on their toxicity. For instance, the MWCNTs 150 nm thick in diameter was found to be less toxic compared to the 50 nm thick ones [79].

The CNTs also show cell type dependent activity since the CNTs can be accumulated mostly in cells with more phagocytic activity. It was found that they can enhance or inhibit adhesion and proliferation of the some cell types. For example, the proliferation of human lung cancer cells (A549) was inhibited while both proliferation and viability of normal human lung cells (BEAS-2B) was inhibited after the exposure to SWCNTs [80].

As seen, the toxicity of CNTs is mostly attributed to metal impurities, aggregation of CNTs inside cells and their thin, longer needle-like structure [81–84]. In addition, the effect of CNTs changes based on cell type, the one that has more endocytosis ability carries more CNTs toxicity risk. After CNTs accumulation inside the cell, ROS production is induced promoting activation of an inflammatory response or DNA damage. The reason of toxicity of CNTs in scaffolds is attributed to the metal and amorphous carbon impurities on the CNTs surface [81]. The use of purified, hydrophilic, and shorter length with larger diameters of CNTs can be solutions to avoid their adverse effects in implants. However, in order to understand the possible adverse effect of CNTs in vivo, their distribution to the organs has been undertaken. The distribution of the CNTs through the organs depends on the combination of the parameters including their size and surface properties. The administration route and concentration also alter their distribution through the organs and tissues. The surface chemistry especially plays an important role in blood circulation. The unmodified SWCNTs were found to accumulate in rabbit liver, lungs, and spleen in a longer time of circulation. In another study, the SWCNTs labeled with an ^{125}I-isotope were hydroxylated. When the unmodified and hydroxylated forms were injected into mice, the hydroxylated ones were quickly distributed to the entire body and accumulated in bone kidney and stomach and eliminated from the body by renal clearance. In a long time period, they mostly accumulated in bone. However, the unmodified ones mostly were found in the liver, lung, and spleen. In another study, the modification of the SWCNTs surface with PEG increased their circulation time in the body and increased the distribution through the organs and tissues. The PEGylated SWCNTs were found in the brain within 1 h after injection into mice. Following a 7-day incubation, their accumulation was observed in the liver, spleen, and skin. The MWCNTs modified with Taurine were found to accumulate in the liver for a 3-month incubation with a low liver toxicity. Diethylenetriaminepentaacetic (DTPA) labeled SWCNTs were not found to accumulate in any of the reticuloendothelial system organs such as liver or spleen. They were eliminated from the body by renal excretion within a 3 h of injection [74]. As seen, various modifications on CNTs have been undertaken to understand their biodistribution, accumulation, and clearance. The surface characteristics mostly determine their interaction with the cells and their uptake.

10.2.3 Boron nitride nanotubes (BNNTs)

Boron nitride in its bulk form has become an indispensable material in many fields through the last century. It is being widely used in cosmetics, lubricants, and abrasives. Its nanosized form, BNNTs, is another class of nanotubes that are structurally similar to CNTs. Unlike the covalent C—C bonds in CNTs, BNNTs carry B—N covalent bonds which provide the structure an ionic character and a more stable ionic property. They were first discovered theoretically in 1994, followed by their actual synthesis a year after [85,86]. Since then, BNNTs have gained growing attention for their superior mechanical strength, chemical, and thermal stability all of which provide properties such as resistance to chemical attack and oxidation, electrical insulation, thermal conductivity, and temperature resistance [87–91]. These properties opened the way up for BNNTs to be applied in a wide range of applications, such as sensors, lubricant additives, hydrogen storage, nanovectors for drug delivery, transducers in neural, osteoblastic and muscle cells, and contrasting agents in boron–neutron capture therapy (BNCT) [92–100].

In implant technology, they have been mostly applied in bone implants since they were shown to improve mechanical strength, promote bone regeneration capability, cell adhesion, and cell spreading [101–105]. Similar to titanium implants, implants containing surface oxides of a boron layer have been shown to form a hydroxide layer which can act as a lubricant between two bones and joints [106]. Moreover, they are preferred as reinforcement additives against corrosion and wear. As mentioned earlier, their ability to act as nanotransducers may also contribute to neural and muscle cell regeneration.

10.2.3.1 Toxicity and biodistribution of BNNTs

BNNT use in implant materials is still in its early stage, thus there is not an extensive amount of research-based knowledge on their toxicity at the moment. However, the field is growing rapidly. There are already three in vivo animal studies on BNNTs: one on rabbits, one on mice, and one on planarians [107–109]. In none of them BNNTs were observed as toxic. Blood parameters, hepatic and renal activity, behavioral patterns were in line with the control group in a rabbit study where the animals were exposed to glycol-chitosan-coated BNNTs [107]. In the mice study, GC-coated BNNTs labeled with ^{99m}Tc were found to rapidly accumulate in the spleen, liver, gut, and bladder which indicated renal excretion even though a rate of free radioisotope could not be excluded from the observed results [108]. In the regenerative capability of a planarian study, the morphology and phototactive behavior were retained as well as mitotic cell number, transcriptional levels of stemness markers, DNA integrity, blastema size, and wound regeneration stayed within normal limits, indicating no toxic effects [109].

Among the many in vitro studies about BNNT toxicity, almost all of them demonstrate BNNTs as a safe material. Up to date, only one study by Horvath *et al.* reported them as toxic where they tested ~50 nm diameter-sized BNNTs of 10 μm length. In this study, they pointed out to a few important parameters that might alter the toxicity results of BNNT-treated cells [110]:

(i) BNNTs are highly hydrophobic, thus they tend to form aggregates in cell culture medium. Once they aggregate, they behave like their bulk counterparts and might not exert toxic effects.
(ii) MTT assays might not be suitable for BNNTs due to interference of MTT formazan with tubular NPs.
(iii) BNNTs are cytotoxic in a cell type-dependent manner; being highly toxic to cells with more endocytic capability.
(iv) The higher the aspect ratio of tubular NPs, the more likely the cells will be affected in a negative way and the shape is another important factor.

Considering the above-mentioned points, BNNTs are still promising candidates for future implants especially as a reinforcement material. The BNNT-containing implant surfaces were shown to adhere to surfaces more efficiently than other materials that are being used. Therefore, it is likely that the wear debris of a BNNT-coated implant will be much less than the other commonly used materials.

10.2.4 Silver nanoparticles (AgNPs)

AgNPs are formed by clustering of the silver atoms with physical, chemical, or biological methods in a 1–100 nm size range with desired shapes. The ability to prepare AgNPs with a variety of sizes and shapes provides them multifunctional properties such as thermal conductivity, catalytic activity, and optical behavior [111–116]. The anti-bacterial and anti-fungal properties also make them valuable in biomedical applications such as in implants and catheters [117–119]. Indeed, the infections and biofilm formation by gram-positive *Staphylococcus aureus* and gram-negative *Escherichia coli* in the implantation side are known as the main problem. They have resistance to antibiotics causing complications, which might result in the removal of the implant or revision of the surgery. The use of AgNPs in implants shows efficient results against the formation of a biofilm without exhibiting resistance [120,121].

The AgNPs has an efficient antibacterial property compared with the ionic silver nitrate since the AgNPs have large surface areas providing a more efficient interaction with bacteria. One of the underlying mechanisms of AgNPs–antibacterial activity is known as the interaction of NPs with the bacterial cell wall. The interaction causes the penetration of AgNPs (if their size is small enough) inside bacterial cells. The intracellular uptake of AgNPs causes intermolecular structural changes and production of free radicals that result in the membrane disruption [122]. Another mechanism is the intracellular release of Ag^+ from AgNPs, which interacts with thiol, sulfur, or phosphorous groups of the intracellular components such as enzymes or DNA, and inactivates the several vital functions such as respiration for energy or DNA replication for proliferation [122–124].

AgNPs containing composites can be fabricated by surface coating or adsorption and cross-linking of modified nanoparticles [125–128]. The AgNPs should have ideal properties to be used in implants. First of all, they should be biocompatible to avoid toxic or inflammatory responses to the host cells. The prolonged effectiveness of a broad spectrum of bacteria is also important for the effectiveness of AgNPs in implants. For this purpose, the important point is the determination of the effective size, shape, and exact concentration of AgNPs in scaffolds providing sustained release of Ag^+. For example, the particle sizes between 5 and 100 nm were tested and 5 nm showed the fastest antibacterial effect against *E. coli* whereas the degree of antibacterial effect of 5, 7, and 10 nm AgNPs was observed to be the same after a while [129]. In another study, the AgNP-embedded calcium phosphate biomaterials were prepared by incubation of sodium citrate and silver nitrate. The silver ion release kinetics and the amount were adjusted by controlling the concentration and the incubation time of the sodium citrate and silver nitrate [130].

10.2.4.1 Toxicity and biodistribution of AgNPs

The potential risk of AgNPs on mammalian cells and organs is a debated issue even though they are currently used in clinics for their antibacterial properties [131]. The exact source of toxicity is still in question since different sizes and shapes of NPs have been prepared with different reducing agents and tested for their toxicity with different cell lines. The toxicity of AgNPs is mostly attributed to the oxidation of

AgNPs, initiating the corrosion following the Ag^+ release [132,133]. The released Ag^+ interferes with the intracellular reactions such as the reactions taking place in the mitochondria or interacts with the cell membrane causing ROS production [134,135]. The amount of Ag^+ released from the AgNPs alters depending on their size, shape, surface chemistry, and colloidal stability [133,136–139].

The size of AgNPs has an inverse relationship with the surface area and Ag^+ release capacity. As the AgNP size gets smaller, the surface area of the particle increases, and dependently the release of Ag^+ increases. Besides the size, the colloidal stability based on the reduction methods and surface capping agent also alters the amount of Ag^+ release [140]. The salt concentration and the protein corona formation influence the stability of AgNPs, thus the released Ag^+ concentration [141]. For instance, the behaviors of citrate and PVP coated 10 nm AgNPs in cell culture media are different even though they have the same size. The weakly bound small citrate ions on AgNPs can easily exchange with the serum proteins influencing the stability of particles resulting in releasing of Ag^+ rapidly to the environment compared to non-charged larger PVP coated ones [142]. It was also reported that the agglomeration and the release of silver ions in the extracellular environment did not influence the cytotoxicity of the 10 nm AgNPs. The main reason was associated with the intracellular silver ion release [142]. The intracellular toxicity of AgNPs is demonstrated on osteoblasts (OBs) and osteoclasts (OCs). Since OBs have more phagocytosing activity, they are more subjected to AgNP-induced toxicity [143]. Besides cell types, the surface functional groups also have a role in their cellular uptake [139].

The adverse effect of implants prepared by AgNPs has not been observed since the amount of AgNPs loaded in implants was very low and the effective concentration of released Ag^+ from the scaffold (e.g. 10% for 7 days), which was sufficient to kill the infection, did not show an adverse effect on host cells. In vivo studies demonstrated the safety of the AgNPs in implants by the evaluation of their acute toxicity and biodistribution. They confirmed that the AgNPs toxicity was based on size and the concentration. The AgNPs with 40 nm particle size were injected to the mice at a 6–15 mg/kg concentration range. The 12 and 15 mg/kg concentration injections caused death of the mice in 30 min. The 9 mg/kg concentration injection caused the death of 1 male and 3 female mice out of 12 mice. The remaining mice continued to live for 14 days. The 7.5 and 6 mg/kg concentrations of AgNP-injected mice lived for 14 days. The tissue sections obtained from the liver, spleen, blood, heart, lung, and kidney showed that the AgNPs were accumulated in liver and spleen in a concentration dependent manner. The 9 mg/kg concentration of AgNP-injected mice induced toxicity to the spleen. The 7.5 and 6 mg/kg concentrations did not cause any adverse effect to the mice [144]. In another study, an evaluation of the biodistribution of AgNPs based on their size shows that 80 and 110 nm AgNPs accumulated in the spleen, liver, and lungs whereas 20 nm AgNPs was found in the liver, kidney, and spleen of rats [145]. The particle size did not play a role for their accumulation in the other organs including heart, brain, and testes. In addition, the concentration of the accumulated 20 nm AgNPs was observed very low since the smaller particles were rapidly eliminated from the body or easily dissolved. Since the released Ag^+ from the implants remained under this concentration, the use of the AgNPs in implants is indicated as safe.

10.2.5 Gold nanoparticles (AuNPs)

AuNPs are commonly synthesized with various size and shapes by the reduction of Au^{3+} to Au^0 in the presence of various small molecules, proteins, surfactants, and polymers. Depending on their size and shape, they exhibit unique optical, electrical, and thermal properties. AuNPs are less toxic to living organisms. They can be used for the various biomedical applications such as drug and gene delivery, photothermal and radiotherapy, contrast agents for the cancer diagnostic, cell imaging, and implant technology. Due to their conductive properties, the implants prepared by the AuNPs favor the differentiation, growth, and proliferation of the neurons and cardiomyocytes by enabling a quick and efficient electrical communication between neighboring cells. AuNPs also alter the topography of the fibers providing more anchorage sites for the cells. For example, both the topographical and conductive properties of the fibers prepared with AuNPs promote the neural growth, axonal elongation, and further formation of complex neuronal networks [146]. Another study shows that, better contraction forces were observed from the cardiac tissues cultured within the AuNP-implants [147,148]. The size and concentration of the AuNPs determine the conductivity and topology of the scaffolds. The higher concentration of AuNPs was found to cause very poor distribution of the electrical signal uniformity due to non-uniform distribution of AuNPs around the scaffold [149]. In addition, AuNP concentration did not influence the tissue growth due to toxicity. However, by the help of increased conductivity the cell differentiation was induced. For example, the higher concentration of AuNPs caused the muscle cells to differentiate into myotubes and decreased the contractile response [149].

10.2.5.1 Toxicity and biodistribution of AuNPs

The AuNPs are known as non-toxic to the living cells. However, they are not completely biocompatible. A study performed with 10 nm AuNPs results showed that the particle exposure did not influence the cell viability but it caused the release of the cytokines. The effects of the AuNPs on living cells may change depending on size, shape, surface chemistry, and concentration as typical for all nanomaterials. A toxicity study based on the different sizes of the AuNPs, ranging from 0.8 to 1.8 nm showed that 1.4 nm-size was 60- to 100-fold more toxic to the cells since their size was similar with the major groove of DNA [150]. Their interaction with DNA was indicated as the major source of the toxicity. The surface charge of AuNPs also plays an important role for their interaction with the cells and intracellular components. For example, the cationic 2 nm AuNPs were found more toxic compared to the anionic 2 nm AuNPs [151].

The shape of the AuNPs can be another parameter of their toxicity. The gold nanorods (AuNRs) are more toxic compared to the spheroid shaped ones [152]. Although the mechanism of the toxicity induced by the AuNRs compared to spheroid ones has not been explored yet, the toxicity of nanorods decreased when the reducing surfactant was replaced with a less toxic one [152]. For example, the use of phosphatidylcholine (PC), poly(acrylic acid), or poly(allylamine hydrochloride) polymers instead of hexadecyltrimethylammonium bromide (CTAB), a toxic surfactant, significantly decreases the toxicity of the AuNPs although the prepared NRs have similar size (30–40 nm in diameter) and surface potential (−20 mV) [152]. This result indicates that the inherent toxicity of the reductant significantly influences the toxicity rather than the particle size and surface potential.

The concentration of AuNPs is another factor altering their effect on the living organism. In an in vitro study, 1 nm sizes of AuNPs were exposed to fibroblast cells from 0.2 to 0.8 mg/mL concentration range. At higher concentration, the accumulation of the AuNPs in vacuoles and damage of the actin fibers were observed [153]. As seen, various physicochemical properties influence the cellular responses, but the cell line used for the evaluation can also affect the cytotoxicity results. The citrate reduced 13 nm AuNP was found to be more toxic to the human carcinoma lung cell line compared to a human liver carcinoma cell line at the same concentration exposure.

Although in vitro studies provide an insight about their possible effects, their potential adverse effect on humans cannot be predicted. Therefore, in most of the studies, their accumulation in organs is linked with their toxicity. The biodistribution of the AuNPs was investigated based on their size, shape and surface chemistry. The accumulation of 10 and 50 nm AuNPs in different organs of rats was investigated for 3 and 7 days after intraperitoneally administration. Although the uptake of 50 nm AuNPs by the cells were known as the highest in in vitro studies, much more accumulation of the 10 nm AuNPs in the heart, liver, and kidney was observed. The accumulation of the particles larger than 10 nm was found in the liver and spleen in mice up to 6 months without any adverse effect. However, there was no information about the complete clearance from the body. The size dependent accumulation of the AuNPs in tissues and organs was demonstrated by intravenously injection of different sizes of AuNPs ranging from 10 to 250 nm into albino mice. The small particles (10–15 nm) were found in blood, liver, lung, spleen, kidney, and heart whereas the larger particles (200–250 nm) resided in the blood, brain, and spleen [153]. The biodistribution of the gold nanostars with 56 nm size was investigated in mouse and their accumulation was observed in the liver inducing moderate toxicity. The behavior of AuNPs may alter depending on the model organism. For example, the influence of 15 nm citrate reduced AuNPs was investigated on one of the model organisms, *Drosophila melanogaster*. The decrease in their life span and the fertility and an increase in DNA fragmentation and stress protein expression were observed after ingesting AuNPs (12 μg/g per day). As shown in many studies, different physicochemical properties of AuNPs influence their distribution through the organ and tissues. The reports basically point out that AuNPs are non-toxic to living organisms.

10.3 Conclusions

Even though only a few types of NMs were mentioned in this chapter, the combination of these NPs is also a promising approach to enhance implant properties. It is possible to draw some general conclusions on the toxicity of the mentioned NMs used in implant technology. First of all, it should be born in mind that the toxicity data of NM-containing implants are still premature and care should be taken while implementing these materials as well as while evaluating their toxicity. This ambiguity mostly comes from the lack of uniformity in the in vitro toxicity studies of NMs. The variations in the preparation, administration, and evaluation of NMs often cause variations in the overall toxicity results. For instance, the metallic or other impurities in NM synthesis, dispersion methods, size and shape variations, exposure periods, dosages, tests to evaluate the toxicity are all parameters that highly impact the study outcomes.

On the other side, implants are designed to be resistant to corrosion and wear. Therefore, in an ideal world an implant is not expected to release its contents to the patient's body. Since NMs mostly exert toxicity if they are taken up by the cells, they can be considered safe as long as they are not released from the implant. In real world scenarios, however, corrosion and wear do exist and can only be optimized to the least. Therefore, the NM with the best mechanical resistance should be chosen for the implants – or at least for their outer surface – receiving high mechanical loads, such as hip, knee, or dental implants. The biocompatibility with the body fluids should also be optimized so that any agglomeration or protein adsorption complications will be minimized.

Apart from resistance and compatibility properties, the characteristics of the released NMs are also required to investigate their possible toxicity. The size and shape as well as the location of the released NMs affect their biodistribution, thus their fate in the body. In the cases where NMs are aggregated and form larger particles, they can safely be excreted from the body and toxicity will not be an issue. However, particles smaller than 10 nm might be positioned in a higher risk group. The coating of the NMs is also important for their biodistribution.

Finally, the advancements in nanomaterials science contribute to improved implants with better biocompatibility and resistance. The design and toxicity evaluation of NM-containing implants should still be cautious paying attention to various parameters including in vitro NM toxicity study outcomes.

References

[1] Mineral Commodity Summaries 2015 – mcs2015.pdf (Internet). (cited 2015 Oct 26). Available from: http://minerals.usgs.gov/minerals/pubs/mcs/2015/mcs2015.pdf

[2] Leventhal GS. 'Titanium, a metal for surgery'. *Journal of Bone & Joint Surgery, American Volume.* 1951;**33**(2):473–474.

[3] Bothe RT, Beaton LE, Davenport HA. 'Reaction of bone to multiple metallic implants'. *Journal of Surgery, Gynecology and Obstetrics.* 1940;**71**(6):598–602.

[4] Brånemark PI, Hansson BO, Adell R, *et al.* 'Osseointegrated implants in the treatment of the edentulous jaw. Experience from a 10-year period'. *Scandinavian Journal of Plastic and Reconstructive Surgery. Supplementum.* 1977;**16**:1–132.

[5] Adell R, Lekholm U, Rockler B, Brånemark PI. 'A 15-year study of osseointegrated implants in the treatment of the edentulous jaw'. *International Journal of Oral Surgery.* 1981;**10**(6):387–416.

[6] Adell R, Eriksson B, Lekholm U, Brånemark PI, Jemt T. 'Long-term follow-up study of osseointegrated implants in the treatment of totally edentulous jaws'. *International Journal of Oral & Maxillofacial Implants.* 1990;**5**(4):347–359.

[7] Yu K-N, Chang S-H, Park SJ, *et al.* 'Titanium dioxide nanoparticles induce endoplasmic reticulum stress-mediated autophagic cell death via mitochondria-associated endoplasmic reticulum membrane disruption in normal lung cells'. *PLoS ONE (Internet).* 2015;**10**(6).

[8] Rahman Q, Lohani M, Dopp E, *et al.* 'Evidence that ultrafine titanium dioxide induces micronuclei and apoptosis in Syrian hamster embryo fibroblasts'. *Environmental Health Perspectives.* 2002;**110**(8):797–800.

[9] Gurr J-R, Wang ASS, Chen C-H, Jan K-Y. 'Ultrafine titanium dioxide particles in the absence of photoactivation can induce oxidative damage to human bronchial epithelial cells'. *Toxicology.* 2005;**213**(1–2):66–73.

[10] Wang J, Liu Y, Jiao F, *et al.* 'Time-dependent translocation and potential impairment on central nervous system by intranasally instilled TiO(2) nanoparticles'. *Toxicology.* 2008;**254**(1–2):82–90.

[11] Liu H, Ma L, Zhao J, *et al.* 'Biochemical toxicity of nano-anatase TiO_2 particles in mice'. *Biological Trace Element Research.* 2009;**129**(1–3):170–180.

[12] Chen J, Dong X, Zhao J, Tang G. 'In vivo acute toxicity of titanium dioxide nanoparticles to mice after intraperitoneal injection'. *Journal of Applied Toxicology: JAT.* 2009;**29**(4):330–337.

[13] Kobayashi N, Naya M, Endoh S, Maru J, Yamamoto K, Nakanishi J. 'Comparative pulmonary toxicity study of nano-TiO(2) particles of different sizes and agglomerations in rats: different short- and long-term post-instillation results'. *Toxicology.* 2009;**264**(1-2):110–118.

[14] Onuma K, Sato Y, Ogawara S, *et al.* 'Nano-scaled particles of titanium dioxide convert benign mouse fibrosarcoma cells into aggressive tumor cells'. *American Journal of Pathology.* 2009;**175**(5):2171–2183.

[15] Ma L, Liu J, Li N, *et al.* 'Oxidative stress in the brain of mice caused by translocated nanoparticulate TiO_2 delivered to the abdominal cavity'. *Biomaterials.* 2010;**31**(1):99–105.

[16] Li N, Duan Y, Hong M, *et al.* 'Spleen injury and apoptotic pathway in mice caused by titanium dioxide nanoparticules'. *Toxicology Letters.* 2010;**195**(2–3):161–168.

[17] Sang X, Zheng L, Sun Q, *et al.* 'The chronic spleen injury of mice following long-term exposure to titanium dioxide nanoparticles'. *Journal of Biomedical Materials Research. Part A.* 2012;**100**(4):894–902.

[18] Alarifi S, Ali D, Al-Doaiss AA, Ali BA, Ahmed M, Al-Khedhairy AA. 'Histologic and apoptotic changes induced by titanium dioxide nanoparticles in the livers of rats'. *International Journal of Nanomedicine.* 2013;**8**:3937–3943.

[19] Sheng L, Wang L, Sang X, *et al.* 'Nano-sized titanium dioxide-induced splenic toxicity: a biological pathway explored using microarray technology'. *Journal of Hazardous Materials.* 2014;**278**:180–188.

[20] Yu X, Zhao X, Ze Y, *et al.* 'Changes of serum parameters of TiO_2 nanoparticle-induced atherosclerosis in mice'. *Journal of Hazardous Materials.* 2014;**280**:364–371.

[21] Sheng L, Ze Y, Wang L, *et al.* 'Mechanisms of TiO_2 nanoparticle-induced neuronal apoptosis in rat primary cultured hippocampal neurons'. *Journal of Biomedical Materials Research. Part A.* 2015;**103**(3):1141–1149.

[22] 2011-160.pdf (Internet). (cited 2015 Oct 26). Available from: http://www.cdc.gov/niosh/docs/2011-160/pdfs/2011-160.pdf

[23] IARC Working Group on the Evaluation of Carcinogenic Risks to Humans. 'Carbon black, titanium dioxide, and talc'. IARC Monogr Eval Carcinog Risks Hum World Health Organ. *International Agency for Research on Cancer*. 2010;93:1–413.

[24] Urban RM, Jacobs JJ, Tomlinson MJ, Gavrilovic J, Black J, Peoc'h M. 'Dissemination of wear particles to the liver, spleen, and abdominal lymph nodes of patients with hip or knee replacement'. *Journal of Bone and Joint Surgery.* 2000;**82**(4):457–476.

[25] Hallam P, Haddad F, Cobb J. 'Pain in the well-fixed, aseptic titanium hip replacement. The role of corrosion'. *Journal of Bone and Joint Surgery. British Volume.* 2004;**86**(1):27–30.

[26] Simka W. 'Preliminary investigations on the anodic oxidation of Ti–13Nb–13Zr alloy in a solution containing calcium and phosphorus'. *Electrochimica Acta.* 2011;**56**(27):9831–9837.

[27] Lorenzetti M, Pellicer E, Sort J, *et al.* 'Improvement to the corrosion resistance of Ti-based implants using hydrothermally synthesized nanostructured anatase coatings'. *Materials.* 2014;**7**(1):180–194.

[28] Wang JY, Wicklund BH, Gustilo RB, Tsukayama DT. 'Titanium, chromium and cobalt ions modulate the release of bone-associated cytokines by human monocytes/macrophages in vitro'. *Biomaterials.* 1996;**17**(23):2233–2240.

[29] Signorello LB, Ye W, Fryzek JP, *et al.* 'Nationwide study of cancer risk among hip replacement patients in Sweden'. *Journal of the National Cancer Institute.* 2001;**93**(18):1405–1410.

[30] Goldacre MJ, Wotton CJ, Seagroatt V, Yeates D. 'Cancer following hip and knee arthroplasty: record linkage study'. *British Journal of Cancer.* 2005;**92**(7):1298–1301.

[31] Wang J, Zhou G, Chen C, *et al.* 'Acute toxicity and biodistribution of different sized titanium dioxide particles in mice after oral administration'. *Toxicology Letters.* 2007;**168**(2):176–185.

[32] Visuri T, Borg H, Pulkkinen P, Paavolainen P, Pukkala E. 'A retrospective comparative study of mortality and causes of death among patients with metal-on-metal and metal-on-polyethylene total hip prostheses in primary osteoarthritis after a long-term follow-up'. *BMC Musculoskeletal Disorders.* 2010;**11**:78.

[33] Smith AJ, Dieppe P, Porter M, Blom AW, National Joint Registry of England and Wales. 'Risk of cancer in first seven years after metal-on-metal hip replacement compared with other bearings and general population: linkage study between the National Joint Registry of England and Wales and hospital episode statistics'. *BMJ (Clinical Research Ed.).* 2012;**344**:e2383.

[34] Migaud H, Putman S, Combes A, *et al.* 'Metal-on-metal bearing: is this the end of the line? We do not think so'. *HSS Journal.* 2012;**8**(3):262–269.

[35] Mäkelä KT, Visuri T, Pulkkinen P, *et al.* 'Cancer incidence and cause-specific mortality in patients with metal-on-metal hip replacements in Finland'. *Acta Orthopaedica.* 2014;**85**(1):32–38.

[36] Reddy A, Caicedo M, Samelko L, Jacobs JJ, Hallab NJ. 'Implant debris particle size affects serum protein adsorption which may contribute to particle size-based bioreactivity differences'. *Journal of Long-Term Effects of Medical Implants.* 2014;**24**(1):77–88.

[37] Iijima S. 'Helical microtubules of graphitic carbon'. *Nature.* 1991;**354** (6348):56–58.

[38] Endo M, Strano MS, Ajayan PM. 'Potential applications of carbon nanotubes'. *Carbon Nanotubes*. Springer; 2008. pp. 13–62.

[39] Iijima S, Ichihashi T. 'Single-shell carbon nanotubes of 1-nm diameter'. *Nature.* 1993;**362**:603–605.

[40] Shah B, Lafleur T, Chen A. 'Carbon nanotube based electrochemical sensor for the sensitive detection of valacyclovir'. *Faraday Discussions.* 2013; **164**:135–146.

[41] Shi X, Hudson JL, Spicer PP, Tour JM, Krishnamoorti R, Mikos AG. 'Injectable nanocomposites of single-walled carbon nanotubes and biodegradable polymers for bone tissue engineering'. *Biomacromolecules.* 2006;**7** (7):2237–2242.

[42] Balani K, Anderson R, Laha T, *et al.* 'Plasma-sprayed carbon nanotube reinforced hydroxyapatite coatings and their interaction with human osteoblasts in vitro'. *Biomaterials.* 2007;**28**(4):618–624.

[43] Han ZJ, Rider AE, Ishaq M, *et al.* 'Carbon nanostructures for hard tissue engineering'. *RSC Advances.* 2013;**3**(28):11058–11072.

[44] Kabiri M, Soleimani M, Shabani I, *et al.* 'Neural differentiation of mouse embryonic stem cells on conductive nanofiber scaffolds'. *Biotechnology Letters.* 2012;**34**(7):1357–1365.

[45] Jin G-Z, Kim M, Shin US, Kim H-W. 'Neurite outgrowth of dorsal root ganglia neurons is enhanced on aligned nanofibrous biopolymer scaffold with carbon nanotube coating'. *Neuroscience Letters.* 2011;**501**(1):10–14.

[46] Hwang J-Y, Shin US, Jang W-C, Hyun JK, Wall IB, Kim H-W. 'Biofunctionalized carbon nanotubes in neural regeneration: a mini-review'. *Nanoscale.* 2013;**5**(2):487–497.

[47] Li X, Gao H, Uo M, *et al.* 'Maturation of osteoblast-like SaoS2 induced by carbon nanotubes'. *Biomedical Materials.* 2009;**4**(1):015005.

[48] Akasaka T, Yokoyama A, Matsuoka M, Hashimoto T, Watari F. 'Thin films of single-walled carbon nanotubes promote human osteoblastic cells (Saos-2)

proliferation in low serum concentrations'. *Materials Science and Engineering: C.* 2010;**30**(3):391–399.

[49] Khang D, Sato M, Price RL, Ribbe AE, Webster TJ. 'Selective adhesion and mineral deposition by osteoblasts on carbon nanofiber patterns'. *International Journal of Nanomedicine.* 2006;**1**(1):65.

[50] Venkatesan J, Ryu B, Sudha P, Kim S-K. 'Preparation and characterization of chitosan–carbon nanotube scaffolds for bone tissue engineering'. *International Journal of Biological Macromolecules.* 2012;**50**(2):393–402.

[51] Supronowicz P, Ajayan P, Ullmann K, Arulanandam B, Metzger D, Bizios R. 'Novel current-conducting composite substrates for exposing osteoblasts to alternating current stimulation'. *Journal of Biomedical Materials Research.* 2002;**59**(3):499–506.

[52] Stout DA, Webster TJ. 'Carbon nanotubes for stem cell control'. *Materials Today.* 2012;**15**(7):312–318.

[53] Zanello LP, Zhao B, Hu H, Haddon RC. 'Bone cell proliferation on carbon nanotubes'. *Nano Letters.* 2006;**6**(3):562–567.

[54] Nayak TR, Jian L, Phua LC, Ho HK, Ren Y, Pastorin G. 'Thin films of functionalized multiwalled carbon nanotubes as suitable scaffold materials for stem cells proliferation and bone formation'. *ACS Nano.* 2010; **4**(12):7717–7725.

[55] Liu D, Yi C, Zhang D, Zhang J, Yang M. 'Inhibition of proliferation and differentiation of mesenchymal stem cells by carboxylated carbon nanotubes'. *ACS Nano.* 2010;**4**(4):2185–2195.

[56] Price RL, Waid MC, Haberstroh KM, Webster TJ. 'Selective bone cell adhesion on formulations containing carbon nanofibers'. *Biomaterials.* 2003;**24**(11):1877–1887.

[57] Webster TJ, Waid MC, McKenzie JL, Price RL, Ejiofor JU. 'Nanobiotechnology: carbon nanofibres as improved neural and orthopaedic implants'. *Nanotechnology.* 2004;**15**(1):48.

[58] Pan L, Pei X, He R, Wan Q, Wang J. 'Multiwall carbon nanotubes/polycaprolactone composites for bone tissue engineering application'. *Colloids and Surfaces B: Biointerfaces.* 2012;**93**:226–234.

[59] Mattioli-Belmonte M, Vozzi G, Whulanza Y, *et al.* 'Tuning polycaprolactone–carbon nanotube composites for bone tissue engineering scaffolds'. *Materials Science and Engineering: C.* 2012;**32**(2):152–159.

[60] Takagi A, Hirose A, Nishimura T, *et al.* 'Induction of mesothelioma in p53+/− mouse by intraperitoneal application of multi-wall carbon nanotube'. *Journal of Toxicological Sciences.* 2008;**33**(1):105–116.

[61] Ellinger-Ziegelbauer H, Pauluhn J. 'Pulmonary toxicity of multi-walled carbon nanotubes (Baytubes®) relative to α-quartz following a single 6 h inhalation exposure of rats and a 3 months post-exposure period'. *Toxicology.* 2009;**266**(1):16–29.

[62] Ma-Hock L, Treumann S, Strauss V, *et al.* 'Inhalation toxicity of multi-wall carbon nanotubes in rats exposed for 3 months'. *Toxicological Sciences.* 2009:kfp146.

[63] Fisher C, Rider AE, Han ZJ, Kumar S, Levchenko I, Ostrikov K. 'Applications and nanotoxicity of carbon nanotubes and graphene in biomedicine'. *Journal of Nanomaterials.* 2012;**2012**:3.

[64] Pulskamp K, Diabaté S, Krug HF. 'Carbon nanotubes show no sign of acute toxicity but induce intracellular reactive oxygen species in dependence on contaminants'. *Toxicology Letters.* 2007;**168**(1):58–74.

[65] Kagan VE, Konduru NV, Feng W, *et al.* 'Carbon nanotubes degraded by neutrophil myeloperoxidase induce less pulmonary inflammation'. *Nature Nanotechnology.* 2010;**5**(5):354–359.

[66] Jia G, Wang H, Yan L, *et al.* 'Cytotoxicity of carbon nanomaterials: single-wall nanotube, multi-wall nanotube, and fullerene'. *Environmental Science & Technology.* 2005;**39**(5):1378–1383.

[67] Dumortier H, Lacotte S, Pastorin G, *et al.* 'Functionalized carbon nanotubes are non-cytotoxic and preserve the functionality of primary immune cells'. *Nano Letters.* 2006;**6**(7):1522–1528.

[68] Fubini B, Ghiazza M, Fenoglio I. 'Physico-chemical features of engineered nanoparticles relevant to their toxicity'. *Nanotoxicology.* 2010;**4**(4): 347–363.

[69] Haniu H, Saito N, Matsuda Y, *et al.* 'Effect of dispersants of multi-walled carbon nanotubes on cellular uptake and biological responses'. *International Journal of Nanomedicine.* 2011;**6**:3295.

[70] Haniu H, Saito N, Matsuda Y, *et al.* 'Elucidation mechanism of different biological responses to multi-walled carbon nanotubes using four cell lines'. *International Journal of Nanomedicine.* 2011;**6**:3487.

[71] Hirano S, Fujitani Y, Furuyama A, Kanno S. 'Uptake and cytotoxic effects of multi-walled carbon nanotubes in human bronchial epithelial cells'. *Toxicology and Applied Pharmacology.* 2010;**249**(1):8–15.

[72] Zhu L, Chang DW, Dai L, Hong Y. 'DNA damage induced by multiwalled carbon nanotubes in mouse embryonic stem cells'. *Nano Letters.* 2007; **7**(12):3592–3597.

[73] Wang H, Wang J, Deng X, *et al.* 'Biodistribution of carbon single-wall carbon nanotubes in mice'. *Journal of Nanoscience and Nanotechnology.* 2004;**4**(8):1019–1024.

[74] Singh R, Pantarotto D, Lacerda L, *et al.* 'Tissue biodistribution and blood clearance rates of intravenously administered carbon nanotube radiotracers'. *Proceedings of the National Academy of Sciences of the United States of America.* 2006;**103**(9):3357–3362.

[75] Allen BL, Kichambare PD, Gou P, *et al.* 'Biodegradation of single-walled carbon nanotubes through enzymatic catalysis'. *Nano Letters.* 2008;**8**(11): 3899–3903.

[76] Liu X, Hurt RH, Kane AB. 'Biodurability of single-walled carbon nanotubes depends on surface functionalization'. *Carbon.* 2010;**48**(7): 1961–1969.

[77] Long N-lCN. 'Asbestos activate the NLRP3 inflammasome through a similar mechanism'. Palomaki. Jaana.

[78] Sato Y, Yokoyama A, Shibata K-i, *et al.* 'Influence of length on cytotoxicity of multi-walled carbon nanotubes against human acute monocytic leukemia cell line THP-1 in vitro and subcutaneous tissue of rats in vivo'. *Molecular BioSystems.* 2005;**1**(2):176–182.

[79] Nagai H, Okazaki Y, Chew SH, *et al.* 'Diameter and rigidity of multiwalled carbon nanotubes are critical factors in mesothelial injury and carcinogenesis'. *Proceedings of the National Academy of Sciences of the United States of America.* 2011;**108**(49):E1330–E1338.

[80] Herzog E, Casey A, Lyng FM, Chambers G, Byrne HJ, Davoren M. 'A new approach to the toxicity testing of carbon-based nanomaterials—the clonogenic assay'. *Toxicology Letters.* 2007;**174**(1):49–60.

[81] Lobo AdO, Antunes E, Machado A, Pacheco-Soares C, Trava-Airoldi V, Corat E. 'Cell viability and adhesion on as grown multi-wall carbon nanotube films'. *Materials Science and Engineering: C.* 2008;**28**(2):264–269.

[82] Lobo A, Corat M, Antunes E, *et al.* 'An evaluation of cell proliferation and adhesion on vertically-aligned multi-walled carbon nanotube films'. *Carbon.* 2010;**48**(1):245–254.

[83] Lobo A, Marciano F, Ramos S, Machado M, Corat E, Corat M. 'Increasing mouse embryonic fibroblast cells adhesion on superhydrophilic vertically aligned carbon nanotube films'. *Materials Science and Engineering: C.* 2011;**31**(7):1505–1511.

[84] Giannona S, Firkowska I, Rojas-Chapana J, Giersig M. 'Vertically aligned carbon nanotubes as cytocompatible material for enhanced adhesion and proliferation of osteoblast-like cells'. *Journal of Nanoscience and Nanotechnology.* 2007;**7**(4–5):1679–1683.

[85] Rubio A, Corkill JL, Cohen ML. 'Theory of graphitic boron nitride nanotubes'. *Physical Review B.* 1994;**49**(7):5081–5084.

[86] Chopra NG, Luyken RJ, Cherrey K, *et al.* 'Boron nitride nanotubes'. *Science.* 1995;**269**, (5226):966–967.

[87] Blase X, Rubio A, Louie SG, Cohen ML. 'Stability and band gap constancy of boron nitride nanotubes'. *EPL Europhysics Letters.* 1994;**28**(5):335.

[88] Chen Y, Zou J, Campbell SJ, Caer GL. 'Boron nitride nanotubes: pronounced resistance to oxidation'. *Applied Physics Letters.* 2004;**84**(13): 2430–2432.

[89] Zettl A, Chang CW, Begtrup G. 'A new look at thermal properties of nanotubes'. *Physica Status Solidi B.* 2007;**244**(11):4181–4183.

[90] Chopra NG, Zettl A. 'Measurement of the elastic modulus of a multi-wall boron nitride nanotube'. *Solid State Communications.* 1998;**105**(5): 297–300.

[91] Chang CW, Han W-Q, Zettl A. 'Thermal conductivity of B–C–N and BN nanotubes'. *Applied Physics Letters.* 2005;**86**(17):173102.

[92] Reddy ALM, Gupta BK, Narayanan TN, Martí AA, Ajayan PM, Walker GC. 'Probing of Ni-encapsulated ferromagnetic boron nitride nanotubes by time-resolved and steady-state photoluminescence spectroscopy'. *Journal of Physical Chemistry C.* 2012;**116**(23):12803–12809.

[93] Raffa V, Ciofani G, Cuschieri A. 'Enhanced low voltage cell electropermeabilization by boron nitride nanotubes'. *Nanotechnology.* 2009; **20**(7):075104.

[94] Pawlak Z, Pai R, Bayraktar E, Kaldonski T, Oloyede A. 'Lamellar lubrication in vivo and vitro: Friction testing of hexagonal boron nitride'. *Biosystems.* 2008;**94**(3):202–208.

[95] Ma R, Bando Y, Zhu H, Sato T, Xu C, Wu D. 'Hydrogen uptake in boron nitride nanotubes at room temperature'. *Journal of the American Chemical Society.* 2002;**124**(26):7672–7673.

[96] Ciofani G, Raffa V. Yu J, *et al.* 'Boron nitride nanotubes: a novel vector for targeted magnetic drug delivery'. *Current Nanoscience.* 2009;**5**(1):33–38.

[97] Danti S, Ciofani G, Moscato S, *et al.* 'Boron nitride nanotubes and primary human osteoblasts: in vitro compatibility and biological interactions under low frequency ultrasound stimulation'. *Nanotechnology.* 2013;**24**(46): 465102.

[98] Ciofani G, Raffa V, Menciassi A, Cuschieri A. 'Folate functionalized boron nitride nanotubes and their selective uptake by glioblastoma multiforme cells: implications for their use as boron carriers in clinical boron neutron capture therapy'. *Nanoscale Research Letters.* 2009;**4**(2):113–121.

[99] Ciofani G, Danti S, D'Alessandro D, *et al.* 'Enhancement of neurite outgrowth in neuronal-like cells following boron nitride nanotube-mediated stimulation.' *ACS Nano.* 2010;**4**(10):6267–6277.

[100] Danti S, Ciofani G, Pertici G, *et al.* 'Boron nitride nanotube-functionalised myoblast/microfibre constructs: a nanotech-assisted tissue-engineered platform for muscle stimulation'. *Journal of Tissue Engineering and Regenerative Medicine.* 2015;**9**(7):847–851.

[101] Lahiri D, Singh V, Benaduce AP, Seal S, Kos L, Agarwal A. 'Boron nitride nanotube reinforced hydroxyapatite composite: Mechanical and tribological performance and in-vitro biocompatibility to osteoblasts'. *Journal of the Mechanical Behavior of Biomedical Materials.* 2011;**4**(1):44–56.

[102] Lahiri D, Rouzaud F, Richard T, *et al.* 'Boron nitride nanotube reinforced polylactide–polycaprolactone copolymer composite: Mechanical properties and cytocompatibility with osteoblasts and macrophages in vitro'. *Acta Biomaterialia.* 2010;**6**(9):3524–3533.

[103] Lahiri D, Singh V, Keshri AK, Seal S, Agarwal A. 'Apatite formability of boron nitride nanotubes'. *Nanotechnology.* 2011;**22**(20):205601.

[104] Li L, Li LH, Ramakrishnan S, *et al.* 'Controlling wettability of boron nitride nanotube films and improved cell proliferation'. *Journal of Physical Chemistry C.* 2012;**116**(34):18334–18339.

[105] Li X, Wang X, Jiang X, *et al.* 'Boron nitride nanotube-enhanced osteogenic differentiation of mesenchymal stem cells'. *Journal of Biomedical Materials Research Part B: Applied Biomaterials.* 2016;**104**(2):323–329.

[106] Klepper CC, Williams JM, Truhan JJ, *et al.* 'Tribo-mechanical properties of thin boron coatings deposited on polished cobalt alloy surfaces for orthopedic applications'. *Thin Solid Films.* 2008;**516**(10):3070–3080.

[107] Ciofani G, Danti S, Genchi GG, *et al.* 'Pilot in vivo toxicological investigation of boron nitride nanotubes'. *International Journal of Nanomedicine.* 2012;**7**:19–24.

[108] Soares DCF, Ferreira TH, Ferreira C de A, Cardoso VN, de Sousa EMB. 'Boron nitride nanotubes radiolabeled with ^{99m}Tc: Preparation, physicochemical characterization, biodistribution study, and scintigraphic imaging in Swiss mice'. *International Journal of Pharmaceutics.* 2012;**423**(2): 489–495.

[109] Salvetti A, Rossi L, Iacopetti P, *et al.* 'In vivo biocompatibility of boron nitride nanotubes: effects on stem cell biology and tissue regeneration in planarians'. *Nanomedicine.* 2015;**10**(12):1911–1922.

[110] Horváth L, Magrez A, Golberg D, *et al.* 'In vitro investigation of the cellular toxicity of boron nitride nanotubes'. *ACS Nano.* 2011;**5**(5): 3800–3810.

[111] Warrier P, Teja A. 'Effect of particle size on the thermal conductivity of nanofluids containing metallic nanoparticles'. *Nanoscale Research Letters.* 2011;**6**(1):1–6.

[112] Chou C-W, Hsu S-h, Chang H, Tseng S-M, Lin H-R. 'Enhanced thermal and mechanical properties and biostability of polyurethane containing silver nanoparticles'. *Polymer Degradation and Stability.* 2006;**91**(5):1017–1024.

[113] Kvitek L, Panáček A, Soukupova J, *et al.* 'Effect of surfactants and polymers on stability and antibacterial activity of silver nanoparticles (NPs)'. *Journal of Physical Chemistry C.* 2008;**112**(15):5825–5834.

[114] Xu R, Wang D, Zhang J, Li Y. 'Shape-dependent catalytic activity of silver nanoparticles for the oxidation of styrene'. *Chemistry – An Asian Journal.* 2006;**1**(6):888–893.

[115] Evanoff DD, Chumanov G. 'Synthesis and optical properties of silver nanoparticles and arrays'. *ChemPhysChem.* 2005;**6**(7):1221–1231.

[116] Vodnik VV, Božanić DK, Bibić N, Šaponjić ZV, Nedeljković JM. 'Optical properties of shaped silver nanoparticles'. *Journal of Nanoscience and Nanotechnology.* 2008;**8**(7):3511–3515.

[117] Ge L, Li Q, Wang M, Ouyang J, Li X, Xing MM. 'Nanosilver particles in medical applications: synthesis, performance, and toxicity'. *International Journal of Nanomedicine.* 2014;**9**:2399.

[118] Jafari SM, Coyle C, Mortazavi SJ, Sharkey PF, Parvizi J. 'Revision hip arthroplasty: infection is the most common cause of failure'. *Clinical Orthopaedics and Related Research.* 2010;**468**(8):2046–2051.

[119] Connaughton A, Childs A, Dylewski S, Sabesan VJ. 'Biofilm disrupting technology for orthopedic implants: what's on the horizon?'. *Frontiers in Medicine.* 2014;**1**.

[120] Cardoso VS, Quelemes PV, Amorin A, *et al.* 'Collagen-based silver nanoparticles for biological applications: synthesis and characterization'. *Journal of Nanobiotechnology.* 2014;**12**(1):36.

[121] Shahverdi AR, Fakhimi A, Shahverdi HR, Minaian S. 'Synthesis and effect of silver nanoparticles on the antibacterial activity of different antibiotics

against *Staphylococcus aureus* and *Escherichia coli*'. *Nanomedicine: Nanotechnology, Biology and Medicine*. 2007;**3**(2):168–171.

[122] García-Barrasa J, López-de-Luzuriaga JM, Monge M. 'Silver nanoparticles: synthesis through chemical methods in solution and biomedical applications'. *Central European Journal of Chemistry*. 2011;**9**(1):7–19.

[123] Lansdown AB. 'A review of the use of silver in wound care: facts and fallacies'. *British Journal of Nursing*. 2004;**13**(Suppl 1):S6–S19.

[124] Ovington LG. 'The truth about silver'. *Ostomy/Wound Management*. 2004;**50**(Suppl 9A):1S–10S.

[125] Mei L, Lu Z, Zhang W, *et al.* 'Bioconjugated nanoparticles for attachment and penetration into pathogenic bacteria'. *Biomaterials*. 2013; **34**(38):10328–10337.

[126] Xu W-P, Zhang L-C, Li J-P, *et al.* 'Facile synthesis of silver@graphene oxide nanocomposites and their enhanced antibacterial properties'. *Journal of Materials Chemistry*. 2011;**21**(12):4593–4597.

[127] Liu Q, Zhou Z, Qiu G, Li J, Xie J, Lee JY. 'A Surface reaction route to increase the loading of antimicrobial Ag nanoparticles in forward osmosis membranes'. *ACS Sustainable Chemistry & Engineering*. 2015;**3**(11): 2959–2966.

[128] Keleştemur S, Kilic E, Uslu Ü, *et al.* 'Wound healing properties of modified silver nanoparticles and their distribution in mouse organs after topical application'. *Nano Biomedicine and Engineering*. 2012;**4**(4):170–176.

[129] Agnihotri S, Mukherji S, Mukherji S. 'Size-controlled silver nanoparticles synthesized over the range 5–100 nm using the same protocol and their antibacterial efficacy'. *RSC Advances*. 2014;**4**(8):3974–3979.

[130] Lee JS, Murphy WL. 'Functionalizing calcium phosphate biomaterials with antibacterial silver particles'. *Advanced Materials*. 2013;**25**(8):1173–1179.

[131] Gosheger G, Gebert C, Ahrens H, Streitbuerger A, Winkelmann W, Hardes J. 'Endoprosthetic reconstruction in 250 patients with sarcoma'. *Clinical Orthopaedics and Related Research*. 2006;**450**:164–167.

[132] Liu J, Sonshine DA, Shervani S, Hurt RH. 'Controlled release of biologically active silver from nanosilver surfaces'. *ACS Nano*. 2010;**4**(11): 6903–6913.

[133] Kittler S, Greulich C, Diendorf J, Koller M, Epple M. 'Toxicity of silver nanoparticles increases during storage because of slow dissolution under release of silver ions'. *Chemistry of Materials*. 2010;**22**(16):4548–4554.

[134] AshaRani P, Low Kah Mun G, Hande MP, Valiyaveettil S. 'Cytotoxicity and genotoxicity of silver nanoparticles in human cells'. *ACS Nano*. 2008; **3**(2):279–290.

[135] Cronholm P, Karlsson HL, Hedberg J, *et al.* 'Intracellular uptake and toxicity of Ag and CuO nanoparticles: a comparison between nanoparticles and their corresponding metal ions'. *Small*. 2013;**9**(7):970–982.

[136] Stoehr LC, Gonzalez E, Stampfl A, *et al.* 'Shape matters: effects of silver nanospheres and wires on human alveolar epithelial cells'. *Particle and Fibre Toxicology*. 2011;**8**(36):3–15.

[137] El Badawy AM, Silva RG, Morris B, Scheckel KG, Suidan MT, Tolaymat TM. 'Surface charge-dependent toxicity of silver nanoparticles'. *Environmental Science & Technology.* 2010;**45**(1):283-7

[138] Sur I, Altunbek M, Kahraman M, Culha M. 'The influence of the surface chemistry of silver nanoparticles on cell death'. *Nanotechnology.* 2012;**23** (37):375102.

[139] Hatipoglu MK, Keleştemur S, Altunbek M, Culha M. 'Source of cytotoxicity in a colloidal silver nanoparticle suspension'. *Nanotechnology.* 2015;**26**(19):195103.

[140] Rivera-Gil P, Jimenez De Aberasturi D, Wulf V, *et al.* 'The challenge to relate the physicochemical properties of colloidal nanoparticles to their cytotoxicity'. *Accounts of Chemical Research.* 2012;**46**(3):743–749.

[141] Monopoli MP, Åberg C, Salvati A, Dawson KA. 'Biomolecular coronas provide the biological identity of nanosized materials'. *Nature Nanotechnology.* 2012;**7**(12):779–786.

[142] Gliga AR, Skoglund S, Wallinder IO, Fadeel B, Karlsson HL. 'Size-dependent cytotoxicity of silver nanoparticles in human lung cells: the role of cellular uptake, agglomeration and Ag release'. *Particle and Fibre Toxicology.* 2014;**11**(11):1–17.

[143] Albers CE, Hofstetter W, Siebenrock KA, Landmann R, Klenke FM. 'In vitro cytotoxicity of silver nanoparticles on osteoblasts and osteoclasts at antibacterial concentrations'. *Nanotoxicology.* 2013;**7**(1):30–36.

[144] Wang H, Cheng M, Hu J, Wang C, Xu S, Han CC. 'Preparation and optimization of silver nanoparticles embedded electrospun membrane for implant associated infections prevention'. *ACS Applied Materials and Interfaces*. 2013;**5**(21):11014–11021.

[145] Lankveld DPK, Oomen AG, Krystek P, *et al.* 'The kinetics of the tissue distribution of silver nanoparticles of different sizes'. *Biomaterials.* 2010;**31**:8350–8361.

[146] Baranes K, Shevach M, Shefi O, Dvir T. 'Gold nanoparticle-decorated scaffolds promote neuronal differentiation and maturation'. *Nano Letters.* 2015;**16** (5):2916–2920.

[147] Fleischer S, Shevach M, Feiner R, Dvir T. 'Coiled fiber scaffolds embedded with gold nanoparticles improve the performance of engineered cardiac tissues'. *Nanoscale.* 2014;**6**(16):9410–9414.

[148] Shevach M, Fleischer S, Shapira A, Dvir T. 'Gold nanoparticle-decellularized matrix hybrids for cardiac tissue engineering'. *Nano Letters.* 2014;**14**(10):5792–5796.

[149] McKeon-Fischer KD, Freeman JW. 'Characterization of electrospun poly (L-lactide) and gold nanoparticle composite scaffolds for skeletal muscle tissue engineering'. *Journal of Tissue Engineering and Regenerative Medicine.* 2011;**5**(7):560–568.

[150] Pan Y, Leifert A, Ruau D, *et al.* 'Gold nanoparticles of diameter 1.4 nm trigger necrosis by oxidative stress and mitochondrial damage'. *Small.* 2009;**5**(18):2067–2076.

[151] Goodman CM, McCusker CD, Yilmaz T, Rotello VM. 'Toxicity of gold nanoparticles functionalized with cationic and anionic side chains'. *Bioconjugate Chemistry*. 2004:**15**:897–900.

[152] Fratoddi I, Venditti I, Cametti C, Russo MV. 'How toxic are gold nanoparticles? The state-of-the-art'. *Nano Research*. 2015;**8**(6):1771–1799.

[153] Sonavane G, Tomoda K, Makino K. 'Biodistribution of colloidal gold nanoparticles after intravenous administration: effect of particle size'. *Colloids and Surfaces B Biointerfaces*. 2008;**66**(2):274–280.

Part IV

Current and future clinical applications of nanomaterials

Chapter 11

Nanotechnology and consumer products: challenges of risk assessment of nanomaterials and current regulatory status

Haiou Qu[1]

Abstract

This chapter discusses nanomaterial's safety in consumer products from a regulatory perspective. The challenges of risk assessment will be illustrated from four basic aspects, including hazard identification, hazard characterization, exposure assessment, and risk characterization. The technical difficulties in performing a comprehensive physical and chemical characterization of nanomaterials in complex product matrices will be presented. The issues in the toxicity test system, particularly in the assessment of long-term toxic effect and the limitations of the applied methodology in exposure assessment will be investigated. This chapter will also give an overview of the regulatory status in the United States, Europe, and China, and the current approaches that each government agency has adopted in identifying and evaluating the risk associated with the use of nanomaterials.

11.1 Introduction

The last two decades have seen the huge boom in the research and development of nanotechnology. According to the National Nanotechnology Initiative, nanotechnology is helping considerably improve, even revolutionize, many technology and industry sectors such as information technology, energy, environmental science, medicine, homeland security, food safety, and transportation, among many others [1]. Currently, there are more than one thousand national and international corporations that are conducting nanotechnology related research, development, designing, and manufacturing business [2]. It is strongly believed that nanotechnology will bring significant impact across almost every field of science and technology and tremendously improve the quality of human life [1]. Among all industrial sectors, medical and health related application is one of the fastest

[1]Department of Chemistry, University of New Orleans, New Orleans, LA, USA

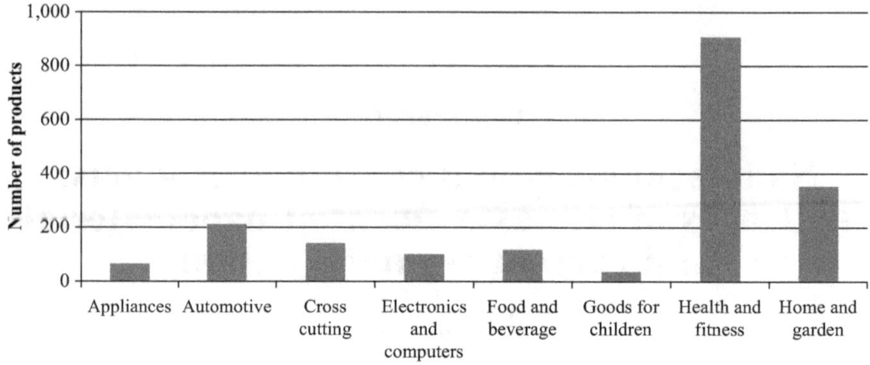

Figure 11.1 Number of nanotechnology based products in each category (Source: PEN [5]; analysis: February 2015)

growing areas in nanotechnology. For example, according to a study in 2012, the global nanomedicine market reached $50.1 billion in 2011 and is expected to grow to $96.9 billion by 2016, which accounts for 19.5% of the global pharmaceuticals market [3]. Nanotechnology has shown strong potentials in a variety of biomedical applications, such as early stage cancer detection and disease diagnosis, controlled target drug delivery, biosensor, and tissue repair. A study from the Project on Emerging Nanotechnologies (PEN) has collected information for 1,800 products on the market containing nanomaterials or using nanotechnology during the manufacturing process, and the number of health and fitness related products is around 900 (Figure 11.1), making up for about 50% of the total products [4].

With the continually growing number of nanotechnology based products, there is an increased public concern of using engineered nanomaterials (ENMs) in consumer products, from both a safety and effectiveness point of view. For example, although the chemical compositions of almost all the nanomaterials that are under research and production have already been chemically identified, the characteristics of the nanomaterials can be significantly different from their bulk counterparts [6]. Many new properties, e.g. surface plasma resonance of gold nanoparticles, can emerge when the dimension of a material is reduced to the nanometer range. Moreover, the extreme large surface area of the nanomaterials may induce much stronger interactions with the biological system, resulting in a completely different toxic profile from the bulk. Therefore, it is a priority to investigate whether it is safe to use nanotechnology based products. In addition to the safety issues, ineffective products should also be prohibited from entering the market. Some manufacturers have taken the advantages of the unfamiliarity of the public with nanotechnology and exaggerated the benefits or even added false health claims to the labels. Taking the results from PEN again, among the 1,200 products in the database, there are less than 4% having verified claims (i.e. additional supporting documents about the nanomaterials in the products are available), whereas the claims from more than 70% of the products are not supported with any evidence [4].

It is becoming clearer that we are at a stage where the regulatory agencies need to step in and properly regulate these products. The risk associated with using nanomaterials in a product should be determined by conducting a thorough risk assessment. The possible route and pattern of the exposure and the subsequent health effect from the exposure need to be carefully evaluated. Both underestimating and overcalculating the risk can cause unnecessary cost to the economy and public health. Although the challenges the regulatory bodies are facing in regulating nanomaterials containing products might not be significantly different from what they have encountered with other chemicals, it is complicated by the "nano effect" [7].

The very first issue associated with nanomaterials is how to define them. Even though nanotechnology has been continually developed for the last 20 years, the scientific community and regulation bodies have not reached a unified definition of the term "nanomaterials". For fundamental research purposes, a material having at least one dimension within the range of 1–100 nm can be called a nanomaterial. However, by looking at the literature, we can find numerous research articles considered the materials in their study as nanomaterials even though all the dimensions are larger than 100 nm. In 2008, the International Organization for Standardization (ISO) first published a document, intending to establish normative terminology for a series of terms commonly used in nanotechnology. According to the document, a size range from approximately 1 to 100 nm was defined as nanoscale and a material with one, two or three external dimensions at the nanoscale was defined as nano-object [8]. A similar definition was also provided for nanoparticle, nanofiber, and nanoplate. In 2009, Health Canada released a policy statement describing their working definition of nanomaterials. In the statement, Health Canada promoted ISO definition on the term "nanoscale", stating that it means 1–100 nm [9]. Any manufactured substance or product and any component material, ingredient, device, or structure is considered to be nanomaterial if:

1. It is at or within the nanoscale in at least one external dimension, or has internal or surface structure at the nanoscale, or;
2. It is smaller or larger than the nanoscale in all dimensions and exhibits one or more nanoscale properties/phenomena.

It can be seen that in addition to the definition by ISO, Health Canada added that a material can still be considered as nanomaterial even its all dimensions are outside the range of 1–100 nm as long as they exhibit nanoscale properties/phenomena. Here, nanoscale properties/phenomena mean properties which are attributable to size and their effects, which are distinguishable from the chemical or physical properties of individual atoms, individual molecules, and bulk material [9].

A reoccurring question when defining nanomaterials is whether there should be a lower limit on what percentage of materials must be in the nanometer range in order to be considered as nanomaterial. Because, it is not uncommon to find materials that are in the nanometer range in either natural occurred or manufactured bulk materials. For example, inorganic pigment particles, such as Fe_2O_3 and TiO_2, are commonly added to cosmetic products for their color effect. While most of the pigment materials have a fairly large grain size (>500 nm), a small portion of the materials may be smaller than 100 nm. The percentage of this portion is highly dependent on the preparation method.

The definition proposed by European Commission (EU) addressed this issue. In 2011, the EU introduced a comprehensive science-based definition of nanomaterials in Union legislation and a recommendation was adopted in 2011 after public consultation [10]. Based on this recommendation a "nanomaterial" means:

1. A natural, incidental, or manufactured material containing particles, in an unbound state or as an aggregate or as an agglomerate and where, for 50% or more of the particles in the number size distribution, one or more external dimensions is in the size range 1–100 nm.
2. In specific cases and where warranted by concerns for the environment, health, safety, or competitiveness, the number size distribution threshold of 50% may be replaced by a threshold between 1% and 50%.

The definition gives a cut off of 100 nm for nanomaterials, which can be problematic when regulatory action needs to be taken on certain materials that fall outside the size range and yet still show nano-related properties. In an extreme case, if a material is 101 nm in size, we would not consider it as nanomaterials since it is large than 100 nm. In addition, none of the definitions above have put a frame on the term "size" when describing the materials. The particle size is meaningless without specification of the type of size [11]. As we know that, there are a set of techniques that can be used to obtain size information of the materials, and the determined size is highly dependent on which aspects of the materials are being measured. Electron microscopy (EM) techniques determine the "dry size" of the material, and in most cases the organic coating layer does not generate sufficient contrast in order to be seen in EM. However, this coating layer could have a drastic effect on the hydrodynamic radius of nanoparticles when DLS is the measuring technique. For example, a magnetite nanoparticle with an obtained diameter of 12 nm can double or even triple its size in a DLS measurement (28–35 nm for nanoparticles with polyacrylic acid coating of different molecular weight) if it is functionalized with polymers [12]. It is highly likely that a particle which is determined to be smaller than 100 nm in one technique appears much larger in size in another type of measurement. From this point, the second term from the definition proposed by Health Canada, stating that as long as the materials exhibit nanoscale phenomenon and properties they are considered as nanomaterials, is absolutely necessary.

Another topic that always creates debate between scientists when discussing the safety and benefit of nanomaterials is that whether we should establish a complete new framework to regulate nanotechnology-based product, which is of course both labor and cost intensive, or try to fit them into the existing system. The later one is an obviously simpler way that would be especially preferred by the regulatory agencies. However, there are still many obstacles in front of us even if we choose the second pathway. The primary problem is that most of the policies in the current regulatory structure are addressing issues associated with the chemical properties of substances. Many properties that are unique to nanomaterials are determined by their physical characteristics [6]. The U.S. Food and Drug

Administration (FDA) is one of the agencies that collectively administer laws that are related to food safety. According to the U.S. Federal Food, Drug and Cosmetic Act, "any substance that is intentionally added to food is a food additive, that is subject to premarket review and approval by FDA, unless the substance is generally recognized, among qualified experts, as having been adequately shown to be safe under the conditions of its intended use, or unless the use of the substance is otherwise excluded from the definition of a food additive". The FDA is currently maintaining a GRAS list that includes a variety of substances. Many substances in this list, such as ferric oxide, zinc oxide, and silica, are common materials that are being investigated in nano-related studies. This list, however, only defines the chemical composition of the substances and does not have information regarding the physical characteristics of the materials. Now the question is that if nanoscale zinc oxide is added to the food, should the FDA require pre-market review and ask manufactures to submit evidence to FDA, demonstrating that no harm will be resulted from the proposed use of an additive? One would argue that such a requirement is not necessary because the chemical composition of the nanomaterials is no different from the one that is currently in the GRAS list. On the other hand, many would favor a "yes" to the question, providing that due to the significant changes in the physical properties, the interaction between the nanomaterials and the biological or environmental system is drastically affected.

Although nanomaterial is a simple word, it brings tens of thousands of materials under its umbrella. A huge number of chemicals and substances that are being manufactured and used everyday can be made into nanomaterials. The situation is further complicated by the fact that even for a single nanomaterial its properties can be affected by its physical state. For example, nanoparticles can be categorized into four categories as surface bound, suspended in liquid, suspended in solid, and free airborne based on the location of the nanomaterials, and an associated risk can vary significantly [13]. It is urgent to establish a system that can efficiently identify, investigate, and analyze problems and challenges associated with nanotechnology and find solutions to address them.

11.2 Toxicity testing of nanomaterials

The first step in a risk assessment is to determine the identity of the hazard and their toxicity (to humans and the environment). Because nanomaterial is the point of concern, it is identified as the potential hazard in the current focus. For many nanotechnology-based materials, their chemical compositions are no different from the known substances. The question is, can we completely rely on the existing knowledge to predict the toxicity of nanomaterials? Because the properties of nanomaterials are different from their bulk or ionic counterpart, the emerged optical, mechanical, and electronic properties at nanoscale may affect the toxicity. For example, it has been demonstrated that at the same concentration, CdTe quantum dots can cause more severe morphological damage than ionic cadmium to an invertebrate freshwater model organism in a time and dose dependent manner [14].

The mechanism that causes this difference can be largely attributed to the distinctive biophysicochemical interaction of nanomaterials with biological systems. When the dimension of a material gradually decreases, its surface area to size ratio grows at a rapid rate, which can lead to strong and complex interactions with surrounding medium and biological substrates [15]. Dozens of factors which belong to the four major components of the nano-bio interaction (i.e. nanomaterials, suspending media, solid–liquid interface, and nano-bio interface) can have influence on a series of forces (e.g. hydrodynamic and electrostatic interaction) that governing the interaction, resulting in specific behaviors of nanomaterials.

With the realization of the unique challenges of nanomaterials, it is essential to carry out independent toxicity studies to address this issue. Currently, most of the toxicology studies on nanomaterials rely on in vitro cell culture tests as the screening method, which is largely based on the existing techniques initially designed for small molecules. While the utilization of a wide variety of commercial available assay kits allows the test to be performed in a timely and cost-effective manner, many researchers questioned the reliability of these assays in assessing the toxicity of nanomaterials, as quite a number of disagreed or even contradictory results were reported in the literature [16]. Due to the strong and complex interaction between nanomaterials and cells, specific biological reaction can be triggered depending on the testing condition. For example, Patra *et al.* reported a study on the cytotoxicity of gold nanoparticles in which they observed a cell selective response. They suggested that gold nanoparticles may not universally target all cell lines [17]. Because many factors in the experimental design such as choice of cell line, dosing range, and duration and the select of end point can potentially affect the outcomes, it is critical to establish harmonized testing conditions to allow meaningful comparison of studies reported by different groups [16]. In 2005, with the aim to identify key components in nanomaterial's toxicity test, the Nanomaterial Toxicity Screening Working Group, established by the International Life Sciences Institute/Risk Science Institute (ILSI/RSI), evaluated a variety of existing methods, and gave recommendations on the type of cell lines, assay, and end points that can be used for initial toxicity screen [18]. The working group also emphasized that the selected cell line should closely represent the potential target in human exposure.

Interferences in the toxicity assay are caused by the interactions of substances with an analyte of interest, which alters the measureable concentration and results in false positive or false negative results. The problem is especially evident in toxicity tests for nanomaterials because of their unique physiochemical properties. Although the exact mechanisms that cause the interference depend on the type of assays and nanomaterials, they can be generally categorized based on four factors.

11.2.1 Optical properties

A wide range of nanomaterials, particularly inorganic nanomaterials including metallic, metal oxide, and semiconductor nanoparticles exhibit certain intrinsic optical properties which can interference with measurements or even mask the true effect. When evaluating the toxic potential on materials, e.g. iron oxide nanoparticles, which have strong absorbance in the UV–vis wavelength range, colorimetric type of assays such as MTT assay can exhibit elevated signals. In addition, materials like gold nanoparticles which have quenching effects on dye molecules can significantly reduce the luminescence signals. Furthermore, any assays that relies on measuring fluorescent signals can be compromised by materials that have strong luminescence properties such as quantum dots. One simple strategy to eliminate or reduce the interference caused by nanomaterials with absorbance is to use a "no cell control". By acquiring the readouts of nanomaterials themselves without adding any cell, it is able to determine whether there is any interference, and obtain a baseline for the assay. For nanomaterials with a quenching effect, the measurement should be collected before and after the addition of nanomaterials in order compensate signal suppression [19].

11.2.2 Chemical reactivity

The interference can also come from a chemical reaction between nanomaterials and substances that are essential in an assay. Nanoparticle surfaces can be highly reactive toward oxidation or reduction which degrade chemicals, resulting in false outcomes. Some nanomaterials also exhibit strong catalytic properties and are able to catalyze reactions in an assay without the addition of certain reagents. For assays that experience this type of issue, readings should be collected before and after the addition of nanomaterials or add a control group where no reagent is spiked in order to probe and assess the potential interference.

11.2.3 Physical properties

One of the characteristics that differs nanomaterials from their bulk counterparts is the extreme large surface area. Both substances in the cell culture medium and added reagents can adsorb onto nanomaterial surfaces, resulting in either elevated or reduced readout. For example, protein molecules can adsorb on nanoparticle surfaces and form a protein corona, resulting in an inhibition of properties of the protein (e.g. catalytic properties of enzymes). For tests that involve the use of proteins such as hemolytic, LAL, LDH, and Cytokine ELISA assays, a control group in which assays are measured with nanomaterials spiked with the protein reagent should be added to minimize interference. In addition to chemicals and substances in the assays, impurities from the nanomaterial preparation and environment contaminants can also adsorb to nanoparticles and lead to toxicity effects that are not directly related to nanomaterials [20]. For this type of interference, it is important to apply a proper purification method, such as repeated washing, if the aim is to evaluate the toxicity of the particular nanomaterials other than the whole formulation.

11.2.4 Aggregation

When nanomaterials are added to the cell culture medium, a range of forces including attractive van del Waals (VDW) force, electrostatic, repulsive steric, solvation interaction, and several other short range forces govern the colloidal condition of nanomaterials. Because the biological medium usually has very high ionic strength (150 mM), it can shield the surface charge and significantly reduce the electrostatic repulsion. Attractive forces will then come into play and cause aggregation and form agglomeration. The formation of agglomeration can either interfere with the measurement (e.g. increase absorbance) or alter the interaction between nanomaterials and cells. Visual examination is a proper approach to detect the presence of any false positive or negative effect, and a no cell control is effective in reducing the effect caused by the increased absorbance of agglomeration.

A good practice when conducting in vitro toxicity assays on nanomaterials is to review literature reports to determine possible interference that may exist in a particular test for certain nanomaterials. Assays that significantly bias the results should be avoided and alternative methods should be used when possible. If interference cannot be eliminated, appropriate assay control must be in place and a second assay, which measures the same parameter based on a different mechanism, is strongly recommended in order to assist the interpretation of the data.

In the late middle age, Paracelsus, the Father of Toxicology, had pointed out that, "all things are poison and solely the dose determines that a thing is not a poison". He pioneered a concept that highlighted the importance to establish a quantitative dose–response relationship to evaluate the toxic potential of a chemical substance. Traditionally, mass concentration is used to describe the dose in almost all toxicology studies. Recent studies have suggested that such measures may not well indicate the level of harm caused by nanomaterials. Metrics such as nanoparticle number concentration and surface area concentration may be more suitable to evaluate the toxicity of nanomaterials. Nevertheless, it is highly recommended that in a toxicity study, responses against all three dose metrics (mass, surface area, and particle number concentration) are collected to allow comparable interpretation of data [18].

When investigating nanomaterial toxicity, an array of attributes can be selected to study their contribution on the toxicity profile. The situation is, however, complicated by the fact that the characteristic of nanomaterials is defined by the summation of a group of physicochemical properties. Although a range of different aspects of nanomaterials such as size, shape, crystal structure, surface area, and charges have been studied to establish correlation with toxicity, it is still unclear which factor contributes the most. With the continual development in nanotoxicology, it is becoming clear that the biophysicochemical properties of nanomaterials intertwine with each other and the toxicity of nanomaterials might be a result of the combined effect of multiple attributes. Because it is impossible to study every single factor, a feasible practice is to rank the importance of all common attributes to the properties of nanomaterials and focus the studies on the most critical ones. For example, because many properties of the nanomaterials exhibit dramatic changes when their sizes are altered, i.e. quantum confinement of quantum dots, size is commonly selected as the attribute to study the toxicity, and many related factors can be reasonably estimated based on the existing information.

Several issues can potentially compromise the reliability of the size based results. Particle aggregation is a very common problem when studying particle toxicity. The colloidal stability of nanoparticles is a result of the balance of multiple interactions including VDW, electrostatic, hydrophobic interactions, and etc., and any change to the surrounding medium or particle itself will disturb the balance. As mentioned repeatedly earlier, many physicochemical or biological properties of nanomaterials are dependent on size. The toxic profile of a material may change if individual particles form large aggregations. Taking iron oxide nanoparticles as an example, ultra-small iron oxide nanoparticles with a hydrodynamic size smaller than 10 nm can evade the reticuloendothelial (RES) system and accumulate in the lymph node whereas larger size particles are taken up by RES and rapidly cleared by liver or spleen [21]. The materials can exhibit distinct toxicity due the different metabolism pathways.

Excessive energy introduced to the sample, e.g. long time of sonication, can potentially change the aggregation state. Therefore, sample preparation conditions should be carefully controlled not to alter the initial state of the materials. When performing size characterization on nanomaterials, an appropriate measuring condition is the key to obtain meaningful results. For example, although dynamic light scattering is one of the most applied techniques for size determination of nanoparticles due to its simplicity and low cost, the results should be carefully interpreted to avoid drawing incorrect size dependent toxicity because not only the hydrodynamic size measured by DLS varies significantly depending on the suspending medium, the particle stability can be affected as well. Nanoparticles that are highly stable in water may form large agglomeration in biological fluids such as serum due to its extreme high ionic strength. It is always recommended that, during the size related characterization, the measuring condition should, if possible, match with what would be used in the subsequent toxic studies.

A unique feature of nanomaterials is that, with the core materials remaining intact, simply modifying the composition of the surface of the material can dramatically change a series of biophysicochemical properties which may or may not influence the toxic profile of the materials. As a result, conclusions drawn from an individual toxicity study cannot be directly used to describe the hazardousness of a class of similar nanomaterials if their surface chemistry is different. For example, a surface coating largely determines the colloidal stability of a nanomaterial. Replacing the low charge density ligand with a high charge density one can effectively increase the zeta potential which improves the stability. Alternatively, many researchers choose coating nanomaterials with a thick layer of polymers such as polyvinylpyrrolidone (PVP) as their strategy to prevent aggregation because VDW attraction depletes rapidly with increasing particle–particle distance. Another purpose of the coating layer is to provide protection to the core. Many inorganic nanomaterials, especially metal oxide and semiconductors are prone to chemical decomposition, such as oxidation, and can gradually release ions into surroundings. The ions themselves can be highly toxic or possess catalytic properties which can induce the formation of free radical species, causing damage to the cells. Coating the materials can effectively separate the core from the environment and enhance their chemical stability. Some studies have suggested that the toxicity of inorganic nanomaterials can be reduced once they were coated with several types of molecules [21]. Once entering a living organism, the surface coating will be the first in contact with the in vivo system, and due to the extreme large surface area of nanomaterials, the interaction between the surface and the biological environment cannot be ignored. In general, naked nanomaterials can be easily recognized and cleared out by phagocytic systems as foreign substances. Studies have demonstrated that functionalization of nanomaterials with polyethylene glycol (PEG) can help nanomaterials avoid the immune system and prolong circulation time by reducing the interaction with biological systems [22]. Many ligand molecules render charges to the surface of nanomaterials which shape the interactions at the nano-bio interface. Evidence has shown that nanoparticles with cationic surfaces are more cytotoxic than their neutral or anionic counterparts by damaging the cell integrity because of the stronger binding with the negatively charge cell membrane. Tuning the surface charge can adjust the particle–cell interaction and control the cellular uptake of nanomaterials. Furthermore, the surface chemistry can also direct the metabolism pathway of a nanomaterial and decide their fate in the body. For example, by carefully manipulating the number of ligands and their charges on the surface, renal clearance rate of quantum dots can be achieved [23,24]. So far, many efforts have been placed to investigate the effect of surface coating on the properties of nanomaterials. The toxicity of the coating materials themselves, however, was often ignored. Detailed characterization such as mass spectrometry for organic matter and elemental analysis (XPS) for inorganic materials should be included during the study to acquire in-depth information on the chemical composition of the surface coating.

The synthesis process of nanomaterials usually involves multiple chemicals and reactions, and a residual level of precursors and byproducts can remain with the materials even after purification. In addition, nanomaterials in the environment often adsorb substances from the surroundings. It is important to assess the contribution of these contaminants to the total toxicity when evaluating the hazardousness of the nanomaterials [25].

In vitro tests are an efficient approach for rapidly screening the toxic potential of nanomaterials. It produces valuable information to assist researchers to elucidate the mechanism of the toxic effect from exposure to nanomaterials. However, the human body is a rather complex entity, and extrapolating the results from a simple cell culture test to a whole, living organisms is problematic. In vitro assays do not provide any insight into basic components in pharmacokinetics, which is how nanomaterials are absorbed, distributed, metabolized, and excreted upon exposure [26]. The toxic effect from an in vivo test can be completely different from in vitro because of the breakdown of nanomaterials by metabolism. The dose response curve may also vary significantly from an in vitro assay due to compartmentalization of nanomaterials in different organs and tissues. As discussed earlier, because an array of factors such as size, shape, charge, and surface coating, such properties governor the interaction at the nano-bio interface, the fate and toxic potential of each type of nanomaterial depends on its unique physicochemical and biological properties. Therefore, it is essential to conduct toxicity tests using an in vivo model to quantitatively evaluate the pharmacokinetic behavior of nanomaterials.

11.3 Exposure assessment of nanomaterials

Exposure assessment is one of the most critical steps in the risk assessment process, which ultimately determines whether there is a risk associated with the hazard.

If there is no exposure, there will be no risk and the mere existence of a hazard does not necessarily lead to harm. Furthermore, not all exposures cause the same level of harm. Nanomaterials that are embedded and fixed in the solid matrix pose no harm to humans unless the matrix is altered [27]. Silver nanoparticles that are added to sport socks would have much less potential to cause harm through dermal contact than a silver particle based dietary supplements which are consumed directly by the consumers. Conducting an accurate and reliable exposure assessment will establish a solid foundation for the subsequent risk characterization step, which largely affects the decision making in risk management.

In a common exposure assessment process, the investigation usually starts with five questions: what, where, when, who, and how [28]. The first step is to determine what the identity of the hazard is? The challenge of this step resides in the fact that very limited information can be obtained from the manufacture regarding the composition and other characteristics of the products. Here we use the data from PEN to illustrate the issue [4]. According to this database, nanomaterial containing products or nanotechnology based products are categorized into five levels based on what we know about the products. Among all the products, about 71% of them do not have specific information provided to the support the nanotechnology related claim and are classified into in the "Unsupported Claim" category. The highest level in the category is called "extensively verified claim" and it requires not only information from the manufacturer to support the nanotechnology claim, this claim must also be verified by an independent source [4]. Solely relying on the information provided by the manufacture, even if it is comprehensive, is risky. The definition further states that the actual product must be tested for nanomaterial or supporting documentation references such product or the product was described in more than one published scientific document in order to be qualified in this category [4]. Unfortunately, only less than 1% of the products are in this level. Without clearly knowing the composition of the products, it will be extremely difficult to perform an exposure assessment. Moreover, manufacturers may use nanotechnology in the process but decided not to advertise it due to the concern of the unknown consequences. Currently regulations, especially in the United States, do not mandate companies to report the use of nanotechnologies during the manufacturing process. It has come to an agreement that, characterization on the products in addition to the information released by the manufacturer is absolutely necessary.

With the knowledge of the composition of a product, "Where are the nanomaterials located?" becomes a critical question to be answered in the next step. Titanium dioxide nanoparticles in the powder sugar on donuts will have different pathway and degree of exposure compared to the ones in sunscreens and other cosmetic products. Acquiring details regarding the location of the nanomaterials can be as challenging as determining the composition of a product because such information is usually not shared by the manufacturers. Conducting independent analysis on the products might be the best solution to the problem. For nanomaterials in consumer products, some researchers suggested to group the exposure into three categories based their location in the products [13]. For example, airborne nanoparticles or nanoparticles in liquid suspension are categorized as "expected to cause exposure" because of the possible direct human contact under the intended use. Group 2, "may cause exposure", includes products that can release a certain amount of nanomaterials by wear and tear, although the products are not designed for such purpose. Products that have nanomaterials fixed in the solid matrix and are resistant to normal use are generally in the "no expected exposure to the consumer" category. It is expected that this grouping strategy can improve the efficiency when evaluating the exposure without largely reducing the reliability of the assessment.

The group of individuals who are at risk of exposure will change at different stages of the nanomaterial's life cycle. Occupational exposure is always considered to be more prevalent due to the direct interaction with the raw nanomaterials. Workers at nanomaterials research and development departments, manufacturing or recycling facilities have the largest opportunities to be exposed to a high level of nanomaterials through inhalation, dermal, ocular, and ingestion pathways. The consumer exposure to nanomaterials has the characteristics of high variety because not only the exposure route is more changeable, the suggested direction for use of a nanomaterial containing products might not be strictly followed [29]. Some products may also target groups of individuals who are more vulnerable such as children and elderly, and special considerations are needed to decide whether these groups are at risk of exposure.

One of the key factors that determines what would an exposure result be the concentration of nanomaterials. Quantification is helpful in improving the reliability of an assessment, and this task can be hindered by the lack of available information on the content of the sample, which is especially true for environmental samples and many consumer products. Although for occupational exposure, we usually have a better knowledge of the identity of the hazard, the actual scenario can be different from place to place and independent analysis might be necessary.

Caution should be taken when collecting samples for the exposure assessment as contradicting results may appear and lead to a deviated conclusion if a standard operating protocol is not strictly followed. Several factors should be considered when perform the sampling activities on nanomaterials exposure assessment. First, the sampling location must be recorded in detail because the concentration may vary significantly depending on the distance to the source of the nanomaterials [29]. For example, at a nano-pigment synthesis facility, sampling the air inside the production room, close to the ventilation system, and outside the facility will probably give largely different readings. In addition to the sampling location, the time a sample is collected also matters. Taking the same facility as the example, the airborne nanoparticle concentrations will be much lower in the morning than the one right after an entire day operation. Therefore, it is recommended that samples should be collected at multiple relevant locations and time periods to obtain a comprehensive data set that can correctly reflect the real situation. If the sample is collecting outside (e.g. nanoparticle generated by the vehicle fuel combustion), weather conditions must be taken into account and time weighted values for several days or longer might be necessary. Second, the metric that is selected to be measured during the sampling can potentially mask the real situation or cause the result deviating from its true value. For instance, due to the high level of background nanoparticle concentration coming from the control burning nearby, the concentration of nano-particle of interest outside a manufacturing facility may be significantly over-estimated if only a condensation particle counter (CPC) is used for the measurement. Because CPC is a number based particle measuring device that does not distinguish the type of the particle, both nanoparticle released by the facility and the ones generate by the burning will be counted together. Furthermore, some techniques require adjusting the instrument setting when measuring different sample. The device sensitivity for a particular sample may be significantly low in a mode which is able to work normally for another analyte [30]. A solution to this problem is the combined use of multiple techniques that measure different metrics of the nanomaterials. If the results show that two metrics (e.g. number based measurement and mass based measurement) do not agree with each other, it possibly is an indication of high background nanoparticles, and further analysis is then warranted.

In nanomaterials related toxicology studies many physicochemical properties (e.g. size, mass, surface area) have been utilized to correlate with the toxicity of the materials. No agreement, however, has been reached so far on which property can ultimately represent the toxic effect. Since it is impossible to monitor every single aspect of the nanomaterials, a few attributes that are deemed more important should be selected for the analysis based on existing evidence or reasonable estimation. Nanomaterials can be first grouped into several categories based a number of characteristics such as size, shape, composition, and heterogeneities. Next, each individual physicochemical property is assigned with a degree of relevance to the toxic profile for a specific category. The characterization and quantification in exposure assessment should be then focused on the high relevant factors. The unique characteristics of nanomaterials also create difficulties in choosing correct metrics to determine the level of exposure. Without sufficient knowledge of which metric is more relevant, either underestimation or over-calculation of the harm can occur. For airborne particles, mass per unit has been used as the metric for many years to indicate the level of exposure. Studies have suggested, however, that exposure assessment based on mass concentration may not fully represent the harm [29]. It has been reported that smaller sized nanoparticles with the same composition tend to travel deeper in the respiratory tract and may cause more harm once they deposit. There are also evidence suggesting that surface area of the nanomaterials can be a good metric for some types of nanoparticles [29]. Other metrics that are potentially useful in establishing a sound dose response relationship with the adverse effect of the nanomaterials are particle number concentration, volume concentration, and surface charge density [18]. Maynard and Aitken have proposed a rather useful grouping system with the aim of reducing the number of metrics that need to be measured for a meaningful exposure assessment [29]. In this system, four types of concentrations (number, length, mass, and surface area) were selected as the main metrics for four individual groups. Other attributes (e.g. shape, composition, crystal structure, porosity, etc.) that are commonly used to represent the characteristics of the nanomaterials are then assigned with level of relevance (high, medium, low, and none) based on their relationship with the main metrics. By applying this strategy, it is possible to measure only a few aspects of the nanomaterials and obtain an estimated contribution of other attributes with enough confidence, which largely decreases the workload.

Life cycle analysis which determines the origin, transport, transformation, and fate of nanomaterials can be beneficial to the exposure assessment. The exposure potential of nanomaterials can change significantly when they migrate from raw materials in the manufacturing facility to the ingredients in products and eventually end up in the environment during disposal. Furthermore, it is known that nanomaterials in the products can undergo transformation in both the chemical and physical form (e.g. silver nanoparticles can release silver ions in solution), which subsequently affected their exposure potential and toxic profile. Therefore, an individual assessment that targets to a specific period in a nanomaterial's lifetime is needed.

11.4 Characterization of nanomaterials

In order to fully understand the physicochemical of nanomaterials and how they interact with the biological system, an extensive characterization of the nanomaterials is critical. Just like any other products, questions such as what type of nanomaterials, how much of them are added to the products, are the contents consistent with the labeling, is there a batch-to-batch variation and are these nanomaterials harmful to humans, should be answered before the products reach consumers. Thus, analytical characterization methods which can provide information on different characteristics of the nanomaterials are needed.

There are various types of techniques that have been routinely used for the characterization of nanomaterials, and they are able to provide relatively accurate determination on certain aspects of the materials. Although most of the techniques have shown fairly well performance when working with neat nanomaterials, i.e. purified nanoparticle solution, their ability to deal with complex matrices are highly technique dependent. Additionally, a robust analytical method should be able to determine both the identity and quantity of the analytes [11]. In case of nanomaterials, the size and chemical identities are answered by the detection step and information of mass or number based fraction is provided by the quantification. A list of techniques that are proposed for the characterization of ENMs in consumer products are presented below, and their advantages and limitation will also be discussed.

11.4.1 Electron microscopy (EM)

Transmission electron microscopy (TEM) is a heavily used technique in nanomaterial related studies. It is widely considered as the gold standard for the characterization of nanomaterials, especially nanoparticles. Because it is a microscopy technique, it is able to provide direct visual information on the size and size distribution, morphology, and aggregation state of the materials being studied. A modern TEM with a LaB_6 filament or a field emission TEM (FETEM) is commonly capable of a sub-nanometer resolution. The fine structure of the nanomaterials (usually inorganic), such as the packing defecting in crystalline nanomaterials including twins, phase boundary, and stacking fault can be easily revealed by high resolution TEM (HRTEM). In addition, electron diffraction can also be a useful tool in identifying the structure of a material, especially when the specimen contains an unknown composition that is crystalline in nature. It is worth mentioning that many ENMs, particularly inorganic nanomaterials, have a protective coating which is commonly made up with organic compounds. This layer serves as a barricade that stabilizes the inner core from aggregation and degradation. Since the contrast in the TEM is largely affected by the atomic number where heavy elements diffracts the beam stronger and appears darker on the image, the coating layer which generally contains hydrogen, carbon, oxygen, and nitrogen cannot be directly observed. Thus, only the core of the materials is studied by TEM. Although negative staining which involves the use of a heavy metal containing compound such as uranyl acetate can increase the contrast of the organic compound, its effect on other aspects of the materials is not easily predicted.

It is generally agreed that TEM is able to provide reliable information regarding the size of an individual nanoparticle. However, whether the generated results can represent the characteristic of the entire sample is questionable, considering the sample area being examined is relatively small. Furthermore, for an ENM, large numbers of individuals need to be counted to get the average size and deviation with sufficient statistical significance. The counting can be very tedious and time consuming. Moreover, if the nanomaterial concentration in the original sample is low, an additional pre-concentration step might be necessary in order avoid time being wasted in searching the entire grid to find enough materials to image. Some image analysis software such as ImageJ can be used to reduce the workload. However, they are more effective in analyzing highly dispersed, in another word, non-aggregated materials, which is rarely the case in a real world situation. Many nanomaterials are highly aggregated, which require additional image processing to define the boundary of an individual unit. Another major drawback of TEM is the complicated sample preparation. For example, in order examine the nanomaterials in a biological sample (e.g. tissue and cells), a multi-step procedure including extraction, fixation, dehydration, embedding, sectioning, and staining must be followed to prepare a sample that is suitable for TEM observation. The procedure is a complicated and low throughput process which requires both skills and patience to obtain a satisfied specimen. The sample preparation for the ENMs in the consumer products (e.g. TiO_2 and ZnO in sunscreens) can also be a headache because most of the materials are embedded in complex matrices, which make a direct TEM examination unlikely. The matrices may block the visualization of the nanomaterials, and the volatile organic compound in the matrix can damage the high vacuum environment in the microscope column. Therefore, it is essential to perform an extraction prior to the TEM analysis. However, a question that almost always accompanies with the extraction is whether such treatment will maintain the original state of the materials, e.g. degree of aggregation. In addition, all the samples being examined must be highly dried, and the drying itself can also alter the sample.

Compared to TEM, SEM is capable of studying materials in a very dynamic size range (nanometers to microns) with a slightly lower resolution. Additionally, the scanning design allows SEM to examine a much larger area than TEM, making it extremely useful in studying 2D and 3D nanostructured materials and investigate the agglomeration state of the nanoparticles in a complex matrix. Although the electron gun, apertures and the configuration of the column are somehow similar to TEM, the imaging mechanism in the SEM is different. SEM commonly utilizes its in-lens detector to capture secondary electrons from inelastic scattering. In this mode, a variety of non-conducing materials such as polymers, minerals or ceramics can be examined. Back-scattered electrons from elastic scattering, which intensity depends on the composition of the compound, can also be collected to reveal small details such as derivatives and phases of the materials. The requirement of sample preparation in SEM is usually less strict than TEM. In TEM, a sample layer has to be thin enough to let the electron beam pass through in order to generate a signal. In SEM, many samples can be simply placed on a silicon wafer or fixed on a carbon tape and directly examined. Because conventional SEM is operated under high vacuum, samples must be sufficiently dried which create restriction on studying wet biological or environmental samples. The development of environmental SEM permits the observation of volatile samples without the need of a drying and coating step, effectively preserving the original state of the materials.

11.4.2 X-ray fluorescence spectrometry (XRF)

Beside size and shape, elemental composition is another fundamental factor that determines the physicochemical characteristics of the nanomaterials. XRF is a class of widely used elemental analysis techniques. When a high energy beam, e.g. X-ray, impinge on the sample, the atom is excited and one or more electrons from the inner orbit are ejected, leaving a vacancy behind which will be subsequently filled by an electron from a higher orbital. The atom emits photons to release energy and returns to the ground state. By measuring the energy released during the electronic transition, the element can be determined. XRF is a non-destructive technique and allows for in-situ analysis for many samples. With the development of portable XRF, this technique has shown a strong application potential in the routine screening for elements of interest in a variety of samples such as food, pharmaceuticals, geological, and environmental samples. It should be noted that, XRF is an elemental analysis technique and it cannot distinguish nanomaterials from bulk materials. Nevertheless, there has been an increasing interests in using XRF to identify the chemical composition for nanomaterials. For example, Linder and coworkers reported a methodology that utilizes portable XRF in quantifying nano-silver in dietary supplements [31]. The results acquired by p-XRF are in good agreement with inductively coupled plasma mass spectrometry (ICP-MS) measurements. There are a number of factors that can compromise the quality of p-XRF data, and sample matrix effect is the most prominent one when analyzing real products. The calibration curve is usually established by measuring standards, which is a lack of matrices that are commonly used in real samples. The matrix may interfere with the measurement and the results may deviate from the true value. Additionally, homogeneity of the sample is also a crucial factor in XRF measurements where heterogeneous samples increase the difficulty in obtaining repeatable data and results from the analyzed portion may not well represent the feature of the entire product if the sample is not uniform [32]. The detection limit of a common XRF is generally above the ppm level which is sufficient for products that have nanomaterials as the main ingredient such as Zn and Ti in sunscreen. However, if a nanomaterial is present in the product as a contaminant, its concentration may be beyond the detection ability of XRF.

Nowadays, many TEM and SEM are equipped with an energy-dispersive spectroscopy (EDS) detector which has the same fundamental principles as regular XRF. On these instruments, instead of an X-ray tube, an electron beam is used as the high energy source, and information on the elemental composition of the nanomaterials in the selected sample area can be acquired by switching from imaging mode to EDS mode. In addition to the regular energy line scanning mode, the X-ray mapping mode is able to collect an elemental distribution of the sample. This is especially important to study the properties of some composite materials which may have distinct element abundance in different areas of the sample. A common limitation of EDS is the peak overlaps of different elements due the X-ray emission corresponding to the different shell transition of the element. Although, EDS is capable of detecting elements that are heavier than beryllium, the accuracy in analyzing the light element is low. Moreover, the energy resolution of EDS (with a Si detection) is in the range of 100–150 eV, indicating that this technique is unable to extract chemical-state information (energy difference < 10 eV) from the spectra.

11.4.3 X-ray diffraction (XRD)

Many physicochemical properties of the nanomaterials such as mechanical, electrical, and magnetic properties are closely related to the crystal structure of the materials. Moreover, some studies have suggested that two materials that have the same chemical composition and only differ in crystalline phases can induce different levels of cytotoxicity [33]. XRD has been demonstrated to be a rapid and nondestructive technique for the characterization of various organic and inorganic ENMs such as lipids, proteins, metal alloy, metal oxide, and semiconductors. It is especially useful for a multicomponent system. By comparing the diffraction pattern with the standards database, it is possible to determine the identity of each individual unit in a highly complex sample. The results from XRD directly reflect the crystallographic structure of the materials. By applying the Scherrer equation, it is possible to estimate the average size of some types of nanomaterials without conducting microscopy studies. It should be noted that the value calculated from the Scherrer equation corresponds to the crystallite size, and this value is close to the nanoparticle size only when an individual particle is single crystalline in nature. In addition, XRD data do not give any indication on the shape, size distribution, and aggregation state of the nanomaterials. Compared to EM techniques, XRD requires larger quantities of materials to be analyzed to get satisfied data. For samples that are precious such as some active pharmaceutical ingredient or have a nature of low concentration of nanomaterials, analysis through XRD can be difficult.

11.4.4 Dynamic light scattering (DLS)

DLS, which is also known as photo correlation spectroscopy, is probably the most popular technique for the study of colloidal nanoparticles due to its simplicity, rapid measurement, and low cost. When a laser beam is irradiated on to the nanoparticles, the scattering light intensity fluctuates due to the Brownian motion of the particles in solution. The characteristic of the fluctuation is dependent on the diffusion coefficient of the particles. DLS is able to process this information through autocorrelation and the size of the particles is calculated according to the Stoke–Einstein equation. Different from EM techniques which look at the "dry size" of the nanomaterials, DLS gives the hydrodynamic radius which is a measure of how fast a particle diffuses in the solution. There are many factors that can cause the results from DLS significantly deviating from the true value. First, DLS cannot distinguish different shapes of particles with similar size. Since the experimental theory of DLS is based on the assumption that particles are spherical shape, a cubic shape nanoparticle that diffuses at the same rate as a spherical shape particle will have the same reported value in DLS. Second, DLS is known for its inaccuracy in analyzing multimodal samples which contain nanoparticles in several size ranges. According to the Rayleigh approximation, the scattering intensity of small particles ($d < \lambda/20$) is proportional to d^6 [34]. When a sample contains a mixture of nanoparticles that have a distinct size, the results will be strongly biased results toward to bigger size because of the intense scattering by the larger particles. Additionally, the relationship of scattering intensity with particle diameter indicates that contaminants such as dust can also ruin the analysis. Therefore, DLS is not recommended for the sample with a polydispersity index higher than 1.5, and cautions must be taken to avoid contamination. Third, since DLS measures the Brownian motion of nanoparticles in solution, solid, or highly viscous samples cannot be directly analyzed by DLS. They are either dispersed or diluted with appropriate solvents. However, this raises a new concern that the treatment can alter the initial state of the nanomaterials. For example, sonication is a common strategy to disperse nanomaterials and beak aggregations. Whether the results from DLS after the treatment can represent the real situation in the product is difficult to conclude. Nanoparticle tracking analysis (NTA) is another light scattering based particle size analysis technique. Similar to DLS, NTA also detects and analyzes the Brownian motion of laser-illuminated nanoparticles. Differently, NTA uses a microscope to track the movement of many individual particles. The hydrodynamic radius is then calculated through image analysis. Because NTA is a particle-by-particle measurement, it has shown better performance in analyzing polydispersed samples. In terms of the types of samples that can be analyzed, NTA has the same limitation as DLS which are restricted to liquid samples. Moreover, for NTA the samples need considerable dilution in order to avoid overcrowding under the microscope during the analysis. The effect of the dilution on the aggregation state of the sample, however, requires further studies. Furthermore, because very small particles do not scatter enough light to be detected by a microscope, the size limitation is around 10 nm for NTA whereas DLS is able to measure as small as 2 nm of nanoparticles.

11.4.5 Centrifugation

Centrifugation based techniques, such as ultracentrifugation, are routinely used in the study of biomacromolecules [35]. Compared to a number of nanoparticle sizing techniques discussed above, centrifugation, although less discussed in the literature, are better in dealing with a polydispersed system. Analytical ultracentrifugation (AUC), e.g. is a powerful tool in the characterization of multimodal nanoparticle mixtures. The separation by AUC is based on the density, shape and hydrodynamic size of the nanomaterials. Thus nanomaterials that have different morphology (e.g. nanorod with nanoparticle), composition (e.g. Ag with Au nanoparticles of the same size) and surface coating (e.g. naked with polymer coated particles) are all potentially separable in AUC. Additionally, the optical detection system in the AUC allows real-time and in situ monitoring of the sedimentation state of the particles. Another advantage of AUC is that the samples can be recovered and their integrity are not disturbed [35]. In additional to ultracentrifugation, other centrifugation techniques such as rate-zonal centrifugation in which the separation depends on the differences in hydrodynamic behavior were also reported [36]. In general, centrifugation based techniques give results with good confidence in a clean medium, and they are suitable for the characterization liquid form of products that contain nanomaterials, such as some environment samples and dietary supplements. When encountering nanomaterials that are embedded in solid or semi-solid matrices, however, the value of centrifugation is limited. Furthermore, centrifugation induced aggregation is also a problem for unstable nanomaterials [37].

11.4.6 Atomic force microscopy (AFM)

In AFM, the probe scans over the surface of the sample, and deflection responses coming from the probe tip interacting with the sample (direct contact or non-contact mode) are measured. The signals were processed into a 3D profile which reveals the topological details of the nanomaterials such as morphology, surface texture, and roughness. Different from TEM and SEM, AFM works at ambient condition, which allows for the study of liquid samples. Biological specimen can be directly studied by AFM without going through multiple dehydrations and embedding process. However, one should be cautious when using contact mode because the probe tip can get contaminated, especially for liquid samples. The oscillating property of the probe will be changed and cause artifacts [38]. This issue may be overcome by using the tapping mode (non-contact). There have been reports of characterization of nanomaterials in sunscreens and AFM was able to directly study the unmodified products (i.e. sample in its original state). However, it is not a facile process to determine the inorganic nanomaterials from the organic matrix with information acquired solely from AFM [39]. Moreover, when the size of the probe tip is larger than the nanoparticles being imaged, the lateral dimension will be overestimated, and only the measurement from the z-axis (i.e. height) should be used for best accuracy [34]. Furthermore, AFM is an imaging technique, which indicates that it has similar limitations as TEM and SEM in obtaining statistical information on the average size and size distribution of the nanomaterials.

Examination of multiple samples and scan of large areas might be necessary to achieve sufficient confidence. Additionally, samples that have a low concentration of nanomaterials in nature may create a challenge for AFM imaging analysis.

11.4.7 Separation techniques

In nanomaterials related studies, separation techniques can not only be considered as sample preparation methods which isolate nanomaterials from their matrices, they also provide valuable information that directly reflects the physicochemical properties of the materials. Many size based separation techniques have been developed to study different aspects of the nanomaterials such as hydrodynamic diameters, diffusion rate, and charge-to-size ratio. In these methods, many types of detectors such as UV–vis, light scattering, laser induced fluorescence, and mass spectrometry can be coupled online to the separation stage and provide real-time information of nanomaterials.

Liquid chromatography (LC)

A series of techniques that are based on chromatography have been used in the day-to-day analysis of numerous compounds for many years. The techniques can also be used for the separation and characterization of nanomaterials. For example, high performance liquid chromatography (HPLC) coupled with ICPMS has been applied in the separation and characterization of synthesized gold nanoparticles [40]. The same group also demonstrated that it is feasible to use reversed-phase HPLC for the speciation of silver ions and silver nanoparticles released from sport socks during washing processes [41]. Hydrodynamic chromatography (HDC) which separates particles based on their hydrodynamic size was also developed for the analysis of gold nanoparticle mixtures [42]. Chromatography methods are sensitive, fast, and low cost compared to EM methods. Nanomaterial samples that are in liquid form can be either directly injected or mixed with solvents prior to the analysis. This treatment, however, has the potential to change the sample environment in the original media. Solid samples must go through an extraction process to disperse the nanomaterials in the solvent before analyzed by LC methods. A universal problem in the column chromatography method is the analytes column interaction, which can result in reduced sample recovery. It is not uncommon to see recoveries in the range of 70–90% when analyzing nanomaterials. In addition, some reports suggested that the LC method is not very effective in separate multi-component nanoparticle mixtures [42]. Further development is needed to improve separation capacity and sample recoveries in order to enhance the accuracy of the LC method in quantitative analysis.

Asymmetric flow field flow fractionation (AF4)

AF4 has been heavily studies for the separation and characterization of many types of nanomaterials. Different from the column chromatography method, AF4 does not have a stationary phase, and the separation is dependent on the diffusion rate of the materials. After the separation by AF4, multiple detectors such as UV–vis, DLS, and ICPMS can be coupled to the system which provide a set of comprehensive information on the physicochemical characteristic of the analyzed nanomaterials [43]. Although AF4 has shown its advantages in separation, the membrane fouling issue strongly limited its application in the quantitative measurement. During the measurement, the injected samples were first concentrated to a very narrow band by apply a focusing flow. This focusing step usually results in an irreversible binding of the nanoparticles to the membrane causing particle loss and leading to reduced recovery. A possible solution to the problem is to functionalize both membrane and the nanoparticle be analyzed. It has been reported that this strategy efficiently increased the recovery of gold nanoparticles to almost 100% [44]. However, whether is it feasible to perform surface functionalization on all types of nanomaterials without affecting their original state is questionable. In AF4, the migration time can be correlated with the size of the particles. One should be very cautious when using the calibration curve established by one type of nanoparticles (e.g. polystyrene) to characterize a different type of particles (e.g. gold), because in addition to the diffusion rate of the particle, particle–membrane interaction also plays an important role in the separation of AF4. Two types of nanomaterials that have similar size may give distinct fractograms due to different profiles of particle–membrane interaction.

Capillary electrophoresis (CE)

CE has been a popular technique for the study of a variety of materials such as DNA, proteins, and macromolecules. In recent years, CE has also been utilized for the size characterization of many nanomaterials including quantum dots, metallic nanoparticles, dendrimers, and fullerenes [45–47]. Several studies have shown the linear relationship of nanoparticle size with migration time and nanoparticle concentration with peak area [46,48]. Because the mechanism of separation in CE is different from chromatography method, it is able to provide complementary results on the analysis of nanomaterials. So far, CE has demonstrated its advantages in high separation capacity, wide range of selectivity, low volume of sample requirement and better sample recoveries over other techniques. One of the major limitations of CE method lies in its lower repeatability compared to chromatography methods. Factors such as injection plug length, siphoning, solute-capillary surface interaction, Joule heating, and electrodispersion can all affect the precision [49]. The limit of detection in CE is usually at least one order higher than liquid chromatography method which is mainly due to the minute volume of sample (nL in CE compared with μl in HPLC). This poses a difficulty when analyzing environmental samples which commonly have very low concentration of nanomaterials. Although CE has shown excellent separation capacity for a broad range of molecules, the analysis of nanomaterials can still be a challenge. Since the electrophoretic mobility of an analyte is determined by its charge to size ratio, in order to separate different sizes of nanoparticles, the running electrolyte needs to be optimized to maximize the surface charge and maintain the stability of nanoparticles simultaneously. Any adjustment which changes the environment inside the capillary has the potential to affect the colloidal stability and alter the aggregation state of the nanomaterials, leading to different degree of artifacts.

11.4.8 Atomic spectrometry

Chemical composition is one of the most crucial factor that decides the toxic profile of a substance. Even though many nanomaterials have shown distinct properties from their bulk counterpart, their toxicities are still largely dependent on the atoms or molecules that constitute the materials. As for inorganic nanomaterials, since they can be partially or completely digested and release ions after entering a living organism, their elemental compositions must be determined in order to better understand the toxicity of the materials. Atomic Spectrometry, including atomic emission spectrometry (AES), atomic absorption spectrometry (AAS), and atomic fluorescence spectrometry (AFS), are three common techniques in the routine analysis of the elemental composition of samples. In recent years, with the continuous development, ICPMS has become more and more popular over other atomic spectrometry techniques in the elemental determination, by offering a better detection limit and sensitivity, capability of monitoring multiple isotopes, and easiness to interface with different separation techniques. It should be noted that atomic spectrometry is designed for the analysis for a variety of samples rather than specializing for nanomaterials. Nonetheless, they are still providing essential information that is closely associated with the physicochemical and biological properties of the nanomaterials.

From 2003 to 2006, Degueldre and coworkers published several papers, laying the theoretical foundation of characterization of colloidal nanoparticle dispersions using single-particle ICPMS (spICPMS) [50–52]. In single-particle analysis, when a solution with sufficient low particle concentration is introduced to the ICPMS, it is possible to create a series of discrete events where only a single particle enters the plasma at a time and gets ionized, generating a package of ions. By using very short integration time (10 ms in spICPMS as compared to 0.1–1 s in regular ICPMS), these transient signals can be monitored and translated into the size information of the nanoparticles. In contrast to traditional ICPMS where the total ion concentration dissolved in the solution was measured, spICPMS is considered a particle-by-particle based technique. Because spICPMS needs to work with an extreme diluted solution, it is an ideal technique for the characterization of nanomaterials at environmental relevant level. Samples such as river water can be directly analyzed by spICPMS without any pre-concentration steps. Several studies have demonstrated the possibility of using this technique for the characterization of Ag nanoparticle as well as dissolved Ag ions in waste water [53,54].

Although spICPMS has shown promising results in the analysis of ENMs, especially for some environmental samples, the development of such techniques is still at its infancy. Many factors such as dwell time and nebulization efficiency can strongly affect the results and appropriate instrument set up and careful data interpretation is critical in obtaining accurate measurements [55]. So far the smallest size that can be measured by spICPMS is around 20 nm for metallic nanoparticles (e.g. gold and silver nanoparticles). Moreover, most of these studies investigated standard solutions which contain purified single element nanoparticles, creating an ideal condition for the instrument to distinguish the ion pulse produced by the nanoparticle from the background signal [56]. In order to increase the detection limit and instrument ability to handle complex samples, improved detection efficiency is needed [55]. In addition to the measurement of small particles, caution should also be taken when analyzing large nanoparticles. For example, it was reported that a nonlinear relationship between the predicted size and signal response can occurred which causes a shift in the calculated size compared to TEM measurements when analyzing gold nanoparticles larger than 70 nm [55]. Under this circumstance, a reduced sensitivity mode may be used to extend the upper detection size range [55]. There are also a few intrinsic limitations associated with the spICPMS technique. For instance, it would be unlikely to perform a spICPMS analysis on a completely unknown sample because the size of the nanoparticle cannot be calculated without prior knowledge of the shape and the elemental composition of the nanoparticles [56]. This is especially true for environmental samples, since making an assumption of what the type of the nanoparticles is and what the geometry is, are not easy and may require assistance from other technique such as TEM. For samples that are at higher concentrations, dilutions are necessary for the analysis. Since colloidal nanoparticle systems are highly sensitive to the change of the environment, when the chemistry of the dispersing medium such as ionic strength, pH, and particle concentration is altered, whether nanoparticles can maintain their original physicochemical characteristics is difficult to predict without an in-depth investigation of the system.

11.4.9 Particle counting techniques

In addition to the characteristics such as size, shape, elemental composition, and crystal structure which mostly reflect the intrinsic properties of the nanomaterials, the number based particle concentration (the word "particle" is not restricted to only spherical shape nanoparticles, nanomaterials with other geometries are also included) is also a very important factor in risk assessment. It is frequently used in toxicity studies and exposure assessment as a metric and to correlate with the potential hazard. As for the determination of this value, spICPMS is a possible way to obtain the number based concentration of nanomaterials for some environmental samples. In addition to several limitations which were discussed above, spICPMS is not suitable for carbon based materials which is the case for many organic nanomaterials (e.g. polymers), and its application for non-liquid samples is also limited.

CPC is a widely used portable instrument in aerosol particle counting. It has proven to a useful device in analyzing airborne particles at the workplace by offering real-time measurements of particle number concentration and a very wide working range (2 nm–1 μm). Because CPC detects and counts particles by a scattering laser, it suffers a common limitation of all light scattering techniques which is the results that can be strongly biased due to the presence of contaminants. If there is a high level of background nanoparticle, the results from CPC can be significantly overestimated. Additionally, because in CPC, particles first grow to a bigger droplet through nucleation in supersaturated vapor before they reach to the detector, it is only capable of counting the particle number, and does not measure the size. Furthermore, overly high particle concentrations can induce heterogeneous nucleation and increase the variation in the results. Nonetheless, as long as other instruments are used to provide complementary information, CPC can still be a rapid and low cost method to obtain nanoparticle number concentration.

11.4.10 Sample handling and preparation

ENMs are seldom in thermodynamic equilibrium, and their size, shape, crystal structure, and even chemical compositions can change when their environment is altered or disturbed. As many conditions, including temperature, concentration, matrix, pH, ionic strength, oxidative or reducing agents can affect the stability of the ENMs, proper handling of the samples is extremely important in acquiring accurate information. As for the characterization of ENMs, although it is always a good idea to keep the sample preparation to a minimum and study them in situ, most of the instruments require different degrees of treatment before the analysis. If sample treatment is unavoidable, then the procedure should be carefully designed and each step needs to be recorded in detail in order to reduce the possible deviation caused by the artifacts.

Since most of the nanomaterials are in a complex matrix, such as food and cosmetic products, isolation of nanomaterials from the matrix is essential for certain techniques. Common strategies are digestion, extraction, purification, dispersion, and dilution. These methods, however, all introduce specific artifacts to the samples. Currently, there is a lack of studies addressing how the ENMs behave during the entire course of the measurement, and whether the properties of the nanomaterials can be preserved during the analysis is largely unknown. A possible way is to conduct a control study. First, pure nanomaterials, which is the same type as in the real sample, is examined in a set of comprehensive characterization, consisting of multiple instruments which measure different properties of the materials. Then a matrix blank is spiked with this type of nanomaterials that undergoes the exact same analysis procedure as for the real sample. Results from before and after the treatment are then compared to investigate whether the procedure will introduce any artifact to the characterization and how much deviation is generated by the procedure. Even a sample preparation procedure is proven to be useful for an analysis on certain type of nanomaterials, one should be very careful in applying the technique to other nanomaterials as their behavior in the same procedure can be significantly different. For instance, gold nanoparticles are relatively inert to pH and oxidation while silver nanoparticles can easily lose their stability and be converted to ionic form at low pH with high levels of dissolved oxygen in the solution [57].

Some aspects of the nanomaterials may also undergo slow changes during the storage. As in liquid form of the product, unstable nanoparticles may gradually aggregate together and form large agglomerations which eventually precipitate. The aggregation and sedimentation rate can be varied significantly depending on the types of the nanomaterials. Generally, nanomaterials are less stable at high temperature or under extreme pH condition, either acidic or alkaline. Some precipitations, which constitute loosely bound nanoparticles, can be redispersed by shaking, vortexing, or sonication while others, resulting from irreversible binding (e.g. fusing of metallic nanoparticles) of the particles are resistant to the redispersing treatment.

It is worth mentioning that although most of the sample preparations do change certain aspect of the nanomaterials, the characterization results can be considered as reflecting the real nature of the materials as long as the specific physical or chemical characteristics being analyzed are preserved during the treatment. A simple example would be acid digestion of the sample for a total elemental composition determination using ICPMS. Many features of the nanomaterials such as size, shape, aggregation state, and crystal structure are completely destroyed while the identity and concentration of a specific element remain unaffected.

In addition to the factors that directly influence the state of the nanomaterials, an important aspect that can be easily ignored is the heterogeneity of the sample itself. Making a conclusion based on the results that is obtained from measuring a heterogeneous sample is risky, as the analyzed portion may completely misrepresent the original sample. For example, physical sunscreens commonly use complicated formulations that contain a blend a multiple organic and inorganic substances. If the product is stored for some time, a separation can occur where the solid part sediment to the bottom, leaving a layer of oily substance floating on the top. If either individual layer is analyzed, the results may not reflect the situation where sunscreen is remixed and applied onto skin. A careful mixing by shaking, gentle stirring, short time vortexing, or even brief sonication can be used to homogenize the sample. However, the energy that will be introduced to the sample should be under control to minimize sample change during the treatment.

Nonetheless, strictly controlled sample preparation is vital to maintain the properties of the nanomaterials and obtain results without deviation. However, the current situation is that there is not a set of protocols that are universally accepted by different government entities as standard approaches for the sample preparation. More efforts are needed to increase cooperation among different regulatory bodies and initiating harmonization on the standard operating protocol is urgent.

11.4.11 Minimum analytical characterization of nanomaterials

From the discussion above, we are aware of the fact that many well-established material characterization techniques are potentially useful for the analysis of ENMs in the products, and they all have different types of limitations. Additionally, an individual method can only provide information on a very limited number of characteristics, which cannot fully describe the properties of the materials [58]. The bottom line is that there is no single method that can determine all aspects of the ENMs, and a comprehensive characterization that uses multiple complementary techniques is a feasible approach for practical purposes. Furthermore, due to the fact that the basic principle of measuring a nanomaterial varies between techniques, it is recommended that for the determination of a given property, at least two characterization techniques are used in order to reduce the possibility of obtaining strongly biased results from one method [58].

11.5 Current regulatory status

11.5.1 U.S. Food and Drug Administration (FDA)

A wide range of products in the public health related area, including food, food packaging materials, veterinarian products, medicines, medical devices, and cosmetics, have seen the strong application of nanotechnologies. Most of these consumer products are under the jurisdiction of the FDA, which is the main agency responsible for protecting and promoting public health. The utilization of nanotechnology in the products created new challenges to the FDA, because the unique physicochemical properties and interactions with the biological system of nanomaterials may pose unexpected harm to humans even though their bulk counterparts were thoroughly investigated. In August 2006, a nanotechnology task force was formed within the FDA with the mission of identifying knowledge and policy gaps related to nanotechnology and giving recommendation on how to address these issues [59]. The task force released a report in 2007, and the general finding is that the regulatory challenges presented by nanotechnology based products are similar to others that also utilize emerging technologies. The situation, however, can be further complicated by the fact that the properties and safety of nanomaterials may vary as size changes within the nanoscale [7]. In 2011, FDA issued a draft guidance on nanotechnology entitled "Considering Whether an FDA-Regulated Product Involves the Application of Nanotechnology" providing industry with the agency current thinking on nanotechnology related issue [60]. In this guidance document, FDA identifies two points to consider when determining whether a product involves the use of nanotechnology: (1) the engineered material has at least one external dimension in the nano-scale (1–100 nm) and (2) the engineer material shows dimension specific properties. The concept discussed in this document is somehow similar to the definition of nanomaterials developed by Health Canada. Two more product specific guidances in food and cosmetic products were issued in 2012 [61,62]. After receiving comments from the public, the final guidance documents on these topics were released in 2014 [63]. The information described in the FDA guidance documents is not regulation and is treated as nonbinding recommendation. The purpose of these guidance documents is to present the agency's thinking on considerations related to nanotechnology [64]. In general, the FDA intends to maintain its product-focused, science-based regulatory policy and regulate nanotechnology based products under its current framework. All nanomaterials related products that require pre-market approval such as new drugs, biological, and food additives will subject to the same procedure as regular products with additional attention to the special effect nanomaterials may bring to the efficacy and safety related questions. For products that do not require pre-market approvals, industry is responsible for ensuring that products are safe and voluntarily submitting any relevant information to the agency.

11.5.2 U.S. Environmental Protection Agency (EPA)

There are multiple environmental laws such as the Clean Air Act (CAA) or Clean Water Act (CWA) that could be potentially used by the EPA to regulate nanomaterials.

However, due to varied limitations in different laws, most of nanomaterials, which are regarded as chemical substances, are regulated under the Toxic Substances Control Act (TSCA) [6]. In January 2008, the EPA released a document entitled "TSCA Inventory Status of Nanoscale Substances – General Approach" which describes how the EPA currently determines whether a nanoscale substance is a "new" or "existing" chemical substance under TSCA [65]. In this document, the EPA reiterates that current inventory approaches will continue to be applied and molecular identity rather than physical attributes will be used to determine whether a chemical substance is considered as a new chemical. In April 2015, the EPA proposed an information gathering rule regarding reporting and recordkeeping requirements for chemical substances when they are manufactured or processed at the nanoscale [66]. This rule will require manufactures to electronically report information including specific chemical identity, production volume, methods of manufacture and processing, exposure and release information, and existing data concerning environmental and health effects to EPA. The purpose of this rule is to facilitate EPA in evaluating and determining its future action on nanomaterials. This rule is currently receiving comments from the public until August 5, 2015. In addition to TSCA, the EPA can also regulate certain nanomaterials under the Federal Fungicide, Insecticides and Rodenticide Act (FIFRA), particularly for products that contain nanoscale silver due to their antimicrobial property [6]. In March 2015, the EPA responded to a petition which was filed in 2008 by the International Center for Technology Assessment (ICTA) and 13 other organizations, requesting the EPA to regulate nanoscale silver as a pesticide and to assess them as new and different from macrosilver products [67]. The EPA granted the request to treat nanoscale silver as pesticide if intended for pesticidal purposes. In fact, in the same month when the response was released, the EPA started to take action and stopped the selling of two nanoscale silver containing products marketed by the company for anti-microorganism purpose because the products contain a unregistered pesticide-silver nanoparticles [68]. However, the EPA denied the request to treat all nano-silver as a pesticide and stated that the agency will continue to apply its statutory and regulatory criteria as to what is a pesticide on a case-by-case basis.

Other U.S. federal regulatory agencies such as Consumer Product Safety Commission (CPSC), U.S. Department of Agriculture (USDA), and Occupational Safety and Health Administration (OSHA) also have jurisdiction over certain scenarios that involves nanomaterials. For example, OSHA is in charge of establishing standards for industry to protect workers from hazards that may cause death or serious physical harm. Because occupational exposures to nanomaterials, especially through inhalation at nanomaterial manufacturing facilities, are usually much more significant than consumer exposure by using nano-products, OSHA is also responsible to prevent the harm that may be caused by the nanomaterials. In general, similarly to FDA and EPA, other government agencies have no intention to create a complete new framework to regulate nanomaterials containing products. Instead, with the acknowledgment of the presences of challenges, most of regulatory bodies will reply on the current statutory for nanomaterial related issues, and carefully determine their actions in each individual case.

11.5.3 European Union (EU)

As early as 2004, the EC determined that nanotechnology has become one the fasted growing field with great promising, and huge impact will bring to various aspects of human life. The Commission decided to start a conversation on nanotechnology related issues among the members and purposed a strategy which described what approach the EU would like to use in promoting the safe and responsible development of nanotechnology [69]. Later, an action plan which provided a series of detailed actions for the strategy purposed by the EU in 2004 was developed from 2005 to 2009 [70]. In the first review on regulatory aspects of nanomaterials, which was prepared in 2008, the committee concluded that current legislation systems are able to cover most aspects of nanomaterial related issues, especially for products that require pre-market approval or registration [71]. For example, under the current Registration, Evaluation, Authorization and Restriction of Chemicals (REACH) system, all chemical substances that are manufactured or imported above 1 ton/year must be registered. Nanomaterials which are within the definition scope of "substance" are also covered by REACH. In case of the form of chemical substances are changed from bulk to nanomaterials, the registration document must be updated. The committee also acknowledged that for products (e.g. cosmetics) that do not go through pre-market procedural, current legislation may need to be modified to include precautionary measures using a different mechanism to address the potential risk that may be posed by nanomaterials. Additionally, the committee also emphasized that the current information on nanomaterials is insufficient and the knowledge gaps related to nanomaterials concentrate in the characterization method and risk assessment method. However, with the large amount of investment in nanotechnology related research our knowledge about nanomaterials is growing at a fast speed and regulatory change may be needed once new evidences are gained. As a results of the continued development of legislation framework for nanotechnology, the EU issued the definition of nanomaterials in 2011 [10]. Currently, Biocidal Products Regulation (BPR) has specific provision for nanomaterials where nano-form of an active substance may require a separated market the approval even if the bulk form was already approved [72]. On December 13, 2014, EU amended its Food labeling Legislation with one key change related to nanomaterials. The new regulation will require "ingredients present in the form of engineered nanomaterials shall be clearly indicated in the list of ingredients. The names of such ingredients shall be followed by the word 'nano' in brackets" [73]. According to the second regulatory review on nanomaterials, the committee concluded that most of current regulations such as REACH and Classification, Labelling and Packaging regulation (CLP) were able to offer a good framework in addressing nanomaterials related issues, and some modifications to the existing provision may be necessary [74].

11.5.4 China

Besides the United States and EU, China is also a major player in the research and development of nanoscience and nanotechnology. China first released its official definition of nanomaterials in 2004. The terminology is similar to what proposed by Health Canada, which defines nanomaterials as a material that has at least one dimension in the nanometer scale (1–100 nm) or the materials constitute nanostructure and exhibit special properties [75]. In the same year, China issued another six nano-related standards which provide a technical definition of four types of nanomaterials that are commonly used in industry (i.e. nano nickel, zinc oxide, calcium carbonate, and titanium dioxide) and two characterization methods for nanomaterials [76]. Later on, the Chinese government recognized the complexity of the issues generated by the growth in both fundamental research and applications in industrial sectors, and determined that the continued development of national standards in nanotechnology is crucial. A series of nano related standards were released in the later years [77]. In 2011, the Standardization Administration of the People's Republic of China (SAC) announced a plan to establish 17 national standards in nanotechnology including "the Standard on the Health and Safety Practices in Occupational Settings Relevant to Nanotechnologies". However, among many of the published standards or standards that are in draft form, most of them are voluntary or categorized as recommendations, and only a handful of mandatory standards are in place to actually regulate nanotechnology related issues.

11.5.5 Other parts of the world

Countries that have a fast development and large investment in nanotechnology such as Japan and Korea also have started to act on related issues by unifying terminologies, proposing strategic plans, drafting guidance documents, and establishing standards. In general, there is a lot of discussion and conversation within or among different relevant parties and much less action is taken to effectively address the problems. Due to the big challenges posed by nanotechnology, not only academia, government agencies and various industrial sectors should work together to establish a more efficient framework to promote research activities in nanotechnology and to identify and tackle risks associated with the technology, international cooperation should also be encouraged and improved. All major global players in nanotechnology should create a system that allow for better communication, collaboration, and information sharing so that every member can benefit from the healthy and efficient development of nanotechnology.

References

[1] http://www.nano.gov, accessed February 2015.

[2] http://www.nanowerk.com, accessed May 5th 2015.

[3] Evers P., *Nanotechnology in Medical Applications: The Global Market*. BCCResearch report HLC069B, 2012.

[4] Project on Emerging Nanotechnologies. *Nanotechnology Consumer Products Inventory*. http://www.nanotechproject.org, accessed May 4th 2015.

[5] The author have noticed that the PEN database has not been updated since 2013. While some of the information may be outdated this database is a

valuable resource for the public to get an good look at nanotechnology based products that are or have been introduced to the market place.

[6] Chatterjee R., 'The challenge of regulating nanomaterials'. *Environmental Science & Technology*. 2008, 42, (2), 339–343.

[7] U.S. Food and Drug Administration. *Nanotechnology Task Force Report*. 2007.

[8] The International Organisation for Standardization 2008. *Nanotechnologies – Terminology and definitions for nano-objects – Nanoparticle, nanofibre and nanoplate*. ISO/TS 27687.

[9] Health Canada (Canada). *Interim Policy Statement on Health Canada's Working Definition for Nanomaterials*. 2009.

[10] European Commission. *Commission Recommendation of 18 October 2011 on the Definition of Nanomaterial*. 2011.

[11] Linsinger T.P.J., Chaudhry Q., Dehalu V., *et al.*, 'Validation of methods for the detection and quantification of engineered nanoparticles in food'. *Food Chemistry*. 2013, 138, (2–3), 1959–1966.

[12] Xu Y., Qin Y., Palchoudhury S., and Bao Y., 'Water-soluble iron oxide nanoparticles with high stability and selective surface functionality'. *Langmuir*. 2011, 27, (14), 8990–8997.

[13] Hansen S., Michelson E., Kamper A., Borling P., Stuer-Lauridsen F., and Baun A., 'Categorization framework to aid exposure assessment of nanomaterials in consumer products'. *Ecotoxicology*. 2008, 17, (5), 438–447.

[14] Ambrosone A., Mattera L., Marchesano V., *et al.*, 'Mechanisms underlying toxicity induced by CdTe quantum dots determined in an invertebrate model organism'. *Biomaterials*. 2012, 33, (7), 1991–2000.

[15] Nel A.E., Madler L., Velegol D., *et al.*, 'Understanding biophysicochemical interactions at the nano-bio interface'. *Nature Materials*. 2009, 8, (7), 543–557.

[16] Lewinski N., Colvin V., and Drezek R., 'Cytotoxicity of nanoparticles'. *Small*. 2008, 4, (1), 26–49.

[17] Patra H.K., Banerjee S., Chaudhuri U., Lahiri P., and Dasgupta A.K., 'Cell selective response to gold nanoparticles'. *Nanomedicine: Nanotechnology, Biology and Medicine*. 2007, 3, (2), 111–119.

[18] Oberdörster G., Maynard A., Donaldson K., *et al.*, and A report from the ILSI Research Foundation/Risk Science Institute Nanomaterial Toxicity Screening Working Group, 'Principles for characterizing the potential human health effects from exposure to nanomaterials: elements of a screening strategy'. *Particle and Fibre Toxicology*. 2005, 2, (8), 1–35.

[19] Kroll A., Pillukat M., Hahn D., and Schnekenburger J., 'Interference of engineered nanoparticles with in vitro toxicity assays'. *Archives of Toxicology*. 2012, 86, (7), 1123–1136.

[20] Lam C.-w., James J.T., McCluskey R., Arepalli S., and Hunter R.L., 'A review of carbon nanotube toxicity and assessment of potential occupational and environmental health risks'. *Critical Reviews in Toxicology*. 2006, 36, (3), 189–217.

[21] Qiao R., Yang C., and Gao M., 'Superparamagnetic iron oxide nanoparticles: from preparations to in vivo MRI applications'. *Journal of Materials Chemistry*. 2009, 19, (35), 6274–6293.

[22] Rivera-Gil P., Jimenez De Aberasturi D., Wulf V., *et al.*, 'The challenge to relate the physicochemical properties of colloidal nanoparticles to their cytotoxicity'. *Accounts of Chemical Research*. 2013, 46, (3), 743–749.

[23] Soo Choi H., Liu W., Misra P., *et al.*, 'Renal clearance of quantum dots'. *Natural Biotechnology*. 2007, 25, (10), 1165–1170.

[24] Choi H.S., Liu W., Liu F., *et al.*, 'Design considerations for tumour-targeted nanoparticles'. *Natura Nano*. 2010, 5, (1), 42–47.

[25] Kagan V.E., Bayir H., and Shvedova A.A., 'Nanomedicine and nanotoxicology: two sides of the same coin'. *Nanomedicine: Nanotechnology, Biology and Medicine*. 2005, 1, (4), 313–316.

[26] Fischer H.C. and Chan W.C.W., 'Nanotoxicity: the growing need for in vivo study'. *Current Opinion in Biotechnology*. 2007, 18, (6), 565–571.

[27] Abbott L.C. and Maynard A.D., 'Exposure assessment approaches for engineered nanomaterials'. *Risk Analysis*. 2010, 30, (11), 1634–1644.

[28] Richards I.S., *Principles and Practice of Toxicology in Public Health*. Sudbury: Jones and Bartlett Publisher; 2007.

[29] Maynard A.D. and Aitken R.J., 'Assessing exposure to airborne nanomaterials: current abilities and future requirements'. *Nanotoxicology*. 2007, 1, (1), 26–41.

[30] Ono-Ogasawara M., Serita F., and Takaya M., 'Distinguishing nanomaterial particles from background airborne particulate matter for quantitative exposure assessment'. *Journal of Nanoparticle Research*. 2009, 11, (7), 1651–1659.

[31] Sánchez-Pomales G., Mudalige T.K., Lim J.-H., and Linder S.W., 'Rapid determination of silver in nanobased liquid dietary supplements using a portable X-ray fluorescence analyzer'. *Journal of Agricultural and Food Chemistry*. 2013, 61, (30), 7250–7257.

[32] U.S. Environmental Protection Agency. *Environmental Technology Verification Report. Field Portable X-ray Fluorescence Analyzer*. EPA/600/R-97/144-EPA/600/R-97/151. 1998.

[33] Yeo M.-K. and Kang M., 'The biological toxicities of two crystalline phases and differential sizes of TiO_2 nanoparticles during zebrafish embryogenesis development'. *Molecular & Cellular Toxicology*. 2012, 8, (4), 317–326.

[34] Hassellöv M., Readman J., Ranville J., and Tiede K., 'Nanoparticle analysis and characterization methodologies in environmental risk assessment of engineered nanoparticles'. *Ecotoxicology*. 2008, 17, (5), 344–361.

[35] Calabretta M., Jamison J.A., Falkner J.C., *et al.*, 'Analytical ultracentrifugation for characterizing nanocrystals and their bioconjugates'. *Nano Letters*. 2005, 5, (5), 963–967.

[36] Akbulut O., Mace C.R., Martinez R.V., *et al.*, 'Separation of nanoparticles in aqueous multiphase systems through centrifugation'. *Nano Letters*. 2012, 12, (8), 4060–4064.

[37] Calzolai L., Gilliland D., and Rossi F., 'Measuring nanoparticles size distribution in food and consumer products: a review'. *Food Additives & Contaminants: Part A*. 2012, 29, (8), 1183–1193.

[38] Tiede K., Boxall A.B.A., Tear S.P., Lewis J., David H., and Hassellöv M., 'Detection and characterization of engineered nanoparticles in food and the environment'. *Food Additives & Contaminants: Part A*. 2008, 25, (7), 795–821.

[39] Tyner K.M., Wokovich A.M., Doub W.H., *et al.*, 'Comparing methods for detecting and characterizing metal oxide nanoparticles in unmodified commercial sunscreens'. *Nanomedicine*. 2009, 4, (2), 145–159.

[40] Helfrich A. and Bettmer J., 'Analysis of gold nanoparticles using ICP-MS-based hyphenated and complementary ESI-MS techniques'. *International Journal of Mass Spectrometry*. 2011, 307, (1–3), 92–98.

[41] Soto-Alvaredo J., Montes-Bayón M., and Bettmer J., 'Speciation of silver nanoparticles and silver(I) by reversed-phase liquid chromatography coupled to ICPMS'. *Analytical Chemistry*. 2013, 85, (3), 1316–1321.

[42] Gray E.P., Bruton T.A., Higgins C.P., Halden R.U., Westerhoff P., and Ranville J.F., 'Analysis of gold nanoparticle mixtures: a comparison of hydrodynamic chromatography (HDC) and asymmetrical flow field-flow fractionation (AF4) coupled to ICP-MS'. *Journal of Analytical Atomic Spectrometry*. 2012, 27, (9), 1532–1539.

[43] Schmidt B., Loeschner K., Hadrup N., *et al.*, 'Quantitative characterization of gold nanoparticles by field-flow fractionation coupled online with light scattering detection and inductively coupled plasma mass spectrometry'. *Analytical Chemistry*. 2011, 83, (7), 2461–2468.

[44] Mudalige T.K., Qu H., Sánchez-Pomales G., Sisco P.N., and Linder S.W., 'Simple functionalization strategies for enhancing nanoparticle separation and recovery with asymmetric flow field flow fractionation'. *Analytical Chemistry*. 2015, 87, (3), 1764–1772.

[45] Carrillo-Carrión C., Moliner-Martínez Y., Simonet B.M., and Valcárcel M., 'Capillary electrophoresis method for the characterization and separation of CdSe quantum dots'. *Analytical Chemistry*. 2011, 83, (7), 2807–2813.

[46] Qu H., Mudalige T.K., and Linder S.W., 'Capillary electrophoresis/inductively-coupled plasma-mass spectrometry: development and optimization of a high resolution analytical tool for the size-based characterization of nanomaterials in dietary supplements'. *Analytical Chemistry*. 2014, 86, (23), 11620–11627.

[47] Chan K.C., Patri A.K., Veenstra T.D., McNeil S.E., and Issaq H.J., 'Analysis of fullerene-based nanomaterial in serum matrix by CE'. *ELECTROPHORESIS*. 2007, 28, (10), 1518–1524.

[48] Liu F.-K. and Wei G.-T., 'Adding sodium dodecylsulfate to the running electrolyte enhances the separation of gold nanoparticles by capillary electrophoresis'. *Analytica Chimica Acta*. 2004, 510, (1), 77–83.

[49] Laure H.H. and Rozing G.P., *High Performance Capillary Electrophoresis*. Agilent Technologies; 2010. 2nd edn.

[50] Degueldre C. and Favarger P.Y., 'Colloid analysis by single particle inductively coupled plasma-mass spectroscopy: a feasibility study'. *Colloids and Surfaces A: Physicochemical and Engineering Aspects*. 2003, 217, (1–3), 137–142.

[51] Degueldre C., Favarger P.Y., and Bitea C., 'Zirconia colloid analysis by single particle inductively coupled plasma–mass spectrometry'. *Analytica Chimica Acta*. 2004, 518, (1–2), 137–142.

[52] Degueldre C., Favarger P.Y., and Wold S., 'Gold colloid analysis by inductively coupled plasma-mass spectrometry in a single particle mode'. *Analytica Chimica Acta*. 2006, 555, (2), 263–268.

[53] Tuoriniemi J., Cornelis G., and Hassellöv M., 'Size discrimination and detection capabilities of single-particle ICPMS for environmental analysis of silver nanoparticles'. *Analytical Chemistry*. 2012, 84, (9), 3965–3972.

[54] Pace H.E., Rogers N.J., Jarolimek C., *et al.*, 'Single particle inductively coupled plasma-mass spectrometry: a performance evaluation and method comparison in the determination of nanoparticle size'. *Environmental Science & Technology*. 2012, 46, (22), 12272–12280.

[55] Liu J., Murphy K.E., MacCuspie R.I., and Winchester M.R., 'Capabilities of single particle inductively coupled plasma mass spectrometry for the size measurement of nanoparticles: a case study on gold nanoparticles'. *Analytical Chemistry*. 2014, 86, (7), 3405–3414.

[56] Pace H.E., Rogers N.J., Jarolimek C., Coleman V.A., Higgins C.P., and Ranville J.F., 'Determining transport efficiency for the purpose of counting and sizing nanoparticles via single particle inductively coupled plasma mass spectrometry'. *Analytical Chemistry*. 2011, 83, (24), 9361–9369.

[57] Liu J. and Hurt R.H., 'Ion release kinetics and particle persistence in aqueous nano-silver colloids'. *Environmental Science & Technology*. 2010, 44, (6), 2169–2175.

[58] Bouwmeester H., Lynch I., Marvin H.J.P., *et al.*, 'Minimal analytical characterization of engineered nanomaterials needed for hazard assessment in biological matrices'. *Nanotoxicology*. 2011, 5, (1), 1–11.

[59] U.S. Food and Drug Administration. *FDA Forms Internal Nanotechnology Task Force*. 2006.

[60] U.S. Food and Drug Administration. *Draft Guidance for Industry: Considering Whether an FDA-Regulated Product Involves the Application of Nanotechnology*. 2011.

[61] U.S. Food and Drug Administration. *Draft Guidance for Industry: Assessing the Effects of Significant Manufacturing Process Changes, including Emerging Technologies, on the Safety and Regulatory Status of Food Ingredients and Food Contact Substances, Including Food Ingredients that Are Color Additives*. 2012.

[62] U.S. Food and Drug Administration. *Draft Guidance for Industry: Safety of Nanomaterials in Cosmetic Products*. 2012.

[63] U.S. Food and Drug Administration. *FDA Issues Guidance to Support the Responsible Development of Nanotechnology Products*. 2014.

[64] U.S. Food and Drug Administration. *Nanotechnology Fact Sheet*. http://www.fda.gov/ScienceResearch/SpecialTopics/Nanotechnology, accessed July 2015.

[65] U.S. Environmental Protection Agency. *TSCA Inventory Status of Nanoscale Substances – General Approach*. 2008.

[66] U.S. Environmental Protection Agency. *Chemical Substances When Manufactured or Processed as Nanoscale Materials: TSCA Reporting and Recordkeeping Requirements*. 2015.

[67] U.S. Environmental Protection Agency. *EPA Responds to Petition on Nanoscale Silver Regulation*. 2015.

[68] U.S. Environmental Protection Agency. *EPA Takes Action to Protect the Public from an Unregistered Pesticide/EPA Issues Order to Stop the Sale of BioStorm and NanoStrike*. March 2015.

[69] European Commission. *Towards a European Strategy for Nanotechnology*. 2004.

[70] European Commission. *Nanosciences and Nanotechnologies: An Action Plan for Europe 2005–2009*. 2005.

[71] European Commission. *Regulatory Aspects of Nanomaterials*. 2008.

[72] Nanomaterials under Biocidal Products Regulation. http://echa.europa.eu/regulations/nanomaterials-under-bpr, accessed June 2015.

[73] European Union. *Regulation (EU) No. 1169/2011 of the European Parliament and of the Council on the Provision of Food Information to Consumers*. 2011.

[74] European Commission. *Nanomaterials in REACH and CLP*. http://ec.europa.eu/environment/chemicals/nanotech/reach-clp/index_en.htm, accessed June 2015.

[75] General Administration of Quality Supervision, Inspection and Quarantine of the People's Republic of China. *Announcement of Newly Approved National Standards of P.R. China: Terminology for Nano Materials*. GB/T 19619-2004. 2004.

[76] General Administration of Quality Supervision, Inspection and Quarantine of the People's Republic of China. *Announcement of Newly Approved National Standards of P.R.China: Notice No. 2004-10*. GB/T 13221-2004; GB/T 19587-2004 to GB/T 19591-2004.

[77] Jarvis D.S. and Richmond N., *Regulation and Governance of Nanotechnology in China: Regulatory Challenges and Effectiveness*. 2011.

Chapter 12

The scientific rationale for using silicon nitride in biomedical implants

Giuseppe Pezzotti[1], *Bryan McEntire*[2], *Wenliang Zhu*[3] *and B. Sonny Bal*[2,4]

Abstract

In order for synthetic materials to be utilized as biomedical implants, their short- or intermediate-term biocompatibility must be validated. However, most biomaterials are not aptly assessed for their ability to successfully interface with the human milieu long-term. To be optimally effective, a biomaterial's state of biocompatibility and tissue integration should increase with time. Yet, the standard for selection of suitable materials today remains primarily bioinertness. In this regard, ceramics are popular because of their perceived stability within the human body. However, complete bioinertness for any material is virtually impossible; and acknowledging this fact is an important first step in the development of better biomedical devices. In this chapter, advanced spectroscopic techniques were employed to delve deep into the performance of several advanced bioceramics used in prosthetic joint implants. In contrast to the conventional notion that these materials are completely bioinert, it was found that surface chemical and structural changes readily occur upon contact with biological fluids. Furthermore, it was discovered that their "non-bioinertness" can be either beneficial or detrimental with respect to long-term reliability. In particular, the peculiar properties of silicon nitride, a non-oxide ceramic, make it a unique candidate for artificial joints and other biomedical devices. In contrast to oxide ceramics, the surface chemistry and mechanical behavior of silicon nitride appear to ideally lead to improved long-term performance.

[1]Ceramic Physics Laboratory, Kyoto Institute of Technology, Sakyo-ku, Matsugasaki, 606-8585 Kyoto, Japan
[2]Amedica Corporation, Salt Lake City, UT, USA
[3]Department of Medical Engineering for Treatment of Bone and Joint Disorders, Osaka University, 2-2 Yamadaoka, Suita, Osaka 565-0854, Japan
[4]Department of Orthopaedic Surgery, University of Missouri, Columbia, MO, USA

12.1 Introduction

Total joint arthroplasty (TJA) has become a commonly applied orthopedic therapy used to relieve pain and improve mobility for patients suffering from arthritis, osteoporosis, congenital deformities, tumors, and similar maladies coterminous with advanced age. The large majority of patients receiving prosthetic devices express satisfaction with their implants in the short- and intermediate-term [1–3]. In fact, some widely used prostheses have demonstrated 10-year survival rates of greater than 90% [4]. However, for the majority of patients, statistics also show that revision surgeries to replace original devices typically occur within 10–15 years of their index surgeries [5]. It is not uncommon for patients to endure two or even three revisions during their lifetime. This is not only a consequence of both younger patients and increased population longevity, but also a result of inherent prosthetic malfunctions including aseptic loosening, osteolysis, infection, dislocation, fracture, and a number of other related failures [6].

Unfortunately, biomaterial deterioration is either directly or indirectly the cause for most surgical revisions and replacements. Other factors include surgical errors, poor implant design, mistakes during implant manufacturing, or recalls. Revision surgeries are more complicated, costly, and have higher comorbidities than their former index procedures. They also typically require highly experienced surgeons to remove and replace the original implant along with debridement of nearby bone and soft tissue. Even "successful" joint replacement surgeries will eventually end in revisions because the patient's lifespan will generally exceed the functional life of the implant. To minimize the potential for revision surgeries, continuous improvements are being sought in joint design, surgical techniques, and biomaterials, with the eventual goal of creating a lifetime implant. Consequently, the main challenge for biomaterial scientists is the development of reliably durable biomaterials that eliminate the possibility of revision surgeries even for "successful" hip implants.

Using standard definitions for bioinertness, traditional ceramic oxides (e.g., alumina, Al_2O_3, and yttria-stabilized zirconia, ZrO_2, or Y-TZP) are among the most suitable materials for joint implants [7,8]. A reasonably new ceramic composite made of Al_2O_3 and ZrO_2 (i.e., zirconia-toughened alumina or ZTA) is reported to be even more bioinert; and currently considered the "gold standard" in ceramic arthroplasty bearings [9]. Nevertheless, the bioinertness of these oxide materials is clearly a matter of definition. Despite their improved durability, implants made of these bioceramics are also routinely removed during revision surgery. Their long-term non-bioinert nature became even more evident when screening for oxygen stability at the atomic level [10–12]. Also, recent studies on advanced polyethylene counterparts have brought attention to the pivotal role of oxidation chemistry on the expected lifetime of these prosthetic components [13,14]. Consequently, the bar must be raised on the definition of bioinertness beyond classic mechanical and tribological properties (i.e., high fracture strength and low wear). While realistically accepting the concept that no material is truly bioinert (cf. an early paper by Williams) [15], a paradigmatic shift in surface engineering is required to increase the durability of bearing components. Of necessity, this change must take into account the pivotal roles of surface reactivity and long-term off-stoichiometry. Bearing couples with adequate physical, mechanical, and tribological performance are necessary but insufficient. To achieve elongated prosthetic implants into the third and fourth decade under the extreme physicochemical conditions of the human body (i.e., contact stress, temperature, and corrosion), a complete understanding of surface tribochemistry and its evolution with time are essential. This information will enable fine tuning of the biomaterial's surface chemistry at the molecular and atomic scale.

In this chapter, new criteria are set-forth for the complex interactions between artificial bearing surfaces and the biological environment. While revisiting concepts on surface chemistry, toughness, and oxidative degradation, emphasis is placed on molecular-scale processes which lead to surface degradation during in vivo exposure. Using experimental data, an improved definition of "bioinertness" will be developed encompassing not only the macroscopic retention of bulk integrity, but also the molecular level resistance to surface degradation. The importance of local physical, mechanical, and chemical properties at the bearing surface will be emphasized. It will be demonstrated that there are marked differences between properties measured at the surface and within the bulk of today's biomaterials. Owing to the ionic nature of their chemical bonds, oxide ceramics profoundly interact with the biological environment and polyethylene counterparts in non-beneficial ways. A non-oxide bioceramic, silicon nitride (Si_3N_4), is highlighted as a potential new material for improving the functional in vivo life of joint prostheses due to its unique surface chemistry and the covalent character of its bonds [16–18]. Although it is also not a bioinert material, its biological interactions appear to be highly beneficial.

12.2 Oxygen activity in the anaerobic environment

Recently, certain physical chemistry aspects of oxide ceramics have been unveiled which may impact their long-term performance as artificial joints [19]; and they are key to understanding the intrinsic lifespan limits of these materials. Their aspects will be presented and discussed in the three following sub-sections. At first glance, the three topics appear independent of each other, but all possess a common element related to oxygen activity in the tribochemical layer in an otherwise anaerobic environment.

12.2.1 Oxidation in polyethylene liners

In 2010, Muratoglu *et al.* published an enlightening analysis of explanted acetabular liners made of ultra-high molecular weight polyethylene (UHMWPE) [13]. Prior to their implantation, these liners were irradiated at 100 kGy and remelted to fully annihilate residual free radicals (Longevity®, Zimmer, Warsaw, IN, USA). The surprising outcome of this study was that in vivo chemical changes resulted in increased free radical formation, a loss of crosslinking, and oxidation. Even after their explantation, the liners continued to oxidize. Their finding contradicts the widely held belief that free radicals are negligible in a melted liner [13]. If the presence of free radicals in these highly cross-linked and annealed UHMWPE liners were indeed eliminated (which is supported by oxidation analyses on long-term shelf-stored liners in air and by real-time ageing in aqueous environments at homeostatic temperatures) [13,20], then a different mechanism must be responsible for their in vivo and ex vivo degradation. Accordingly, a chemical rather than mechanical origin was postulated for the observed free-radical formation and ensuing oxidation. It was reasoned that the number of loading cycles in vivo would have been too limited to explain oxidation as a loading fatigue-related phenomenon. It was therefore hypothesized that lipids (e.g., squalene) absorbed from the synovial fluid were the source of the free radicals and ensuing oxidation. It is known that UHMWPE absorbs lipids from synovial fluid in vivo [14]. These lipids will start to oxidize as soon as they come into contact with oxygen molecules either in vivo or ex vivo (e.g., on the shelf in presence of air after removal from the patient in the case of the report by Muratoglu *et al.* [13]). In support of this hypothesis, it was observed that even very short durations (e.g., 2 months) of in vivo exposure led to high levels of oxidation after long-term ex vivo storage [13]. The unsaturated C=C bonds in the lipid molecules (e.g., cholesterol, hexadecanoic and octadecanoic esters of cholesterol-like cholesteryl stearate, squalene, or other unsaturated precursors in cholesterol synthesis) [21] tend to react with the saturated polyethylene molecules and initiate degradation of the host polymer (cf. Step 1 in Figure 12.1). Note that peroxidation of squalene can be initiated by different chemical species, such as metal ions, reactive oxygen species, or hydroxyl free radicals present in the human body [22]. Formation of peroxy free radicals in the lipid molecules occurs because of their reaction with oxygen, which in turn leads to the formation of hydroperoxides by abstracting hydrogen atoms from the surrounding polyethylene chains (cf. Step 2 in Figure 12.1). This attack by lipids forms new free radicals in polyethylene, preferentially in its amorphous phase, and their high mobility in the amorphous phase quickly propagates oxidation. This results in a significant increase in crystallinity, chain scission, formation of a lower molecular weight polyethylene, and recrystallization. In fact, this is the physical circumstance that allows non-destructive oxidation measurements of UHMWPE by Raman spectroscopy [23]. Our analyses of retrieved irradiated and melted liners are in full agreement with the findings of Muratoglu *et al.* [13].

As an example, Figure 12.2(a) shows the infrared spectroscopy protocol for analyzing in-depth oxidation profiles in an irradiated/melted UHMWPE liner retrieved after medium-term in vivo exposure of 4.1 years and a post-explantation shelf period of 4.2 years. The analyzed liner articulated against a CoCr femoral head (i.e., the same implant as investigated by Muratoglu *et al.*) [13]. In Figure 12.2(b), infrared spectroscopy data were used to plot the oxidation index (OI) as a function of the in-depth abscissa, z, along the liner thickness at various locations in the main-wear and non-wear zones. These data clearly show that high levels of oxidation (i.e., up to $\mathrm{OI} = 6$) can be found in irradiated and melted liners. Moreover, data collected in the main-wear zone showed a pronounced sub-surface peak of oxidation (at around $z = 600$ μm) when compared to the non-wear zone (cf. locations 1, 2, 3, and 4 with locations 5 and 6). This difference demonstrates that

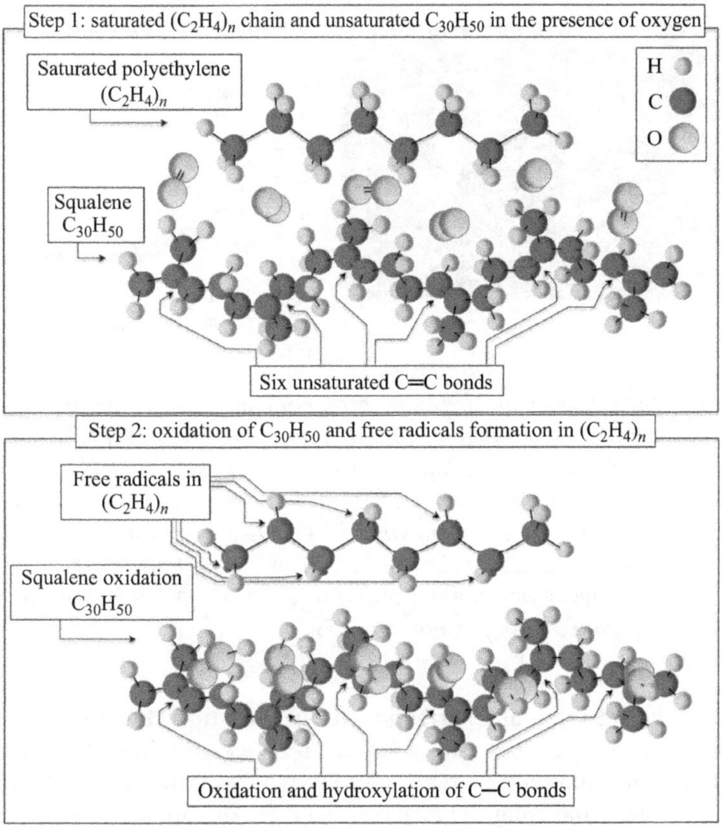

Figure 12.1 Chemical mechanisms involved with polyethylene oxidation, which initiates with the absorption of lipids from the synovial fluid. (© 2015 Elsevier, Reprinted with permission [179])

in vivo loading actually exacerbates free radical formation. The pioneering findings by Muratoglu *et al.* [13] are of fundamental importance in future developments of joint arthroplasty not only because they revealed a previously overlooked mechanism for free radical formation in UHMWPE (which greatly decreases its oxidative stability), but also because they triggered a new line of research into antioxidant vitamin E doped UHMWPE liners [24–26]. It is believed that these important findings will also prompt new research in the ceramic field as well.

Our recent spectroscopic analyses fully support oxidation of identical liners as investigated by Muratoglu *et al.* [13]. However, unlike their interpretation, the dependence of the OI on exposure time was not erratic. Our experimental results indicated that the OI was primarily determined by in vivo implantation time,

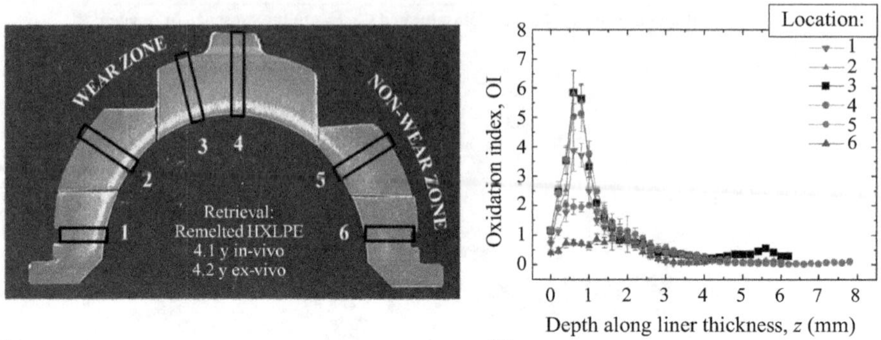

(a) (b)

Figure 12.2 (a) Infrared spectroscopy protocol employed for analyzing in-depth profiles of oxidation in an irradiated/melted UHMWPE liner retrieved after medium-term in vivo exposure of 4.1 years and a post-explantation on shelf period of 4.2 years; (b) infrared spectroscopy data showing the OI, as a function of the in-depth abscissa, z, along the liner thickness at various locations in the main wear and non-wear zones (cf. (a)). (© 2015 Elsevier, Reprinted with permission [179])

although we cannot rule out contributions from shelf time. Figure 12.3(a) and (b) are graphs of the OI parameter for a set of 11 Longevity®liners for short-term (7 retrievals) and medium-term (4 retrievals) in vivo exposure, respectively. Elapsed shelf-time before completing the analyses, with maximum and average OI values for the main-wear and non-wear zones is also provided. OI values were found to consistently increase with cumulative in vivo exposure, while they were erratic with respect to on-shelf exposure. The overall picture for the data shown in Figure 12.3 suggests that free radicals are actually formed in the UHMWPE structure during in vivo exposure. Their formation is cumulative and more pronounced in the main-wear zone. Free radicals readily turn to oxidation sites as soon as oxygen becomes available. Whether or not oxygen availability occurs during or after implantation is a matter of speculation, although it is conceivable to expect a hypoxic environment in joint articulation.

Figure 12.4(a) and (b) are plots of the degree of oxidation as a function of the in-depth abscissa, *z*, for two different types of highly cross-linked polyethylene liners: (i) gamma-irradiated and melted (Longevity® Zimmer, Warsaw, IN, USA) and (ii) gamma-irradiated with a dose of 100 kGy, vitamin E infused, and annealed (E1® Biomet, Inc, Warsaw, IN, USA), respectively. The pristine samples were treated in different ways prior to infrared spectroscopy measurements, including accelerated ageing (100% O_2, under 5 kPa, at 70°C, for 28 days), gamma-irradiation (25 kGy) followed by accelerated ageing (same as above), and accelerated ageing while soaked in squalene. Infusion of vitamin E into the E1® liners was performed to counteract free radical formation. Vitamin E's anti-oxidant property arises from

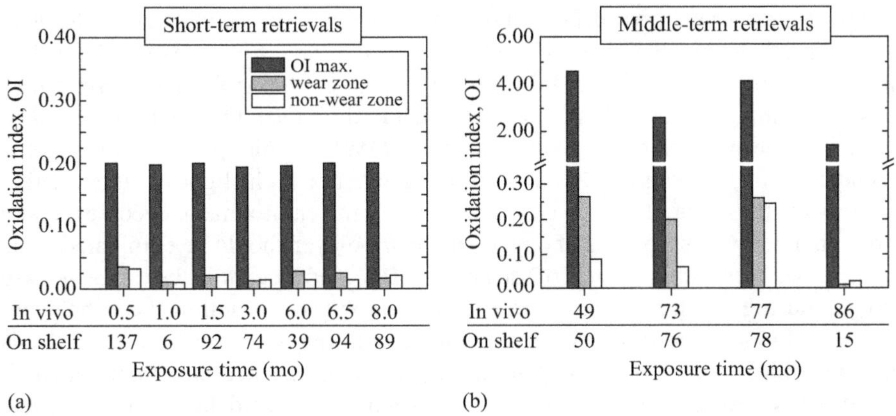

Figure 12.3 Plots of OI as obtained by infrared spectroscopy on a set of 7 short-term retrievals (a) and a set of 4 medium-term retrievals (b). Average data for main-wear and non-wear zones are compared with the maximum oxidation recorded in mapping the retrievals according to the protocol shown in Figure 12.2(a)

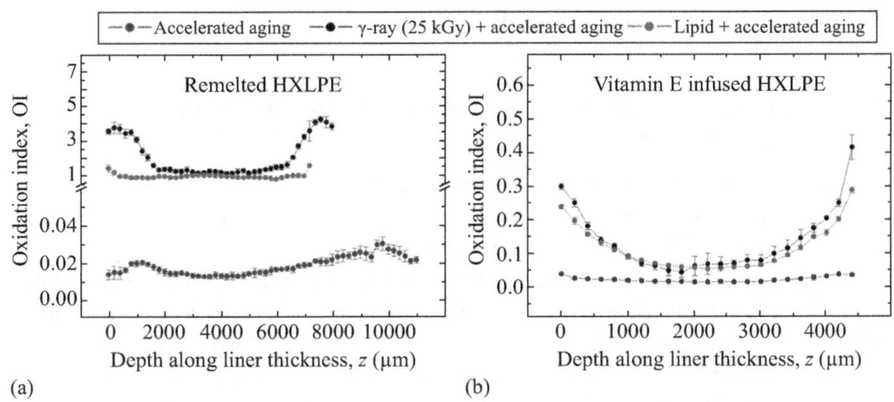

Figure 12.4 Plots of OI as a function of in-depth abscissa, z, for two different types of highly cross-linked polyethylene liners (Longevity® manufactured by Zimmer (a) and E1®, manufactured by Biomet, Inc). The pristine samples were treated in different ways before performing infrared spectroscopy measurements, as shown by symbols in the inset

the presence of an alcoholic hydroxyl group attached to the chroman ring of its structure. The alcoholic hydroxyl has a tendency to give away its hydrogen and become a stabilized phenoxy radical. The free hydrogen quickly annihilates one free radical in the UHMWPE, according to the so-called peroxyl radical trapping mechanism. The surface spectroscopic data in Figure 12.4 clearly show that an in vitro hydrothermal treatment with the concomitant presence of squalene leads to oxidation of vitamin E infused UHMWPE liners, although to a lesser extent than undoped remelted liners [27,28]. Improved in vivo oxidation for short-term retrievals of vitamin E-infused liners has been reported [29,30]. Our own in vitro data show that the propensity for oxidation of UHMWPE in the presence of lipids is attenuated but not completely suppressed by vitamin E. In light of its oxidative propensity, the amount of oxygen within the joint environment becomes a key issue. In other words, sources of oxygen in the tribolayer should be eliminated. This presents an interesting series of questions: Do oxide ceramic bearings release oxygen during service? If so, where does the oxygen go within the tribolayer? There is obviously no direct toxicity involved with the release of oxygen; but this does not mean there are no long-term negative consequences for joint function. UHMWPEs strong intrinsic affinity for oxygen is triggered by the unavoidable presence of free radicals in its structure. Consequently, bearing surfaces prone to release oxygen can hardly be considered "polyethylene friendly."

In conclusion, given the propensity for all UHMWPE liners to oxidize during their second or third decade of in vivo service, it is critical to minimize all sources of oxygen within the tribolayer of an otherwise anaerobic human joint. Note that this is a direct consequence of the findings by Muratoglu *et al.* [13]. Accordingly, judging the oxidative stability of UHMWPE liners based solely on the concentration of preexisting free radicals needs to be reconsidered.

12.2.2 Oxygen vacancies and sub-valent cations in the alumina lattice

The nascent wear surfaces of artificial joints are chemically active, having a wide variety of energetically favorable tribochemical reactions [31,32]. In the case of Al_2O_3 bearings, the hydrothermal environment induces lattice dissociation at its surface [33]. This phenomenon is exacerbated by its polycrystalline nature, which incorporates an internal network of off-stoichiometric and crystallographically faulted grain boundaries. Dissociation of Al_2O_3 produces hydroxyls and proton radicals, which in turn promote formation of surface vacancies to preserve electrical neutrality [11,34]. In fact, Elam *et al.* [35] demonstrated the existence of oxygen vacancies through modeling hydroxyl coverage of a (0001) sapphire surface. Rigorously understanding the "intrinsic" surface properties of alumina ceramics requires correctly interpreting the ubiquitous presence of hydrogen at the molecular level [36–47]. Molecular H_2O and surface hydroxyl groups are generally present because of the adsorption of water from the environment [35,47–60]. The low coordination of Al atoms makes these sites strong electron acceptors. Acceptor sites readily adsorb and dissociate H_2O molecules. While adsorption also occurs in a metastable molecular binding mode, dissociative adsorption is more favorable than molecular adsorption and is primarily heterolytic in nature. This means that adsorbed H_2O can be viewed as splitting into H^+ and OH^-, with the proton transferred to a nearby surface oxygen within the lattice [61]. During frictional sliding of alumina surfaces, even under ideal conditions (i.e., without including third-body wear or microseparation processes) [62,63], continuous formation, dissociative adsorption, and removal of hydroxylated layers repeatedly create and annihilate oxygen vacancies in the outer surface of the alumina lattice. This is a mechanochemically activated process. In simulator studies, the hydrophilicity of sliding Al_2O_3 counterparts minimizes the surface adsorption, denaturation, and irreversible unfolding of proteins [64–66]. However, in vivo, the lack of synovial fluid in the hip joint likely creates an extremely severe thermomechanical environment [67], in which modifications of the alumina lattice (i.e., especially in the vicinity of grain boundaries) increasingly take place. Protein by-products (e.g., hydrocarbons) [68] and/or released ions sub-valent with respect to Al^{3+} (e.g., Ca^{2+}, Mg^{2+}, and Na^{+}) induce irreversible stoichiometric alterations of the alumina lattice. For example, it is known that dilution of sodium hyaluronate (with the release of Na^{+} ions) takes place in patients affected by rheumatoid arthritis, as systematically detected by Dahl *et al.* [69]. Although the effect of these tribochemical factors on the wear of alumina bearings has not been discussed in detail, evidence for vacancy formation and for other interactions between the biological environment and the alumina lattice has been found. In the long-run, these tribochemically driven events lead to extensive modifications of the alumina lattice with concomitant alterations to local mechanical properties and wear resistance. In vitro and in vivo experimental data have been obtained using the high spatial resolution of a cathodoluminescence probe at the nanometer scale [10,11]. The use of this shallow probe illuminates the complex cascade of chemical and crystallographic events affecting in vivo wear of alumina (i.e., from the formation of point defects to plastic flow within the lattice). The fact that Al_2O_3 tends to release oxygen from its surface is the main consequence of its hydroxylation. The formation of an increasing concentration of oxygen vacancies with progressive hydrothermal exposure in vitro is shown in Figure 12.5(a-c). The cathodoluminescence spectrum emitted by Al_2O_3 contains important information about the stoichiometry of its oxygen sub-lattice [70–72]. In the ultra-violet region, a doublet consisting of partly overlapping bands is observed (cf. Figure 12.5(a)). The predominant band (at around 325 nm or 3.8 eV), referred to as F^{+}-center band, is emitted from a single oxygen vacancy trapping one electron. The other weaker band of the doublet is the F-center band. It is located at around 350 nm (3.0 eV) and is generated by a single oxygen vacancy trapping two electrons. The presence of the F^{+}/F oxygen vacancy doublet in the CL spectra is common to all Al_2O_3 surfaces. Given identical irradiation conditions, changes in observed intensities are due to the different vacancy concentrations. Higher intensities directly correlate with a higher concentration of vacancies. The CL spectrum of an alumina femoral head (BIOLOX®forte, CeramTec GmbH, Plochingen, Germany) already displays the F^{+}/F doublet in its as-received state. These are Schottky defects that serve as electric charge compensators. Figure 12.5(b) and (c) shows the change in CL intensity (i.e., increase in oxygen-vacancy concentration) as a function of autoclave exposure (H_2O at 121°C and 2 bar pressure, for up to 300 h) for an as-received Al_2O_3 femoral head (BIOLOX®forte). Figure 12.5(c) actually demonstrates that alumina is not bioinert, and that hydroxylation-induced vacancy formation actually occurred at its surface, leading to an

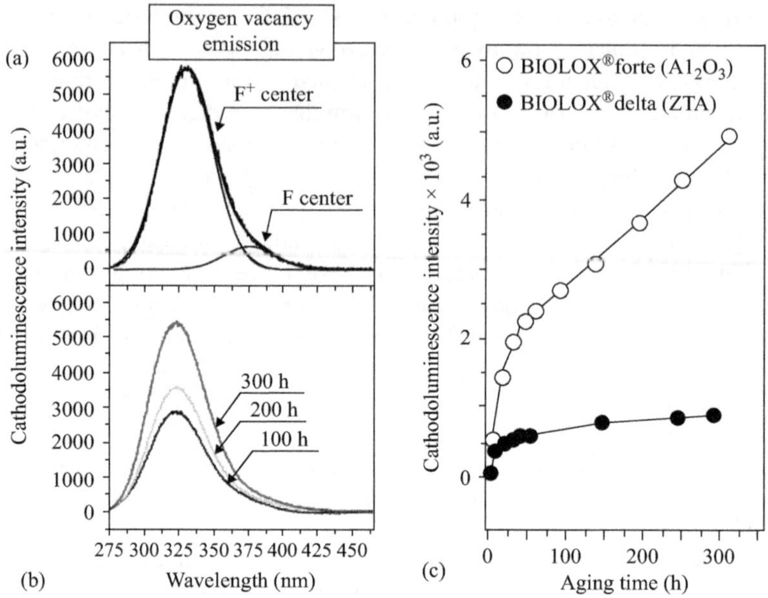

Figure 12.5 (a) The cathodoluminescence spectrum emitted by Al_2O_3 bioceramics in the ultra-violet region deconvoluted into a doublet consisting of partly overlapping bands; (b) evolution of CL intensity (i.e., oxygen-vacancy concentration) upon exposure to a hydrothermal environment for as-received Al_2O_3 femoral heads (BIOLOX®forte). (c) Ageing experiments were conducted in an autoclave operating at 121°C (under 2 bar pressure) in water vapor environment for Al_2O_3 (BIOLOX® forte) and ZTA (BIOLOX® delta) femoral heads

increasingly off-stoichiometric state. Under in vivo tribochemical loading, substantial off-stoichiometry then triggers incorporation of sub-valent cations (i.e., Ca^{2+}, Mg^{2+}, and Na^{+}) from the environment into the lattice structure. Divalent cations substitute for Al^{3+} ions, forming a complex F_c-center, namely a F^+-center adjacent to a divalent cation site. F_c-centers of divalent cations emit a distinct cathodoluminescence spectrum located within a relatively narrow wavelength at around 300 nm [72]. Moreover, a doublet emission in the wavelength interval 450–600 nm corresponds to the formation of a $MgAl_2O_4$ spinel phase with higher ductility than Al_2O_3 [73]. Indeed, these bands are systematically found in retrieved Al_2O_3 femoral heads. As an example, a wear map collected on a self-mated BIOLOX®forte femoral head, exposed in vivo for 7.7 years, is shown in Figure 12.6(a). Its deconvoluted spectrum (Figure 12.6(b)) reveals its surface off-stoichiometry features. In comparing spectra collected on different wear zones (Figure 12.6(c)) with that of an unexposed reference sample (cf. Figure 12.5(a)), complex F_c-center defects (including divalent cations) and the $MgAl_2O_4$ spinel phase were found to form in vivo. Note that their phase fractions increase with the severity of wear (cf. Figure 12.6(a) and (c), where grade 1 to grade 5 indicate mild

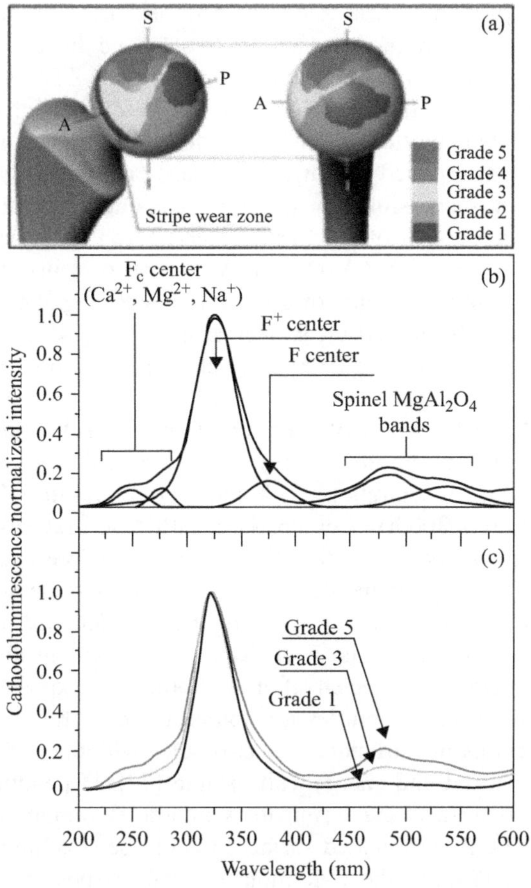

Figure 12.6 (a) Wear map collected on a self-mating BIOLOX®forte femoral head, in vivo exposed for 7.7 years; (b) deconvoluted cathodoluminescence spectrum from the Al_2O_3 retrieval; and (c) comparison among cathodoluminescence spectra collected in zones of the retrieval in (a) with different wear damage

to severe wear, respectively) [74]. This evidence suggests that the alumina lattice is intrinsically limited in its tribochemical resistance and is prone to microscopic degradative wear mechanisms. The demonstration that free oxygen emanates from Al_2O_3 under tribochemical wear suggests that there are few physical reasons to hope for extended lifetimes (i.e., to the second to third decade) from these self-mated bearings.

12.2.3 Chemical triggers to polymorphic transformation of zirconia

The latest technological development in bioceramics for arthroplasty applications is ZTA composites. From the viewpoint of material design, they merge biocompatibility and phase stability of alumina with the enhanced mechanical properties of zirconia. This is unique among ceramic oxides and is achieved by exploiting a peculiar mechanism known as transformation toughening [75,76]. ZTA composites possess improved mechanical properties when compared to biomedical grades of monolithic Al_2O_3. One composite in particular (BIOLOX®delta, CeramTec GmbH, Plochingen, Germany) has gained unrivaled worldwide popularity in the bioceramic market with more than 10 years of successful clinical history. The microstructural composition of this ZTA includes a fraction of 17 vol% of sub-micron ZrO_2 particles which are partially stabilized with ~0.6 wt% Y_2O_3. The Al_2O_3 matrix is mixed with a fraction of 0.3 wt% Cr_2O_3 and a minor amount of SrO. During sintering, the elements of Y and Cr dissolve into the ZrO_2 and Al_2O_3 lattices, respectively, while Sr is incorporated into a distinct third phase, $SrAl_{12-x}Cr_xO_{19}$, which represents about 3 vol% of the final sintered body.

From a stoichiometric point of view, addition of Cr_2O_3 improves the surface stability of the composite. Figure 12.5(c) shows the evolution of its CL intensity (i.e., oxygen-vacancy concentration) with autoclave exposure (H_2O at 121°C and 2 bar pressure, for up to 300 h). A comparison with Al_2O_3 reveals the fundamental role that Cr plays in hampering surface hydroxylation. Its presence strengthens the oxygen sub-lattice by keeping the off-stoichiometry level at the biomaterial surface much lower than that of monolithic Al_2O_3. This is demonstrated by the cathodoluminescence intensities of the F^+/F band, which remain about one order of magnitude lower than Al_2O_3, even after hydrothermal exposures lasting up to 300 h. A range of oxygen stabilities are found among different crystallographic planes within the corundum structure (i.e., α-Al_2O_3), with the (0001) surfaces being the most densely packed and energetically stable [77]. Depending on preparation conditions, α-Al_2O_3 surfaces can terminate with both aluminum and oxygen atoms [78]. However, while Al-terminated surfaces tend to be stoichiometric and remain energetically stable [79,80], the O-terminated surfaces possess the highest reactivity and become stabilized only upon hydroxylation [81,82]. The free surfaces of polycrystalline α-Al_2O_3 artificial joints possess a random orientation of crystallographic planes. Accordingly, the level of hydroxylation will vary depending on local grain orientation and surface roughness. Consequently, measuring the average degree of hydroxylation by monitoring oxygen vacancies using CL spectroscopy requires a statistically meaningful number of measurements (e.g., $>10^4$) at different random locations.

When the hydroxylated surface is subjected to a hydrothermal cycle, neighboring (thermally activated) O—H bonds of hydroxylated molecules start to interact with each other and tend to form new water molecules. Accordingly, hydroxyl groups are removed from the alumina surface and coordinated unsaturated anions (i.e., oxygen) and cations (exposed Al^{3+} and anion vacancies) are formed. In other words, removal of hydroxyl groups (i.e., dehydroxylation) from the surface occurs concurrently with the formation of oxygen vacancies, as schematically shown in Figure 12.7. Several different configurations are possible for the oxide surface at its interface with the biological fluid including: (i) A terminal OH

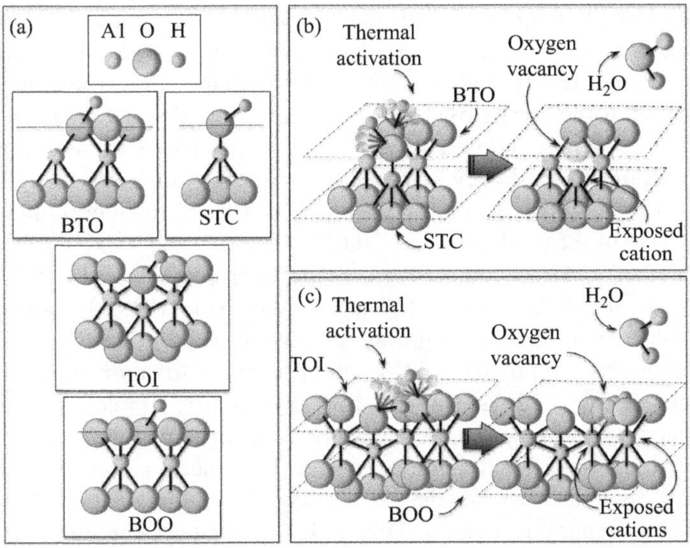

Figure 12.7 *(a) Types of hydroxyl group that quickly develop on the surface of Al_2O_3 in a hydrothermal environment (abbreviations defined in text). During dehydroxylation, (b) STC and BTO nearest neighboring sites recombine to generate a water molecule, concurrently with the formation of one anion vacancy and the exposure of one threefold coordinated Al^{3+} cation; and (c) BOO and TOI nearest neighboring sites recombine to generate a water molecule while producing one anion vacancy and two exposed fivefold coordinated Al^{3+} cations at tetrahedral sites*

group coordinated to a single tetrahedral Al^{3+} cation (referred to as STC); (ii) A bridging OH group linking a tetrahedral and an octahedral cation (BTO); (iii) A bridging OH group linking two cations in octahedral positions (BOO); and (iv) A bridging OH group coordinated to three cations in octahedral interstices (TOI). Illustrations of these different molecular configurations coexisting on the alumina surface are provided in Figure 12.7(a). During dehydroxylation, two main structural rearrangements take place: (a) STC and BTO nearest neighboring sites recombine to generate a water molecule, an anion vacancy, and one threefold coordinated Al^{3+} cation. This latter cation then becomes exposed in a tetrahedral interstice together with an oxygen bridging a tetrahedral and an octahedral cation and (b) BOO and TOI nearest neighboring sites recombine to generate a water molecule while producing one anion vacancy and two exposed fivefold coordinated Al^{3+} cations at tetrahedral sites. The pristine and the recombined structures of the two types (a) and (b) of the dehydroxylated site are schematically shown in Figure 12.7(b) and (c), respectively. In this context, the ease of OH group removal parallels the OH site basicity. This is due to the fact that the net positive charge at the anion vacancy (i.e., hole) is lower than the net negative charge of the removed OH group. Proton acidity and the ease of OH group removal are the two main factors governing the dehydroxylation process. They play a fundamental role in determining the "degree of bioinertness" of oxide surfaces. Cr ions are isovalent to Al but, according to Rashkeev *et al.* [83], differences in bond strength between Cr—O/Cr—H as compared to Al—O/Al—H are crucial in determining the dehydroxylation rate of the oxide surface. $CrO_2 \rightarrow CrO + O$ and $AlO \rightarrow Al + O$ require bond energies of ~5.50 eV and 4.98 eV, respectively; and Cr—H and Al—H have bond energies of ~4.70 eV and 2.95 eV, respectively [84,85]. Note that these marked differences in bond strength are commonly exploited in the field of catalytic dehydrogenation processes. Chromium doping to alumina is adopted to facilitate abstracting hydrogen atoms from hydrocarbon molecules because the energy gained in forming the new Cr—H bond is larger than the energy lost in breaking the C—H bond. Similarly, this difference in bond strength is less energetically favored when isovalent Cr^{3+} enters the alumina lattice. It also results in less oxygen vacancy formation. This is the physical reason why a lower population of F^+/F centers oxygen vacancy defects are observed on the surface of BIOLOX®delta doped with Cr_2O_3 as compared to monolithic undoped BIOLOX®forte. Note that this is a strong point in support of improved durability for this Cr^{3+}-doped ZTA composite. Yet, despite this fundamental improvement, degradation is still observed on the surface of BIOLOX®delta components both in hip simulation and in vivo [86–89]. Indeed, early experimental in vitro data confirmed this degradative phenomena in ZTA [90]. Moreover, recent research on BIOLOX®delta retrievals demonstrates an exacerbation of the in vivo degradative trend observed in vitro as a consequence of tribochemical effects at articulation surfaces. As discussed in the section below, metastability of Y-TZP is important for toughening and strengthening [91]; but, this material suffers from the presence of chemical triggers to its polymorphic transformation. Unfortunately, the metastability of BIOLOX®delta was neglected by researchers until experimental evidence became available in the published literature [90]. Regardless of successive statements by the manufacturer emphasizing their ability to fully control the $t \rightarrow m$ phase transformation [92], data on pristine BIOLOX®delta components from independent researchers show large fluctuations in monoclinic content depending on their year of fabrication and, presumably, on the production lot [89]. Our own analysis of pristine ZTA heads produced in 2014 is in line with the findings of these previous authors. Average histograms of the monoclinic fraction before and after in vitro hydrothermal exposure (200 h at 121°C, 1 bar water-vapor pressure) are shown in Figure 12.8(a). The volumetric percentage of the monoclinic phase was higher than reported by the manufacturer [92] (cf. histogram in Figure 12.8(a)) along with a decreasing trend in its content. Note that the amount of the monoclinic phase doubled after long-term exposure within the autoclave. In rationally considering the observed compositional fluctuations of BIOLOX®delta, it should be noted that fabrication of fine ceramics is certainly not easy; but, in the case of ZrO_2-containing materials, stoichiometry issues are crucial due to the high

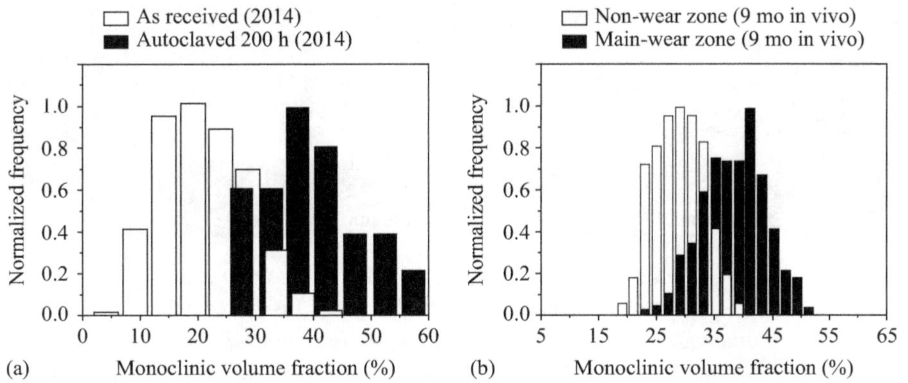

Figure 12.8 (a) Average histograms of monoclinic fraction for femoral heads BIOLOX®delta produced in 2014, before and after in vitro exposure to a hydrothermal environment for 200 h at 121°C (under 1 bar water-vapor pressure) and (b) Average histograms of the monoclinic fraction for a BIOLOX®delta femoral head, from a ceramic-on-polyethylene tripolar implant that was implanted for only 2 months and retrieved for dislocation. Comparison is made between data obtained from the main-wear and the non-wear zones

propensity for local distortion of its oxygen sublattice [93,94]. According to previously published activation energy data for this ZTA material [95], several hundred hours in an autoclave under the above conditions should correspond to more than a few human lifetimes at 37°C in the body, provided that hydrothermal loading is the only trigger for the polymorphic transformation. However, significantly higher monoclinic fractions were systematically found when examining in vivo retrievals of BIOLOX®delta femoral heads [88,89]. As a typical example, Figure 12.8(b) shows the surface phase stability of a retrieved BIOLOX®delta femoral head after only 2 months in vivo. The head was explanted due to dislocation. A statistical comparison made between the main-wear and non-wear zones found a phase composition comparable to the 200-h in vitro autoclaved sample shown in Figure 12.8(a) [95]. Obviously, hydrothermal loading cannot be the only trigger for Y-TZP phase instability; other in vivo chemical causes are crucially important. Note that this explanted femoral head came from a ceramic-on-polyethylene tripolar prosthesis (Active Articulation™E1®Dual Mobility Hip System; Biomet®Orthopedics, Warsaw, IN, USA). The metal cup (M^2a-Magnum™) was an as-cast high carbon cobalt chrome molybdenum (CoCrMo) alloy with a press-fit design. The inner geometry of the cup was highly polished; while a titanium porous plasma spray (PPS®) layer covered the outer surface. The PPS® coating consisted of irregularly shaped titanium particles, which were formed from the impaction and solidification of molten droplets on the substrate surface. As shown in Figure 12.9(a), a cursory visual inspection did not reveal any major damage, which seems reasonable given that the ceramic

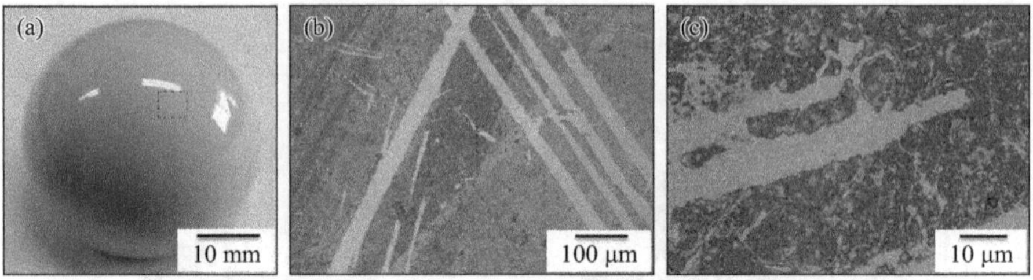

Figure 12.9 *(a) Retrieved BIOLOX®delta femoral head from a ceramic-on-polyethylene tripolar implant; (b) and (c) represent laser micrographs of the head surface, which showed contamination by titanium debris detached from the porous titanium (PPS®) external layer of the CoCrMo metal cup. Ti debris particles remained entrapped between the ceramic/polyethylene sliding counterparts. Micrograph (b) gives an enlargement of the squared area in (a), while (c) enlarges the squared area in (b)*

articulates against a soft counterpart. However, microscopic imaging revealed Ti metal contamination throughout the entire head surface (cf. Figure 12.9(b) and (c)). While the unfortunate migration of Ti debris from the PPS® coating (as plastically deformed "soft" inclusions) to the ceramic–polyethylene interface was easily observed, the destabilizing chemical interaction between Ti and the zirconia lattice was not so obvious. Destabilization of tetragonal ZrO_2 in the presence of Ti is a well-known phenomenon [96]. Tetragonal-to-monoclinic transformation of Y-TZP is actually accelerated by titania doping. Its destabilizing effect has been attributed to a reduced binding energy between the Ti-dopant and oxygen vacancies as compared to Zr sites [96]. The negative consequences of this uncontrolled polymorphic transformation at the bearing surface of ZTA, which results in surface embrittlement and wear damage, will be presented later in this chapter.

In conclusion, our spectroscopic studies revealed significant fluctuations in the content of the *m*-ZrO_2 polymorph for as-manufactured ZTA heads, which points to the inherent difficulty in consistently controlling its surface phase composition. Unexpectedly, high amounts of the monoclinic phase were also found after very short-term exposures in vivo, which were comparable with hundreds of hours of in vitro hydrothermal exposure. Conversely, the oxygen sub-lattice at the surface of ZTA appears "locked" with respect to the Al_2O_3 matrix phase due to incorporation of Cr^{3+}; and yet this bioceramic is still vulnerable to the in vivo environment because of the Y-doped zirconia phase. Consequently, "full control" over phase instability in ZTA bearings appears far from being realized. Moreover, the complex tribochemical events occurring in vivo were demonstrated to be multifaceted, with little attention paid to chemical triggers. Clearly the in vivo behavior of ZTA bioceramics lacks adequate and predictive knowledge.

12.3 Potential long-term advantages of non-oxide ceramics

From basic principles of physical chemistry, it is known that water dissolves many ionic compounds because of its molecular polarity [97]. The mechanism behind this phenomenon is an electrical charge interaction between ionic bonds and the water dipole. On the one hand, this interaction may lead to breaking of the ionic molecule by hydroxylation or dissociative chemisorption as seen within hydrothermal environments. On the other hand, non-polar covalent compounds cannot directly form hydrogen bonds and are therefore expected to have little or no interaction with water. According to this reasoning, use of covalently bonded ceramics in principle appears to be a straightforward solution for improving the bioinertness of orthopedic bearings. In the remainder of this chapter, the characteristics and behavior of non-oxide ceramics are explored, which will inevitably point to Si_3N_4 as a primary candidate among covalent-bonded biomaterials for arthroplasty applications [16–18].

12.3.1 Silicon nitride: a polyethylene-friendly ceramic

There are several intriguing interactive aspects between Si_3N_4 surfaces and the biological environment which may directly impact both the suitability and performance of this biomaterial in the human body. For instance, the tribochemical interaction between silicon nitride and polyethylene favorably differs from oxide ceramics and polyethylene. Silicon nitride surfaces are found to terminate with a nearly monolayer of silicon oxide due to atmospheric exposure [98]. The initial O_2 chemisorption rate is relatively fast and saturation occurs quickly. Chemisorption of H_2O on clean Si_3N_4 surfaces is slower than oxygen, but equally leads to the formation of chemically stable oxide or oxynitride layers in which the surface H:O ratio has been reported to be 1:1 [99]. Tsuruk and Bliznyuk [100] reported strong adhesive and frictional force dependencies for chemically modified silicon nitride surfaces, depending on whether the surface terminal groups were CH_3, NH_2, or SO_3H. These products directly form upon chemisorption of silane-based compounds. Upon immersion in water, the surface silica layer tends to develop a negative charge due to adsorption of hydroxyl radicals. Experimental evidence for improved water lubrication of ceramics at lower sliding velocities was reported in an intriguing study by Hibi and Enomoto [101], where silane coupling agents were used during self-mated sliding of Si_3N_4 and Al_2O_3 in water. The silane agents reduced friction in water both for Si_3N_4 and Al_2O_3 couples. However, unlike Al_2O_3, Si_3N_4 generally showed significant wear reductions, except for the isolated case of a silane coupling-agent containing an epoxy group. The interaction between the silane coupling-agents and the ceramic surfaces provides fundamental information about the roles of silanol groups and polysiloxane film formation on the wetting characteristics of these ceramic surfaces [102]. The amphoteric silica formed on Si_3N_4 acts as an Arrhenius acid with water being the corresponding Arrhenius base. Its surface charge depends on the fluid's pH. With a point of zero charge ranging between 1.2 and 4 [103–105], silica is considerably more acidic than water. It is unlike Al_2O_3, which has a point of zero charge between 8 and 8.5

(i.e., amphoteric to slightly basic). Accordingly, the silica layer formed at the H_2O-chemisorbed surface of Si_3N_4 tends to easily dissolve in water – its solubility being ~100 times larger than that of Al_2O_3 [106,107]. Its dissolution, which is accelerated by friction, leads to formation of silicic acid ($Si(OH)_4$) and the concurrent release of ammonia (NH_3). The elution of silicic acid uniquely reduces self-mated Si_3N_4 frictional forces in water, and the ammonia may serve as a disinfectant.

The interaction of silicon nitride with lipids and proteins has also been the object of interesting studies; but it is still poorly understood. Xu *et al.* [108] described measurements of the force between a silicon nitride scanning tip and a lipid coated silicon nitride surface in the presence of aqueous electrolytes. A long-range repulsive force was found when the surface was immersed in an electrolyte, which was attributed to dissociation of surface molecular groups and/or from the adsorption of ions. Diao *et al.* [109] demonstrated that the surface of silicon nitride could be modified to enhance protein-protein interactions. Modifications included silanization of the native Si_3N_4 surface, which had been previously treated with concentrated hydrofluoric acid to improve surface homogeneity. Then, the surface was oxidized to form a hydrophilic layer, and finally functionalized with amino-propyltriethoxysilane to form amine groups which were activated by the introduction of an aldehyde layer. Senden and Drummond [110] discussed the chemical reactions that could charge a Si_3N_4 surface. Silanols and silylamine surface groups were found to be the main contributors for surface charging in an aqueous medium. In the absence of a background electrolyte, the Si_3N_4 surface is nearly electrically neutral at a pH ranging from 6 to 8.5, meaning that silanol and silylamine surface groups are present in approximately equal concentrations.

The capacity of attracting oxygen to its surface in a hydrothermal environment (rather than expelling it) could play a role in preventing oxidation of polyethylene within a hip joint. In this regard, Si_3N_4 can be classified as an "oxygen cleanser." Conversely, oxide ceramics can be classified as "oxygen polluters" due to their tendency to release oxygen into the tribolayer. The possibility of tailoring the friction coefficient and raising the pH during articulation by fine tuning surface treatments could provide a new dimension in designing the microstructure of joint bioceramics. The chemical peculiarities of Si_3N_4 play important roles in its biomedical applications including bacteriostasis, osseointegration, and tribology.

With these ideas in mind, an applicative experiment using hard-on-soft hip-simulation was performed to examine whether oxygen actually "preferred" the Si_3N_4 surface rather than the polyethylene. Acetabular liners (Apex-Link Poly™, manufactured by OMNI Life Science, East Taunton, MA, USA) were tested in a hip simulator against Ø28 mm Si_3N_4 femoral heads (manufactured by Amedica Corporation, Salt Lake City, UT, USA) and their phase composition was compared to the pristine material. The investigated UHMWPE belonged to the first-generation of annealed liners. Fabricated from GUR1050 resin, they were gamma-irradiated to 78.5 kGy in inert gas and subsequently annealed in nitrogen at 85–90°C for 24 h. Final sterilization was performed in gaseous ethylene oxide. The acetabular liners were artificially aged prior to wear testing according to ASTM F2003-02. The hip simulation study was performed using an AMTI 12-station hip simulator

(AMTI, Watertown, MA) at a rate of 1 Hz. Tests were carried out to a total of 5×10^6 cycles (5 Mc) of simulated gait. The kinematics included a standard walking gait cycle as specified by ISO 14242-1. Crystallinity profiles were non-destructively collected by Raman spectroscopy [109,110] on two-dimensional maps at different focal depths along the liner subsurface. Two-tailed Student *t*-tests (95% confidence level, $p < 0.05$) were used to compare the population means at different depths in the pristine, wear-zones and the non-wear zones of these in vitro tested liners. Figure 12.10 shows the average crystallinity profiles along the in-depth direction, *z*, of the liner thickness in the main-wear zone of the simulator-tested UHMWPE liners in comparison to data collected on the pristine liners. In both plots, the crystalline phase fraction markedly increased within the first few microns of the subsurface leading to a plateau. This trend is typical to a number of commercially available UHMWPE liners, in which the prevalent amorphous state is induced by mechanical machining of the liner surface [111,112]. The important finding is that crystallinity in the worn area of the liner only increased by about 3% within the first 20 μm of subsurface as compared to the pristine sample. Also, there were only marginal differences in crystallinity between worn and unworn regions, especially in the depth of the liners. Overall, independent of measurement location, the increase in crystalline phase fractions for UHMWPE liners tested against Si_3N_4 was always $\leq$ 3% as compared to samples in the pristine state.

As previously discussed in section 12.2.1, the increase in crystallinity which occurred during frictional interaction between head and liner could be ascribed to the formation of oxidized species in the polymer's microstructure. Residual free radicals

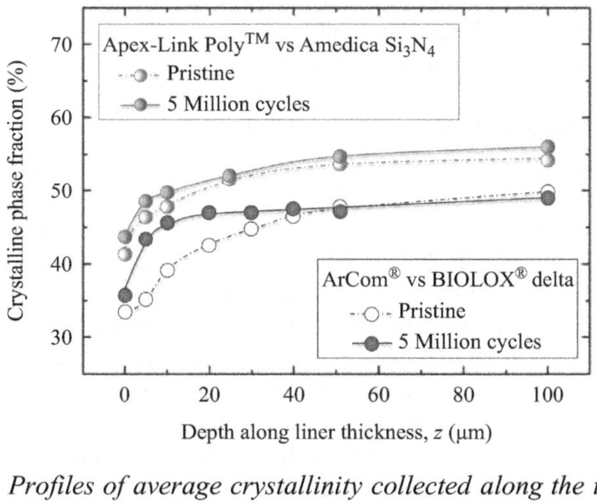

Figure 12.10 Profiles of average crystallinity collected along the in-depth direction, z, of the liner thickness in the main wear zone of the 5 million cycle simulator-tested UHMWPE liners, in comparison with data on pristine liners. The profile recorded for Apex-Link Poly™ liners tested against Amedica Si_3N_4 femoral heads is compared to that of ArCom® liners tested against BIOLOX® delta

tend to preferentially generate in the amorphous phase; they easily react with oxygen and lead to the formation of hydroperoxides (i.e., the first product of oxidation). An oxidative cascade leads to the formation of species containing carbonyl groups, such as ketones, esters, and carboxyl acids. The last compound is responsible for the breakage of polyethylene chains, increased molecular mobility, and subsequent crystallization [113–117]. However, a negligible change in the crystallization of identical UHMWPE liners was observed when tested against Si_3N_4 femoral heads (cf. Figure 12.10). Note also, that according to the reported literature [118–120], crystallization is favored in the main-wear zone because plastic deformation contributes to the formation of free radicals by stretching and weakening of the crosslink sites. Again, this effect was found to be negligible in UHMWPE/Si_3N_4 couples as confirmed by the minor difference between in-depth crystallization profiles collected from main-wear and non-wear zones. Finally, in γ-irradiated polyethylene, ageing before wear testing (i.e., a conservative condition purposely superimposed for simulation in the ASTM standard) should also favor oxidation/crystallization during hip-simulation testing. Moreover, the limited increase in crystallinity with the sub-surface of the liner, similar in both wear and non-wear zones, suggests that most of the observed crystallization happened during the preliminary ageing test.

In order to confirm and strengthen the above results, the hip simulator behavior of UHMWPE/Si_3N_4 couples were compared with UHMWPE/ZTA couples under exactly the same conditions. Figure 12.10 also shows the average crystallinity profiles collected in the main-wear zone of UHMWPE liners (ArCom® polyethylene liner, Biomet. Inc Warsaw, IN, USA) tested against ZTA femoral heads (BIOLOX®delta, manufactured by CeramTec GmbH, Plochingen, Germany), in comparison with data collected on pristine liners of the same type. ArCom®polyethylene is also fabricated from GUR1050 resin and only gamma-irradiated to 33 kGy in Ar gas for final sterilization. As shown in Figure 12.10, the amount of crystallization in ArCom®polyethylene liners after hip simulation was up to 21% higher in the first 30 μm of sub-surface than in the pristine samples.

The volumetric material loss during hip simulation has been reported to be low for both ceramic oxides and non-oxides in articulation against polyethylene [121]. In the case of the present experiments, ArCom® (vs BIOLOX®delta) was in average of twofold more worn than Apex-Link Poly™ (vs Amedica Si_3N_4) (i.e., ~220 vs ~100 mm^3) after 5 Mc. Moreover, the data shown in Figure 12.10 demonstrate, as far as conventional polyethylene structures are concerned, that crystallization by the ZTA material was larger than induced by Si_3N_4 under the same testing conditions. A confirmation of the tribochemical phenomena taking place at the sliding interface between the Si_3N_4 femoral heads and the UHMWPE liners was obtained by cathodoluminescence spectroscopy (i.e., the same spectroscopic tool that unveiled the formation of oxygen vacancies in oxide ceramics, cf. section 12.2.2). Figure 12.11(a) and (b) shows laser micrographs of the bearing surfaces of Si_3N_4 femoral heads before and after testing for 5 Mc in hip simulation, respectively. As seen, the surface of the pristine head contained a small fraction of open porosity from the manufacturing process (Figure 12.11(a)), which became almost completely filled during simulation testing (Figure 12.11(b)). The chemical origin of this

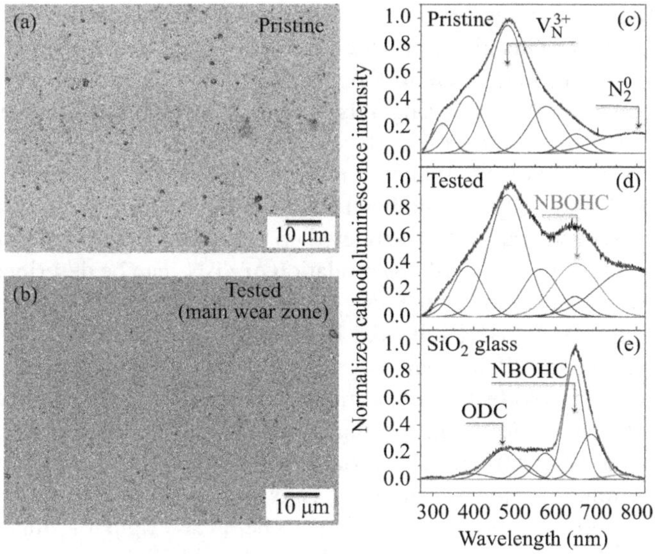

Figure 12.11 *Laser micrograph of the surface of the Si_3N_4 femoral head before (a) and after (b) testing in hip simulator. The outputs of cathodoluminescence analyses on pristine and tested Si_3N_4 head samples are reported in (c) and (d), respectively, in comparison with the cathodoluminescence spectrum of pure silica (SiO_2) glass (e). Labels are explained in the text*

surface modification can be ascertained by comparing the cathodoluminescence spectra of the pristine and tested Si_3N_4 head along with the spectrum of pure silica (SiO_2) recorded under the same experimental conditions (Figure 12.11(e)). The spectra display complex features, which have been discussed in detail in a previously published report [122]. A complete analysis of the spectra is beyond the scope of this chapter, but the main emission bands from the spectra are labeled in Figure 12.11(c)–(e). Of interest are two main features recorded in the tested sample: (i) the pronounced shoulder at around 630 nm and (ii) an enhancement of intensity at around 800 nm. These spectroscopic features were clearly more evident in Si_3N_4 after testing in comparison with the pristine samples (cf. arrows in inset to Figure 12.11(c) and (d)). The above spectral characteristic (i) points to the presence of an oxidized layer of silica-rich glass, whose main emission band is located at around 630 nm and coincides with the location of the shoulder recorded in the tested Si_3N_4 sample. This emission represents oxygen rich sites (non-bridging oxygen hole centers, NBOHC) [123–126]. Note that the other main but weaker band emitted from silica (i.e., arising from oxygen deficient centers labeled ODC in Figure 12.11(e)) strongly overlapped other Si_3N_4 bands and could not be deconvoluted. The different sub-bands were attributed to the stoichiometry state of the Si_3N_4 lattice, among which the most prominent is the emission from nitrogen vacancies (V_N^{3+}). These are located at around 490 nm (cf. spectra in Figure 12.11(c) and (d)), and remained unchanged after testing. One exception to this trend was the band located at around 800 nm, which represents nitrogen-dangling bonds, N_2^0. This latter emission, which was stronger after testing, is due to the presence of aliovalent cations in the Si_3N_4 lattice (e.g., Al^{3+} and Y^{3+} sintering aids in Si_3N_4). Its higher intensity represents a flux of these elements toward the surface during testing.

An understanding of the oxidation reactions and surface chemistry of Si_3N_4 is essential in order to fully comprehend its potential beneficial in vivo behavior. Omitting any second-order effects related to the presence of sintering aids, the sequence of events involved with the oxidation of Si_3N_4 can be described according to the following cascade of elementary phenomena: (i) transfer of oxygen molecules to the surface of Si_3N_4 followed by their physisorption; (ii) dissociation of oxygen molecules and their chemisorption; (iii) thermally activated diffusion of oxygen through the oxide layer toward the SiO_2/Si_3N_4 interface; (iv) chemical reactions producing oxide and nitrogen gas; and (v) gas diffusion through the oxide layer towards the outer surface. This sequence of reactions leads to the formation of various zwitterionic species with multiple isoelectric points at the surface of Si_3N_4. These species indeed represent the physicochemical origin of important Si_3N_4 peculiarities, since surface charge and ionic strength have a strong impact on hydrophilicity, bacterial adhesion, and osteoblast activity [127–131]. The surface of Si_3N_4 is strongly hydrophilic and selective to protein adsorption, both of which are important in resisting bacterial adhesion. This is a consequence of the presence of amine groups which produce a clear antibacterial effect (i.e., comparable to chitosan, a known bactericidal polymer coating for orthopedic implants) [132] and improved osteoinductive properties [131,133]. The rationale for this behavior consists in the electrical interaction of negatively charged bacteria with positively charged amine groups, which induces bacterial membrane disruption [134]. These very same surface characteristics also contribute to pronounced attachment and proliferation of osteoblasts. The results of the Raman and cathodoluminescence experiments fit with the findings of previous biological studies; and, in addition, provide new evidence that the antibacterial effect could also be pronounced in the tribochemical environment of artificial joints.

In conclusion, we have demonstrated that Si_3N_4 is a polyethylene-friendly biomaterial. It is definitely friendlier than oxide ceramics because of its unique ability to scavenge oxygen from the tribolayer. According to the same microscopic mechanisms and independent of the quality of the UHMWPE liner employed, a delay in crystallization/oxidation is also expected to occur in vivo. This peculiar surface chemistry, which was previously reported as the decisive factor regulating the unique antibacterial properties of Si_3N_4 spinal implants [133], might also provide a breakthrough in elongating the lifespan of artificial joints into their second and third decade.

12.3.2 Establishing the concept of "positive" non-bioinertness

In the previous sub-section, the surface of Si_3N_4 was identified as consisting of a oxidized layer with a graded oxynitride film [135]. Infrared measurements have shown that oxidized silicon nitride contains mainly two types of surface groups, namely silanols (Si–OH) and secondary amine groups (Si_2–NH) [136–138]. In fact, it is the competitive presence of these two adsorbates that determines many of its important surface characteristics [139]. For example, the isoelectric point of silicon nitride powder has been reported to be tunable within a wide range between $pH_{ie} = 3–9$ [140–143]; and it may change with ageing and pH. This is clearly an advantage in implant design, because the possibility of controlling the isoelectric point provides a useful parameter in fine tuning its properties by tailored surface treatments [144].

In this sub-section, Si_3N_4 surface phenomena are investigated from a different perspective, namely its chemical evolution with a biological medium. As a solid material, Si_3N_4 has the ability to quickly increase pH in its surrounding environment – from strongly acidic to markedly alkaline. Experiments were conducted using a pH microscope (SCHEM-110; Horiba, Kyoto, Japan) capable of measuring and mapping local pH at the surface of solids with a spatial resolution on the order of tens of microns. pH mapping experiments were performed by embedding a Si_3N_4 sample into an acidic gel consisting of artificial saliva, KCl, and agar (a jelly-like substance obtained from polysaccharide agarose contained in algae agarophytes) [145]. This pH-imaging device consisted of a flat semiconductor plate with a sensing area of 2.5 cm $\times$ 2.5 cm. The highest spatial resolution and the pH sensitivity of the equipment were 100 μm and 0.1, respectively. It was equipped with a light addressable potentiometric sensor, capable of distinguishing protons within the electrolyte. A light beam was directed from the backside of the sensor with a bias voltage applied between the electrolyte and the backside. Characterization of the AC photocurrent depended on the amount of protons at the sensor surface. The pH could be determined to a high degree of precision by measuring the local value of electric current [146]. The detected current signals were converted into a colored visual map, with each pixel arranged to indicate its pH using image analysis software (Image Pro Plus, Media Cybernetics, Rockville, MD, USA).

Figure 12.12 shows the room-temperature pH evolution around an as-received (mm-sized and polished) Si_3N_4 sample (Amedica Corporation, Salt Lake City, UT, USA) as a function of time. The pH-buffering action at the surface of the Si_3N_4 sample was evident by the gradual conversion of the strongly acidic gel (i.e., pH $\approx$ 4.5) to alkaline values (pH $\approx$ 8.5). The gradual spread of the alkaline environment around the sample occurred in tens of minutes; however, the initial change in alkalinity was rather rapid, which made it difficult to obtain a map at time zero. Moreover, the pH buffering effect was found to be independent of the sample size. The same effect was also observed by repeating the experiment with a smaller sample ($\approx$ 400 $\times$ 300 μm in size; data not shown) from the same Si_3N_4 material. Based on this experimental evidence, Figure 12.13 presents schematic illustrations of the sequence of chemical events leading to the observed pH buffering phenomenon. A layer of amphoteric SiO_2, which gradually incorporates increasing amounts of N ions toward the Si_3N_4 core, covers the surface of the polished sample (Figure 12.13(a)). The initial reaction in the acidic media (Step 1; Figure 12.13(b)) consists of dissolution of amphoteric silica through detachment of silanol adsorbates,

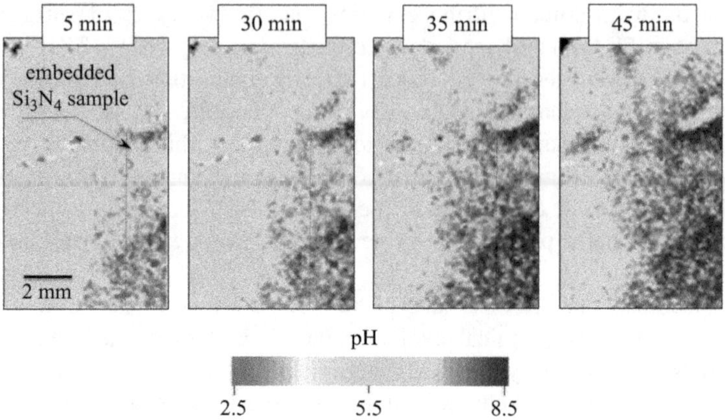

Figure 12.12 Room-temperature evolution of pH around an as-received and polished Amedica Si_3N_4 sample as a function of time. The pH-buffering property of the Si_3N_4 sample enables a gradual increase of pH in wider zones within the surrounding acidic gel (average pH of the unperturbed gel = 4.5). (© 2016 American Chemical Society, Reprinted with permission [180])

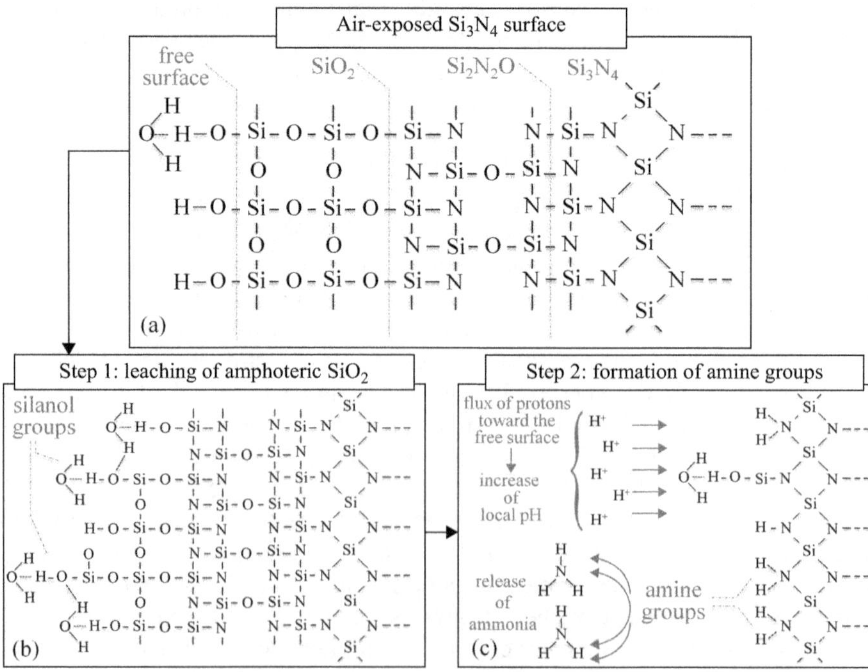

Figure 12.13 Schematic illustration of the chemical events leading to the observed phenomenon of pH buffering by the Amedica Si_3N_4 bioceramic. (© 2016 American Chemical Society, Reprinted with permission [180])

Si–OH, in the form of silicic acid, $Si(OH)_4$. However, $Si(OH)_4$ is unstable in water and it readily condenses to form water and polymeric chains [147,148]. With progressing dissolution of the amphoteric silica layer, amine adsorbates, Si_2–NH (and a minor fraction of Si–NH), are generated on the Si_3N_4 surface (Step 2; Figure 12.13(c)) and fully replace the eluting silanols. Unlike silanols, the amine adsorbates are strongly bonded to the surface [139], but tend to attract protons to form ammonia, NH_3. The strong driving force to produce ammonia induces a massive flux of protons toward the adsorbed amine species. This flux depletes the solution of protons and raises the local pH at the solid/liquid interface. This is exactly what was observed in these pH mapping experiments. Note also that this pH-buffering effect was similarly observed at the surface of silicon nitride immersed in water (at various equilibrium acidic pH values) by means of Raman spectroscopy [149].

The pH-buffering phenomenon found in Si_3N_4 is an important property for a number of biomedical applications. In the field of joint arthroplasty, pH-buffering could play an innovative role. Articular hyaline cartilage consists of a hydrated, proteoglycan-rich matrix in which a network of type II collagen fibers is embedded. In healthy tissue, the hydrated proteoglycan matrix has a high swelling pressure, counteracted by the stiff collagen network, which gives unique biomechanical properties to the articular hyaline cartilage. Degradation of type II collagen is the crucial event in the pathogenesis of osteoarthritis [150]. Indeed, not only neutral, but also acidic cysteine endoproteinases (e.g., cathepsin K) are activated and efficiently degrade collagen at pH values between 4.5 and 6.0 [151,152]. Cathepsin K extracted from healthy articular hyaline cartilage with neutral pH is in the latent, high molecular weight pro form, whereas cathepsin K in osteoarthritic cartilage with low pH is in its active, proteolytically processed, low molecular weight form. Note that acidic endoproteinases become inactivated at neutral pH. In other words, a low pH in the joint promotes degeneration of articular hyaline cartilage (i.e., pH $\approx$ 5.5) as compared to healthy cartilage (pH $\approx$ 7.1) [153]. As a matter of fact, patients suffering rheumatoid arthritis possess quite acidic synovial fluid [154]. Decreased synovial fluid pH correlates strongly with radiographic evidence of polyethylene degradation in rheumatoid patients and, as also discussed in section 12.2.2, might accelerate the off-stoichiometry observed on the surface of oxide bioceramics post-operatively [19]. The capacity of silicon nitride in restoring an alkaline pH environment in the joints could be important not only after arthroplasty surgery, but also in preventive joint therapy.

Other biomedical fields could also benefit from the unique pH-buffering properties of Si_3N_4 bioceramics. In dentistry, for example, it is well known that a healthy mouth should be kept in a non-acidic, neutral, or alkaline environment with a pH $\geq$ 7.0. In fact, tooth roots begin to dissolve as soon as the pH drops even a tiny bit below neutral (i.e., pH = 6.5). Moreover, when acidity levels lower to pH $\leq$ 5.5, tooth enamel will erode, become discolored, and experience a high risk of caries. Acidity always weakens teeth, but it is the amount of time that acids are in contact with teeth that determines the severity of the damage. As erosion and decay worsen upon sipping acidic drinks and nibbling acidic foods, exposure time to acidic environments must be kept to a minimum. Since teeth remineralize when the mouth is alkaline at a pH $\geq$ 7.5, the pH-buffering property of Si_3N_4 could quickly reestablish local pH and alkalinity within the mouth. This might be accomplished by using Si_3N_4 either in the form of powder dispersed in a gel or toothpaste, or as a solid mouthpiece. A rapid and uniform reestablishment of the alkaline environment in the mouth, together with the antibacterial properties of Si_3N_4 [155], could also have therapeutic effects on periodontal diseases [156,157].

In conclusion, if the definition of bioinertness were expanded to include physical, chemical, and mechanical surface modifications at the molecular scale, then Si_3N_4 seems to possess the "right chemistry" to be a beneficial bioceramic. Its surface properties, which include its unique antibacterial, pH buffering, osteointegrative, and tribological advantages, help establish a new definition of bioinertness, one which encompasses a concept of "positive" non-bioinertness. This concept represents the intrinsic capacity of an implanted biomaterial to simulate or facilitate specific biologic functions through suitable chemical and structural in vivo changes.

12.3.3 Surface embrittlement or long-lasting toughening?

Excellent survivability and exceptionally low wear rates with virtually no local adverse tissue reactions have been tagged as main expectations for current ceramic-on-ceramic artificial joints [158,159]. Use of hard-on-hard bearings obviously eliminates polyethylene wear debris, and with it, the cause of long-term implant failure due to osteolysis. However, wear of self-mated bioceramic couples has proven to be complex. It is affected by surface micro-fracture, chemical dissolution, and local heating at the contact interface [160,161]. Moreover, edge loading due to translational mal-positioning under microseparation has been shown to significantly increase the otherwise low wear rate of ceramic-on-ceramic bearings [86,87,162]. The "stiffness" of an all-ceramic hip joint exacerbates even minimal angular misfits which can cause impingement. Several types of wear behavior have been observed for hard-on-hard couples due to differences in their surface properties and their dynamic response to frictional heating, chemical reactions (including biofilm formation), microseparation, and off-stoichiometry drifts [163–165]. However, wear of hard-on-hard bearings generally occurs from a combination of factors, not a single mechanism [166]. Our present understanding of wear in ceramic-on-ceramic joints is incomplete, but even more limited are notions that link phenomenological wear behavior to the mechanical, tribochemical, thermal, and crystallographic properties of the material's surface. In previous studies [10–12,90], local spectroscopic probes (i.e., confocal lasers and electrons) were used to demonstrate the effects of wear on crystallographic, chemical, and off-stoichiometry alterations occurring at the surfaces of oxide bioceramics. These studies revealed direct links between off-stoichiometry, phase stability, and residual stress [10,11]. Chemistry-independent damage was also observed on retrievals due to hard-on-hard micro-shock events associated with microseparation and stripe-wear [11,165]. The extent of this type of wear damage strongly depends more on the surface toughness of the bioceramic than its surface chemistry. Unlike bulk fracture toughness, surface toughness is prone to environmentally driven degradation. Articular surfaces are damaged by repeated shock, indentation, or by third-body abrasive particles. During these events, wear debris are generated by the propagation and coalescence of surface microcracks which results in grain detachment. Consequently, the wear rate of brittle materials is strongly dependent on surface fracture toughness [167–170]. In this latter context, ZTA biomaterials have previously been shown to undergo polymorphic transformation in a hydrothermal environment resulting in local embrittlement of the material's surface [95,171]. Therefore, surface fracture toughness should be clearly distinguished from bulk toughness and separately assessed.

In section 12.2.3, it was also demonstrated that environmental factors triggered crystallographic instability in ZTA, a phenomenon which occurred much quicker in vivo than predicted from in vitro hydrothermal experiments. This led to the transformation of a significant fraction of the initial tetragonal (toughening) phase into its monoclinic polymorph. This raises a natural question: What are the consequences of this transformation on fracture toughness of the bearing surface? Mechanical data reported by the manufacturers of ZTA typically report only bulk toughness. In fact, there is a profound lack of structural property information for the surfaces of ceramic bearings. Consequently, Raman microprobe-assisted indentation studies, coupled with crack opening displacement (COD) measurements [172,173], were used to determine the surface toughness of a number of bioceramics currently employed in joint arthroplasty. An explanatory illustration of this procedure is given in Figure 12.14. It was demonstrated that this improved micromechanics technique can be quantitatively applied to evaluate the surface fracture toughness of several ceramic materials, and to effectively monitor their change in toughness with time, especially after in vitro hydrothermal ageing. Since this procedure correlates two key factors in the durability of hard-on-hard bearings, namely micromechanics and tribochemistry, it may also establish new criteria for material design. Its output consisted of determining the threshold stress intensity for crack initiation, K_{IO}, and the crack resistance parameter, K_R (also usually referred to as rising R-curve behavior) [174]. Common values for K_{IO} in ceramic materials are low (i.e., 1–2 $MPa{\cdot}m^{1/2}$); but K_R can be up to an order of magnitude greater, and it is therefore the important toughness parameter to be monitored. Large R-curve contributions to fracture toughness arrest microcracks and induce stable crack growth [175,176]. In fact, steeply rising K_R values with increasing crack lengths result in high toughness materials. With this background information, comparative plots of these two parameters (i.e., K_{IO} and K_R) are provided in Figure 12.15(a) and (b) for three commercial bioceramic materials, both before and after hydrothermal ageing, respectively. These two plots summarize the rising R-curve behavior for Al_2O_3 (BIOLOX®forte), ZTA (BIOLOX®delta, CeramTec GmbH, Pochingen, Germany), and Si_3N_4 (MC$^{2®}$, Amedica Corporation, Salt Lake City, UT, USA) before and after the long-term autoclaving cycle, respectively. Each material's total crack resistance, $K_{R(\Delta a)} = K_{IO} + \Delta K_{(\Delta a)}$, was obtained at different crack lengths. As previously explained, the K_{IO} value represents the threshold toughness for crack initiation, while the crack-length dependent term, $\Delta K_{(\Delta a)}$, represents the rising R-curve behavior of the material. Environmentally driven effects were found to be negligible on the toughening behavior of both Al_2O_3 and Si_3N_4 biomaterials (cf. Figure 12.15(a) and (b)).

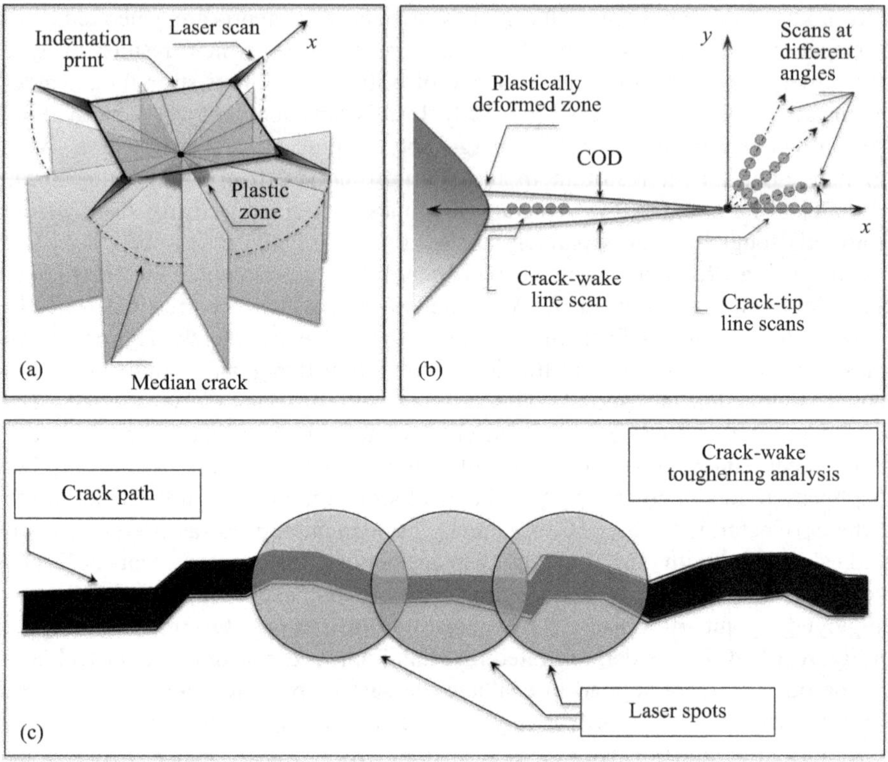

Figure 12.14 Schematic illustration showing the Raman-spectroscopy-assisted procedure used in assessing the near-tip rising R-curve behavior on microcracks propagated from the corners of a Vickers indentation print: (a) the Vickers indentation print with half-penny-shaped (median) microcracks departed from its corners in the neighborhood of the sample surface; (b) analysis of COD along the crack path and stress analyses by Raman spectroscopy both ahead and behind the propagated crack tip; and (c) Raman spectroscopic analysis of crack-wake stress fields for the quantification of crack-wake toughening mechanisms. (© 2016, Elsevier, Reprinted with permission [181])

Their K_{IO} values were similarly low (1.5 and 1.8 MPa·m$^{1/2}$, respectively). However, the toughness of the Al_2O_3 was more than threefold lower than the Si_3N_4 and showed essentially no rising R-curve behavior. Conversely, ZTA clearly underwent surface embrittlement due to the autoclaving cycle (cf. Figure 12.15(a) and (b)). Its toughening mechanism and rising R-curve behavior were inactivated during hydrothermal ageing. In fact, its toughness was effectively reduced to that of monolithic Al_2O_3. Conversely, this detrimental reduction in fracture toughness was not observed in either Al_2O_3 or Si_3N_4 ceramics. In particular, the Si_3N_4 material

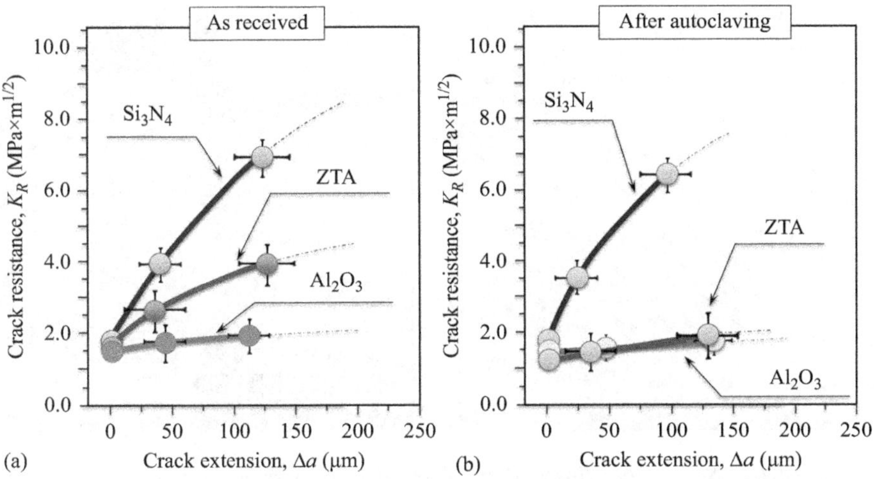

Figure 12.15 Near-tip rising R-curve behaviors of three types of bioceramic used in femoral head components: monolithic Al_2O_3 (BIOLOX®forte), ZTA (BIOLOX®delta), and Si_3N_4 (Amedica). Toughening behaviors before and after autoclaving for 200 h in hydrothermal environment at 121°C under 1 bar in water vapor are shown in (a) and (b), respectively. (© 2016, Elsevier, Reprinted with permission [182])

retained its steeply rising R-curve behavior even after long-term exposure in the autoclave (i.e., up to 200 h).

Explanatory illustrations of the dominant toughening mechanisms in Si_3N_4 and ZTA are provided in Figure 12.16(a) and (b), respectively. Crack-face bridging and transformation toughening are the physical origins for the observed R-curves in these two bioceramics. Crack-face bridging occurs in the presence of acicular grains, which eventually remain unbroken in the wake of a propagating crack, becoming a "bridge" between the crack faces. These bridges oppose crack opening and propagation [177]. This mechanism leads to a steeply rising R-curve for the Si_3N_4 (i.e., $\Delta K \approx 5$ MPa·m$^{1/2}$) within the first 100 μm of crack extension (cf. Figure 12.16(a)). Conversely, toughening in ZTA arises from a substantial lattice expansion (~4 vol%) during the tetragonal to monoclinic polymorphic transformation. The tensile stress field developed at the tip of the propagating crack triggers the transformation. However, volume expansion around the crack path is inhibited by the constraint of the surrounding untransformed material. This constraint in turn results in a compressive closure of the crack faces. Note that both crack-face bridging and transformation toughening mechanisms are cumulative with respect to an increase in crack length, (cf. Figure 12.15(a)). For this reason, they both show a monotonic increase in toughness, ΔK, as a function of crack length, Δa.

In ZTA, the fraction of transforming ZrO_2 grains is limited when compared to a monolithic Y-TZP material. Therefore, the toughening effect on short cracks

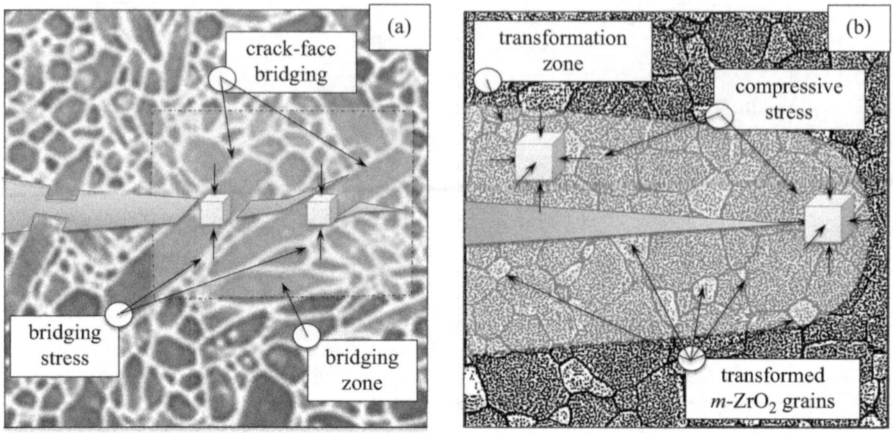

Figure 12.16 Schematic illustrations of toughening mechanisms as occurring in the neighborhood of an advancing crack: (a) crack-face bridging in the acicular microstructure of polycrystalline Si_3N_4 and (b) transformation toughening in the composite ZTA microstructure containing metastable tetragonal-ZrO_2 dispersoids. (© 2016, Elsevier, Reprinted with permission [182])

cannot be as substantial as for monolithic zirconia. Only $\Delta K \approx 2$ MPa·m$^{1/2}$ was observed for ZTA in the as-received state (cf. Figure 12.15(a)). Moreover, data presented in Figure 12.15(b) clearly show that prolonged exposure to a hydrothermal environment completely annihilated this toughening effect in ZTA. On the contrary, hydrothermal exposure leaves the surface fracture behavior of Si_3N_4 almost unaffected. The rationale for the observed difference in toughening between Si_3N_4 and ZTA is directly related to their different toughening mechanisms. Upon hydrothermal exposure, the surface of ZTA spontaneously undergoes polymorphic transformation, triggered by both chemical and off-stoichiometry effects. Accordingly, with no more available t-ZrO_2 to transform around the crack path, no further toughening can occur and ZTA becomes as brittle as a monolithic Al_2O_3 (cf. Figure 12.15(b)). Conversely, the intrinsic strength of the acicular bridging sites in Si_3N_4 is unaffected by the hydrothermal environment, and therefore the toughness of the surface before and after exposure remains durable and unaffected by an aggressive environment.

12.4 Conclusions

In the first part of this chapter, crucial basic aspects associated with biomaterial stability in vivo were discussed (i.e., those having a direct impact on the long-term performance of artificial joints). An updated approach to hip-joint tribology was also presented including newly discovered physical chemistry aspects. Evidence was provided for tribochemical interactions which occur between the biologic environment and lattice cells of oxide ceramics. The off-stoichiometry drift observed in Al_2O_3 due to environmental exposure was shown to play a role in shortening its lifespan as a biomedical device. This phenomenon has been almost completely neglected because there was no efficient analytical methodology to visualize it. In this context, cathodoluminescence spectroscopy located and measured the concentration of point defects and other compositional modifications which demonstrated a continuous drift in surface chemical instability. This degradation is intrinsic to ionic bonds in oxide materials. Their interaction with the water dipole leads to surface hydroxylation and dehydroxylation. It was shown that alumina's propensity to give away oxygen from its surface in the biological environment can be delayed by adding appropriate dopants (e.g., Cr^{3+}), but not completely eliminated. The oxygen released from the oxide surface into the tribolayer promptly diffuses toward the soft UHMWPE counterpart and, in the long-term, results in its oxidative degradation. It was confirmed that polyethylene oxidation can be limited, but not completely obviated, by doping liners with vitamin E. Physical chemistry arguments opposed the conventional wisdom that oxide ceramics are fully bioinert. Indeed, careful spectroscopic experiments suggested that they are not immune to degradation, and, in fact, play a negative role in joint deterioration, serving as oxygen polluters. Also, ZTA, with its improved ionic strength over Al_2O_3 (i.e., obtained through Cr^{3+} doping) suffers from a different kind of surface instability, which is related to the polymorphic transformation of ZrO_2. This phenomenon results from the chemically driven distortion and subsequent instability of the oxygen sub-lattice in t-ZrO_2. It was shown that the occurrence of this surface instability could be much shorter (i.e., on the order of a few months) in vivo than what was actually predicted by standard hydrothermal tests in vitro. Consequently, achieving implants with two-to-three decades of standard lifetime might necessarily require moving away from oxide bioceramics.

In the second part of this chapter, the positive aspects of non-oxide bioceramics were discussed. It was shown that Si_3N_4 appears to possess the "right chemistry" to satisfy synergistic interactions between its solid surface and the biological environment. Oxygen affinity is an important property of Si_3N_4, which leads to its oxygen cleansing behavior. This property represents the physical origin of its "polyethylene friendly" performance as compared to oxide biomaterials (including ZTA). Use of a ceramic that scavenges oxygen might be effective in arresting polyethylene oxidation similar to remelting or vitamin E doping. In this context, the use of Si_3N_4 as a bearing counterpart to polyethylene could yield benefits similar to that of a "passivation layer" preventing corrosion of metals [178]. Advanced spectroscopic procedures have enabled visualization of the changes in surface stoichiometry at the molecular level, which is believed to be an important element in designing bioceramics with elongated lifetimes. The peculiar formation of amine adsorbates on the Si_3N_4 surface lies at the origin of a stream of positive interactions between the solid surface and the biological environment. These include a pronounced pH-buffering effect by attraction of protons to the Si_3N_4 surface, the local formation of ammonia which is interlaced with antibacterial activity, and osteoconductivity arising from the concurrent attachment and proliferation of osteoblasts.

While no synthetic biomaterial can be fully bioinert, Si_3N_4 appears to offer considerable promise in "down-regulating" some of the deleterious oxygen-exchange interactions associated with oxide bioceramics in joint arthroplasty. The surface of this non-oxide material naturally drifts towards protective action with respect to oxidation of UHMWPE soft counterparts. Unlike ZTA, Si_3N_4 bases its toughening characteristics on a microscopic mechanism (crack-face bridging), which remains unaffected by chemical interactions occurring during tribological loading. This peculiar characteristic greatly increases its potentiality for long-lasting hard-on-hard bearings.

In conclusion, presented and discussed in this chapter were unique analytical methods for examining the surface chemistry and mechanical properties of bioceramics. Based on this research, an updated definition of bioinertness was provided, one which suggests both "positive" and "negative" non-bioinertness. Furthermore, this research points to the possibility of entering a new era in joint tribology, one which moves from oxide- to nitride-based bioceramics. Accordingly, a full exploration of the improved potentialities that Si_3N_4 bioceramics offer in arthroplasty is strongly advocated based on the sound scientific rationale and the clinical need for extending the lifespan of joint replacement bearings to coincide with the lifetime of their recipient patients.

References

[1] I.D. Learmonth, C. Young, and C. Rorabeck, "The operation of the century: total hip replacement," *Lancet*, **370** 1508–1519 (2007).

[2] J.-P. Courpied and J.H. Caton, "Total hip arthroplasty, state of the art for the 21st century," *Int. Orthop.*, **35** [2] 149–50 (2011).

[3] P. Räsänen, P. Paavolainen, H. Sintonen, A.-M. Koivisto, M. Blom, O.-P. Ryynänen, and R.P. Roine, "Effectiveness of hip or knee replacement surgery in terms of quality-adjusted life years and costs," *Acta Orthop.*, **78** [1] 108–115 (2007).

[4] S. Stea, B. Bordini, A. Sudanese, and A. Toni, "Registration of hip prostheses at the rizzoli institute: 11 years' experience," *Acta Orthop.*, **73** [s305] 40–44 (2002).

[5] S. Kurtz, K. Ong, E. Lau, F. Mowat, and M. Halpern, "Projections of primary and revision hip and knee arthroplasty," *J. Bone Joint Surg.*, **89-Am** [4] 780–785 (2007).

[6] J.N. Katz, J. Wright, E.A. Wright, and E. Losina, "Failures of total hip replacement: a population-based perspective," *Orthop. J. Harvard Med. Sch.*, **9** 101–106 (2009).

[7] B.R. Schlenker, *Introduction to Materials Science,* SI Edition. J. Wiley & Sons, Milton, 1974.

[8] B. Basu and M. Kalin, *Tribology of Ceramics and Composites: Materials Science Perspective.* John Wiley & Sons, Hoboken, NJ USA, 2011.

[9] F. Di Puccio and L. Mattei, "Biotribology of artificial hip joints," *World J. Orthop.*, **6** [1] 77–94 (2015).

[10] G. Pezzotti, M.C. Munisso, A.A. Porporati, and K. Lessnau, "On the Role of oxygen vacancies and lattice strain in the tetragonal to monoclinic transformation in alumina/zirconia composites and improved environmental stability," *Biomaterials*, **31** [27] 6901–6908 (2010).

[11] Y. Takahashi, W. Zhu, N. Sugano, and G. Pezzotti, "On the role of oxygen vacancies, aliovalent ions and lattice strain in the in vivo wear behavior of alumina hip joints," *J. Mech. Behav. Biomed. Mater.*, **4** [7] 993–1003 (2011).

[12] K. Fukatsu, A. Leto, W. Zhu, N. Sugano, and G. Pezzotti, "Kinetics and the role of off-stoichiometry in the environmentally driven phase transformation of commercially available zirconia femoral heads," *Acta Biomater.*, **8** [4] 1639–1647 (2012).

[13] O.K. Muratoglu, K.K. Wannomae, S.L. Rowell, B.R. Micheli, and H. Malchau, "Ex vivo stability loss of irradiated and melted ultra-high molecular weight polyethylene," *J. Bone Joint Surg.*, **92** [17] 2809–2816 (2010).

[14] L. Costa, P. Bracco, E. Brach del Prever, M.P. Luda, and L. Trossarelli, "Analysis of products diffused into UHMWPE prosthetic components in vivo," *Biomaterials*, **22** [4] 307–315 (2001).

[15] D.F. Williams, "Bioinertness: an outdated principle;" in *Tissue Eng. Prosthet. Vasc. Graft*. Edited by P. Zilla and H.P. Greisler. R. G. Landes Company, Georgetown, TX, USA, 1999.

[16] B.S. Bal and M.N. Rahaman, "Orthopedic applications of silicon nitride ceramics," *Acta Biomater.*, **8** [8] 2889–2898 (2012).

[17] B.S. Bal, A. Khandkar, R. Lakshminarayanan, I. Clarke, A.A. Hofmann, and M.N. Rahaman, "Fabrication and testing of silicon nitride bearings in total hip arthroplasty," *J. Arthroplasty*, **24** 110–116 (2009).

[18] B.S. Bal, A. Khandkar, R. Lakshminarayanan, I. Clarke, A.A. Hofmann, and M.N. Rahaman, "Testing of silicon nitride ceramic bearings for total hip arthroplasty," *J. Biomed. Mater. Res. Part B Appl. Biomater.*, **87** [2] 447–454 (2008).

[19] G. Pezzotti, "Bioceramics for hip joints: the physical chemistry viewpoint," *Materials (Basel).*, **7** 4367–4410 (2014).

[20] K.K. Wannomae, S.D. Christensen, A.A. Freiberg, S. Bhattacharyya, W.H. Harris, and O.K. Muratoglu, "The effect of real-time aging on the oxidation and wear of highly cross-linked UHMWPE acetabular liners," *Biomaterials*, **27** [9] 1980–1987 (2006).

[21] M.L. Eidinoff, J.E. Knoll, B.J. Marano, E. Kvamme, R.S. Rosenfeld, and L. Hellman, "Cholesterol biosynthesis: studies related to the metabolic role of squalene," *J. Clin. Invest.*, **37** [5] 655–659 (1958).

[22] K. Imaeda, K. Ohsawa, and T. Yokota, "Valuation on stability of squalene by use of oxygen determination," *Bunseki Kagaku*, **32** [7] 406–410 (1983).

[23] G. Pezzotti, T. Kumakura, K. Yamada, *et al.*, "Confocal Raman spectroscopic analysis of cross-linked ultra-high molecular weight polyethylene for application in artificial hip joints," *J. Biomed. Opt.*, **12** [1] 14011–14014 (2007).

[24] P. Bracco and E. Oral, "Vitamin E-stabilized UHMWPE for total joint implants: a review," *Clin. Orthop. Relat. Res.*, **469** [8] 2286–2293 (2011).

[25] E. Oral and O.K. Muratoglu, "Vitamin E diffused, highly crosslinked UHMWPE: a review," *Int. Orthop.*, **35** [2] 215–223 (2011).

[26] E. Oral, S.D. Christensen, A.S. Malhi, K.K. Wannomae, and O.K. Muratoglu, "Wear resistance and mechanical properties of highly cross-linked, ultrahigh-molecular weight polyethylene doped with vitamin E," *J. Arthroplasty*, **21** [4] 580–591 (2006).

[27] L. Puppulin, A. Leto, M. Hasegawa, and G. Pezzotti, "A comparative microstructural study of Vitamin-E blended and infused highly crosslinked UHMWPE for total knee arthroplasty," *J. Mech. Behav. Biomed. Mater.*, **39** 247–256 (2014).

[28] Y. Takahashi, K. Yamamoto, and G. Pezzotti, "Effects of Vitamin E blending on plastic deformation mechanisms of highly crosslinked ultrahigh molecular weight polyethylene (HXL-UHMWPE) in total hip arthroplasty," *Acta Biomater.*, **15** 227–236 (2015).

[29] S.L. Rowell, J. Yi, B.R. Micheli, K.K. Wannomae, H. Malchau, and O.K. Muratoglu, "In vivo performance of second generation highly cross-linked UHMWPE;" pp. 4–7 in *Present. as a Poster Exhib. Annu. Meet. Orthop. Res. Soc.* 2012.

[30] S. Rowell and O. Muratoglu, "Short-term follow-up retrieval study on irradiated and Vitamin E-diffused UHMWPE bearings," *Bone Joint J.*, **96-B** [Suppl 11] 203 (2014).

[31] J. Ming-Chang and Y. Li-Yung, "Environmental effects on wear behaviour of alumina," *Wear*, **161** [1–2] 111–119 (1993).

[32] S. Mori, "The lubrication of ceramics—adsorption and chemical reaction," *Tribol. J. Jpn. Soc. Tribol.*, **36** 130–134 (1991).

[33] V. Shapovalov and T.N. Truong, "Ab initio study of water adsorption on α-Al_2O_3(0001) crystal surface," *J. Phys. Chem. B*, **104** [42] 9859–9863 (2000).

[34] E.M. Fernández, R.I. Eglitis, G. Borstel, and L.C. Balbás, "Ab initio calculations of H_2O and O_2 adsorption on Al_2O_3 substrates," *Comput. Mater. Sci.*, **39** [3] 587–592 (2007).

[35] J.W. Elam, C.E. Nelson, M.A. Cameron, M.A. Tolbert, and S.M. George, "Adsorption of H_2O on a single-crystal α-Al_2O_3(0001) surface," *J. Phys. Chem. B*, **102** [36] 7008–7015 (1998).

[36] H.-J. Freund, H. Kuhlenbeck, and V. Staemmler, "Oxide surfaces," *Reports Prog. Phys.*, **59** [3] 283 (1996).

[37] B.C. Lippens and J.J. Steggerda, "Active alumina;" pp. 171–211 in *Phys. Chem. Asp. Adsorbents Catal.* Edited by B.G. Linsen. Academic Press, London, UK, 1970.

[38] H. Knözinger and P. Ratnasamy, "Catalytic aluminas: surface models and characterization of surface sites," *Catal. Rev.*, **17** [1] 31–70 (1978).

[39] C. Morterra, G. Ghiotti, E. Garrone, and F. Boccuzzi, "Infrared spectroscopic characterization of the α-alumina surface," *J. Chem. Soc. Faraday Trans. 1 Phys. Chem. Condens. Phases*, **72** 2722–2734 (1976).

[40] C. Morterra, G. Magnacca, G. Cerrato, N. Del Favero, F. Filippi, and C. V. Folonari, "X-ray diffraction, high-resolution transmission electron

microscopy and Fourier-transform infrared study of Ca-doped Al_2O_3," *J. Chem. Soc. Faraday Trans.*, **89** [1] 135–150 (1993).
[41] C. Morterra and G. Magnacca, "A case study: surface chemistry and surface structure of catalytic aluminas, as studied by vibrational spectroscopy of adsorbed species," *Catal. Today*, **27** [3–4] 497–532 (1996).
[42] A.A. Tsyganenko and P.P. Mardilovich, "Structure of alumina surfaces," *J. Chem. Soc. Faraday Trans.*, **92** [23] 4843–4852 (1996).
[43] A.A. Tsyganenko, K.S. Smirnov, A.M. Rzhevskij, and P.P. Mardilovich, "Infrared spectroscopic evidence for the structural OH groups of spinel alumina modifications," *Mater. Chem. Phys.*, **26** [1] 35–46 (1990).
[44] X. Liu and R.E. Truitt, "DRFT-IR studies of the surface of γ-alumina," *J. Am. Chem. Soc.*, **119** [41] 9856–9860 (1997).
[45] J. Ahn and J.W. Rabalais, "Composition and structure of the $Al_2O_3\{0001\}$-(1×1) surface," *Surf. Sci.*, **388** [1–3] 121–131 (1997).
[46] M.A. Nygren, D.H. Gay, C. Richard, and A. Catlow, "Hydroxylation of the surface of the corundum basal plane," *Surf. Sci.*, **380** [1] 113–123 (1997).
[47] X.-G. Wang, A. Chaka, and M. Scheffler, "Effect of the environment on α-$Al_2O_3(0001)$ surface structures," *Phys. Rev. Lett.*, **84** [16] 3650–3653 (2000).
[48] W.H. Wade and N. Hackerman, "Heats of immersion. IV. The alumina-water system-variations with particle size and outgassing temperature," *J. Phys. Chem.*, **64** [9] 1196–1199 (1960).
[49] R.L. Venable, W.H. Wade, and N. Hackerman, "Heats of immersion. VIII. Differential heats of adsorption as a function of particle size for the alumina-water system," *J. Phys. Chem.*, **69** [1] 317–321 (1965).
[50] J.B. Peri and R.B. Hannan, "Surface hydroxyl groups on γ-alumina," *J. Phys. Chem.*, **64** [10] 1526–1530 (1960).
[51] J.B. Peri, "Infrared and gravimetric study of the surface hydration of γ-alumina," *J. Phys. Chem.*, **69** 211–219 (1965).
[52] Y.-F.Y. Yao, "Adsorption of polar molecules on metal oxide single crystals," *J. Phys. Chem.*, **69** [11] 3930–3941 (1965).
[53] B. Hendriksen, "Heats of adsorption of water on α- and γ-alumina," *J. Catal.*, **24** [1] 82–87 (1972).
[54] G. Della Gatta, B. Fubini, and L. Stradella, "Energies of different surface rehydration processes on 'eta', 'theta' and 'alpha' aluminas," *J. Chem. Soc. Faraday Trans. 2 Mol. Chem. Phys.*, **73** [7] 1040–1049 (1977).
[55] J.G. Chen, J.E. Crowell, and J.T. Yates Jr, "Assignment of a surface vibrational mode by chemical means: modification of the lattice modes of Al_2O_3 by a surface reaction with H_2O," *J. Chem. Phys.*, **84** [10] 5906–5909 (1986).
[56] P.A. Thiel and T.E. Madey, "The interaction of water with solid surfaces: fundamental aspects," *Surf. Sci. Rep.*, **7** [6–8] 211–385 (1987).
[57] B.G. Frederick, G. Apai, and T.N. Rhodin, "Electronic and vibrational properties of hydroxylated and dehydroxylated thin Al_2O_3 films," *Surf. Sci.*, **244** [1–2] 67–80 (1991).

[58] P.F. Rossi, G. Oliveri, and M. Bassoli, "Heat of water chemisorption on α-Al_2O_3 at 200–400°C," *J. Chem. Soc. Faraday Trans.*, **90** [2] 363–367 (1994).

[59] J.M. McHale, A. Navrotsky, and A.J. Perrotta, "Effects of increased surface area and chemisorbed H_2O on the relative stability of nanocrystalline γ-Al_2O_3 and α-Al_2O_3," *J. Phys. Chem. B*, **101** [4] 603–613 (1997).

[60] J.M. McHale, A. Auroux, A.J. Perrotta, and A. Navrotsky, "Surface energies and thermodynamic phase stability in nanocrystalline aluminas," *Science*, **277** [5327] 788–791 (1997).

[61] K.C. Hass, W.F. Schneider, A. Curioni, and W. Andreoni, "The chemistry of water on alumina surfaces: reaction dynamics from first principles," *Science*, **282** [5387] 265–268 (1998).

[62] J. Nevelos, E. Ingham, C. Doyle, *et al.*, "Microseparation of the centers of alumina-alumina artificial hip joints during simulator testing produces clinically relevant wear rates and patterns," *J. Arthroplasty*, **15** [6] 793–795 (2000).

[63] M. Manaka, I.C. Clarke, K. Yamamoto, T. Shishido, A. Gustafson, and A. Imakiire, "Stripe wear rates in alumina THR-comparison of microseparation simulator study with retrieved implants," *J. Biomed. Mater. Res. Part B Appl. Biomater.*, **69** [2] 149–157 (2004).

[64] H. Zeng, K.K. Chittur, and W.R. Lacefield, "Analysis of bovine serum albumin adsorption on calcium phosphate and titanium surfaces," *Biomaterials*, **20** [4] 377–384 (1999).

[65] Q. Garrett, H.J. Griesser, B.K. Milthorpe, and R.W. Garrett, "Irreversible adsorption of human serum albumin to hydrogel contact lenses: a study using electron spin resonance spectroscopy," *Biomaterials*, **20** [14] 1345–1356 (1999).

[66] H. Mishina and M. Kojima, "Changes in human serum albumin on arthroplasty frictional surfaces," *Wear*, **265** [5–6] 655–663 (2008).

[67] J.E. Nevelos, F. Prudhommeaux, M. Hamadouche, *et al.*, "Comparative analysis of two different types of alumina-alumina hip prosthesis retrieved for aseptic loosening," *J. Bone Joint Surg.*, **83-B** [4] 598–603 (2001).

[68] D.E. Deckman, S. Jahanmir, and S.M. Hsu, "Wear mechanisms of α-alumina lubricated with a paraffin oil," *Wear*, **149** [1–2] 155–168 (1991).

[69] L.B. Dahl, I.M. Dahl, A. Engström-Laurent, and K. Granath, "Concentration and molecular weight of sodium hyaluronate in synovial fluid from patients with rheumatoid arthritis and other arthropathies," *Ann. Rheum. Dis.*, **44** [12] 817–822 (1985).

[70] K.H. Lee and J.H. Crawford, "Luminescence of the F center in sapphire," *Phys. Rev. B*, **19** [6] 3217–3221 (1979).

[71] B.G. Draeger and G.P. Summers, "Defects in unirradiated α-Al_2O_3," *Phys. Rev. B*, **19** [2] 1172–1177 (1979).

[72] M. Ghamnia, C. Jardin, and M. Bouslama, "Luminescent centres F and F+ in α-alumina detected by cathodoluminescence technique," *J. Electron Spectros. Relat. Phenomena*, **133** [1–3] 55–63 (2003).

[73] G. Brandt and M. Mikus, "An electron microprobe and cathodoluminescence study of chemical reactions between tool and workpiece when turning steel with alumina-based ceramics," *Wear*, **115** [3] 243–263 (1987).

[74] T. Shishido, I.C. Clarke, P. Williams, *et al.*, "Clinical and simulator wear study of alumina ceramic THR to 17 years and beyond," *J. Biomed. Mater. Res. Part B, Appl. Biomaterials*, **67** [1] 638–647 (2003).

[75] J. Chevalier, L. Gremillard, A. V. Virkar, and D.R. Clarke, "The tetragonal-monoclinic transformation in zirconia: lessons learned and future trends," *J. Am. Ceram. Soc.*, **92** [9] 1901–1920 (2009).

[76] W. Burger and H.G. Richter, "High strength and toughness alumina matrix composites by transformation toughening and 'in situ' platelet reinforcement (ZPTA) – the new generation of bioceramics," *Key Eng. Mater.*, **192–195** 545–548 (2001).

[77] G.E. Brown, V.E. Henrich, W.H. Casey, *et al.*, "Metal oxide surfaces and their interactions with aqueous solutions and microbial organisms," *Chem. Rev.*, **99** [1] 77–174 (1999).

[78] R. Baxter, P. Reinhardt, N. López, and F. Illas, "The extent of relaxation of the α-Al_2O_3 (0001) surface and the reliability of empirical potentials," *Surf. Sci.*, **445** [2–3] 448–460 (2000).

[79] J. Guo, D.E. Ellis, and D.J. Lam, "Electronic structure and energetics of sapphire (0001) and (1102) surfaces," *Phys. Rev. B*, **45** [23] 13647–13656 (1992).

[80] V.E. Puchin, J. D. Gale, A.L. Shluger, *et al.*, "Atomic structure of the (0001) corundum surface;" pp. 633–636 in *Mater. Sci. Forum.* Trans Tech Publ, 1996.

[81] Y.F. Zhukovskii, E.A. Kotomin, B. Herschend, K. Hermansson, and P.W.M. Jacobs, "The adhesion properties of the Ag/α-Al_2O_3(0001) interface: an ab initio study," *Surf. Sci.*, **513** [2] 343–358 (2002).

[82] Z. Łodziana, J.K. Nųrskov, and P. Stoltze, "The stability of the hydroxylated (0001) surface of α-Al_2O_3," *J. Chem. Phys.*, **118** [24] 11179–11188 (2003).

[83] S.N. Rashkeev, K. Sohlberg, M. V. Glazoff, J. Novak, S.J. Pennycook, and S.T. Pantelides, "Transition metal atoms on different alumina phases: the role of subsurface sites on catalytic activity," *Phys. Rev. B*, **67** [11] 115414 (2003).

[84] B. deB. Darwent, "Bond dissociation energies in simple molecules," *NIST Spec. Publ.*, **NSRDS-NBS** [CODEN: NSRDA] 1–52 (1970).

[85] R.H. Petrucci, W.S. Harwood, G.F. Herring, and J.D. Madura, *General Chemistry: Principles and Modern Applications*. Prentice Hall, Saddle River, NJ, USA, 2011.

[86] I.C. Clarke, G. Pezzotti, D.D. Green, H. Shirasu, and T. Donaldson, "Severe simulation test for run-in wear of all-alumina compared to alumina composite THR;" pp. 11–20 in *Bioceram. Altern. Bear. Joint Arthroplast.* Edited by J.A. D'Antonio and M. Dietrich. Steinkopff Verlag, Darmstadt, 2006.

[87] I. C. Clarke, D.D. Green, P.S. Williams, *et al.*, "Hip-simulator wear studies of an alumina-matrix composite (AMC) ceramic compared to retrieval studies of AMC balls with 1–7 years follow-up," *Wear*, **267** [5–8] 702–709 (2009).

[88] F.J. Medel, P. Shah, and S.M. Kurtz, "Retrieval analysis of contemporary alternative femoral head materials: oxinium and Biolox Delta;" Poster No. 2300 in *55th Annu. Meet. Orthop. Res. Soc.* Orthopaedic Research Society, Las Vegas, NV, USA, 2009.

[89] S. Affatato, E. Modena, A. Toni, and P. Taddei, "Retrieval analysis of three generations of Biolox® femoral heads: spectroscopic and SEM characterisation," *J. Mech. Behav. Biomed. Mater.*, **13** 118–28 (2012).

[90] G. Pezzotti, K. Yamada, S. Sakakura, and R.P. Pitto, "Raman spectroscopic analysis of advanced ceramic composite for hip prosthesis," *J. Am. Ceram. Soc.*, **91** [4] 1199–1206 (2008).

[91] G. Pezzotti, K. Yamada, A.A. Porporati, M. Kuntz, and K. Yamamoto, "Fracture toughness analysis of advanced ceramic composite for hip prosthesis," *J. Am. Ceram. Soc.*, **92** [8] 1817–1822 (2009).

[92] M. Kuntz, N. Shneider, and R. Heros, "Controlled zirconia phase transformation in BIOLOX®delta – a feature of safety;" pp. 79–84 in *Bioceram. Altern. Bear. Joint Arthroplast.* Steinkopff, New York, 2005.

[93] S. Fabris, "A stabilization mechanism of zirconia based on oxygen vacancies only," *Acta Mater.*, **50** [20] 5171–5178 (2002).

[94] L.M. Acuña, R.O. Fuentes, M.C.A. Fantini, and D.G. Lamas, "Relation between distortions in the oxygen sublattice and the local order of Zr in nanostructured ZrO_2–CeO_2 mixed oxides," *J. Phys. Chem. C*, **118** [21] 11445–11453 (2014).

[95] J. Chevalier, S. Grandjean, M. Kuntz, and G. Pezzotti, "On the kinetics and impact of tetragonal to monoclinic transformation in an alumina/zirconia composite for arthroplasty applications," *Biomaterials*, **30** [29] 5279–5282 (2009).

[96] Y. Takigawa, Y. Naka, and K. Higashi, "Effect of titania doping on phase stability of zirconia bioceramics in hot water," *Mater. Trans.*, **48** [3] 332–336 (2007).

[97] J.T. Edsall and J. Wyman, *Biophysical Chemistry. Vol. 1, Thermodynamics, Electrostatics, and the Biological Significance of the Properties of Matter*, 2nd Edition. Academic Press, New York, NY, USA, 1958.

[98] D.S. Fox, E.J. Opila, Q.N. Nguyen, D.L. Humphrey, and S.M. Lewton, "Paralinear oxidation of silicon nitride in a water-vapor/oxygen environment," *J. Am. Ceram. Soc.*, **86** [8] 1256–1261 (2003).

[99] Y. Jo, J.A. Schultz, S. Tachi, S. Contarini, and J.W. Rabalais, "Surface stoichiometry, structure, and chemisoprtion on silicon nitride studied by direct recoiling, X-ray photoelectron spectroscopy, and Auger electron spectroscopy," *J. Appl. Phys.*, **60** [7] 2564–2572 (1986).

[100] V. V. Tsukruk and V.N. Bliznyuk, "Adhesive and friction forces between chemically modified silicon and silicon nitride surfaces," *Langmuir*, **14** [2] 446–455 (1998).

[101] Y. Hibi and Y. Enomoto, "Lubrication of Si_3N_4 and Al_2O_3 in water with and without addition of silane coupling agents in the range of 0.05–0.10 mol/l," *Tribol. Int.*, **28** [2] 97–105 (1995).

[102] B. Arkles, "Hydrophobicity, hydrophilicity and silanes," *Paint Coatings Ind. Mag.*, [Oct 2006] 114–131 (2006).

[103] G.A. Parks and P.L. de Bruyn, "The zero point of charge of oxides," *J. Phys. Chem.*, **66** [6] 967–973 (1962).

[104] G.A. Parks, "The isoelectric points of solid oxides, solid hydroxides, and aqueous hydroxo complex systems," *Chem. Rev.*, **65** [2] 177–198 (1965).

[105] G.A. Parks, "Aqueous surface chemistry of oxides and complex oxide minerals," *Adv. Chem. Ser.*, **67** 121–160 (1967).

[106] R. Fournier and J. Rowe, "The solubility of amorphous and high pressures silica in water at high temperatures," *Am. Mineral.*, **62** 1052–1056 (1977).

[107] Y. Zhang, S. Zheng, H. Du, S. Wang, and Y. Zhang, "Solubility of Al_2O_3 in the $Na_2O-Al_2O_3-H_2O-CH_3OH$ system at (30 and 60) °C," *J. Chem. Eng. Data*, **55** [3] 1237–1240 (2010).

[108] W. Xu, B.L. Blackford, J.G. Cordes, *et al.*, "Atomic force microscope measurements of long-range forces near lipid-coated surfaces in electrolytes," *Biophys. J.*, **72** [3] 1404–1413 (1997).

[109] J. Diao, D. Ren, J.R. Engstrom, and K.H. Lee, "A surface modification strategy on silicon nitride for developing biosensors," *Anal. Biochem.*, **343** [2] 322–328 (2005).

[110] T.J. Senden and C.J. Drummond, "Surface chemistry and tip-sample interactions in atomic force microscopy," *Colloids Surfaces A Physicochem. Eng. Asp.*, **94** [1] 29–51 (1995).

[111] Y. Takahashi, K. Yamamoto, T. Shishido, *et al.*, "Strain-induced microstructural rearrangement in ultra-high molecular weight polyethylene for hip joints: a comparison between conventional and Vitamin E-infused highly-crosslinked liners," *J. Mech. Behav. Biomed. Mater.*, **31** 31–44 (2014).

[112] Y. Takahashi, T. Masaoka, K. Yamamoto, *et al.*, "Vitamin-E blended and infused highly cross-linked polyethylene for total hip arthroplasty: a comparison of three-dimensional crystalline morphology and strain recovery behavior," *J. Mech. Behav. Biomed. Mater.*, **36** 59–70 (2014).

[113] P. Bracco, E.M. Brach del Prever, M. Cannas, M.P. Luda, and L. Costa, "Oxidation behaviour in prosthetic UHMWPE components sterilised with high energy radiation in a low-oxygen environment," *Polym. Degrad. Stab.*, **91** [9] 2030–2038 (2006).

[114] P. Bracco, V. Brunella, M.P. Luda, E.M. Brach del Prever, M. Zanetti, and L. Costa, "Oxidation behaviour in prosthetic UHMWPE components sterilised with high-energy radiation in the presence of oxygen," *Polym. Degrad. Stab.*, **91** [12] 3057–3064 (2006).

[115] L. Costa, M.P. Luda, and L. Trossarelli, "Ultra high molecular weight polyethylene—II. Thermal- and photo-oxidation," *Polym. Degrad. Stab.*, **58** [1–2] 41–54 (1997).

[116] K.K. Wannomae, S. Bhattacharyya, A. Freiberg, D. Estok, W.H. Harris, and O. Muratoglu, "In vivo oxidation of retrieved cross-linked ultra-high-molecular-weight polyethylene acetabular components with residual free radicals," *J. Arthroplasty*, **21** [7] 1005–1011 (2006).

[117] S.K. Bhateja, E.H. Andrews, and S.M. Yarbrough, "Radiation induced crystallinity changes in linear polyethylenes: long term aging effects," *Polym. J.*, **21** [9] 739–750 (1989).
[118] V. Premnath, W.H. Harris, M. Jasty, and E.W. Merrill, "Gamma sterilization of UHMWPE articular implants: an analysis of the oxidation problem," *Biomaterials*, **17** [18] 1741–1753 (1996).
[119] B. Yeom, Y.-J. Yu, H.A. McKellop, and R. Salovey, "Profile of oxidation in irradiated polyethylene," *J. Polym. Sci. Part A Polym. Chem.*, **36** [2] 329–339 (1998).
[120] F.J. Medel, C.M. Rimnac, and S.M. Kurtz, "On the assessment of oxidative and microstructural changes after in vivo degradation of historical UHMWPE knee components by means of vibrational spectroscopies and nanoindentation spectroscopies and nanoindentation," *J. Biomed. Mater. Res.*, **89A** [2] 530–538 (2009).
[121] J.G. Bowsher and I.C. Clarke, "Thermal conductivity of femoral ball strongly influenced UHMWPE wear in a hip simulator;" p. 0278 in *Trans. 53rd Annu. Meet. Orthop. Res. Soc.* Orthopaedic Research Society, San Diego, CA, 2007.
[122] W. Zhu, B.J. McEntire, Y. Enomoto, M. Boffelli, and G. Pezzotti, "Point-defect populations as induced by cation/anion substitution in β-Si_3N_4 lattice – a cathodoluminescence study," *J. Phys. Chem. C*, **119** [6] 3279–3287 (2015).
[123] L. Skuja, "Direct singlet-to-triplet optical absorption and luminescence excitation band of the twofold-coordinated silicon center in oxygen-deficient glassy SiO_2," *J. Non. Cryst. Solids*, **167** [3] 229–238 (1994).
[124] M. Goldberg, H.-J. Fitting, and A. Trukhin, "Cathodoluminescence and cathodoelectroluminescence of amorphous SiO_2 films," *J. Non. Cryst. Solids*, **220** [1] 69–77 (1997).
[125] V.A. Gritsenko, Y.G. Shavalgin, P.A. Pundur, H. Wong, and W.M. Lau, "Cathodoluminescence and photoluminescence of amorphous silicon oxynitride," *Microelectron. Reliab.*, **39** [5] 715–718 (1999).
[126] H.-J. Fitting, T. Barfels, A.N. Trukhin, and B. Schmidt, "Cathodoluminescence of crystalline and amorphous SiO_2 and GeO_2," *J. Non. Cryst. Solids*, **279** [1] 51–59 (2001).
[127] R. Kue, A. Sohrabi, D. Nagle, C. Frondoza, and D. Hungerford, "Enhanced proliferation and osteocalcin production by human osteoblast-like MG63 cells on silicon nitride ceramic discs," *Biomaterials*, **20** [13] 1195–201 (1999).
[128] M. Amaral, M.A. Lopes, J.D. Santos, and R.F. Silva, "Wettability and surface charge of Si_3N_4-bioglass composites in contact with simulated physiological liquids," *Biomaterials*, **23** [20] 4123–4129 (2002).
[129] M. Amaral, M.A. Lopes, R.F. Silva, and J.D. Santos, "Densification route and mechanical properties of Si_3N_4-bioglass biocomposites," *Biomaterials*, **23** [3] 857–862 (2002).

[130] M. Amaral, M.A. Costa, M.A. Lopes, R.F. Silva, J.D. Santos, and M.H. Fernandes, "Si_3N_4-bioglass composites stimulate the proliferation of MG63 osteoblast-like cells and support the osteogenic differentiation of human bone marrow cells," *Biomaterials*, **23** [24] 4897–4906 (2002).

[131] C.R. Howlett, E. McCartney, and W. Ching, "The effect of silicon nitride ceramic on rabbit skeletal cells and tissue," *Clin. Orthop. Relat. Res.*, **244** 293–304 (1989).

[132] C.-H. Jou, L. Yuan, and S.M. Lin, "Biocompatibility and antibacterial activity of chitosan and hyaluronic acid immobilized polyester fibers," *J. Appl. Polym. Sci.*, **104** 220–225 (2007).

[133] T.J. Webster, A.A. Patel, M.N. Rahaman, and B.S. Bal, "Anti-infective and osteointegration properties of silicon nitride, poly (ether ether ketone), and titanium implants," *Acta Biomater.*, **8** [12] 4447–4454 (2012).

[134] T. Ikeda, H. Hirayama, H. Yamaguchi, S. Tazuke, and M. Watanabe, "Polycationic biocides with pendant active groups: molecular weight dependence of antibacterial activity," *Antimicrob. Agents Chemother.*, **30** [1] 132–136 (1986).

[135] S.I. Raider, R. Flitsch, J.A. Aboaf, and W.A. Pliskin, "Surface oxidation of silicon nitride films," *J. Electrochem. Soc.*, **123** [4] 560–565 (1976).

[136] G. Busca, V. Lorenzelli, M.I. Baraton, P. Quintard, and R. Marchand, "FT-IR characterization of silicon nitride Si_3N_4 and silicon oxynitride Si_2ON_2 surfaces," *J. Mol. Struct.*, **143** 525–528 (1986).

[137] R. Stine, C.L. Cole, K.M. Ainslie, S.P. Mulvaney, and L.J. Whitman, "Formation of primary amines on silicon nitride surfaces: a direct , plasma-based pathway to functionalization," *Langmuir*, **23** 4400–4404 (2007).

[138] M. Rahaman, Y. Boiteux, and L. DeJonghe, "Surface characterization of silicon nitride and silicon carbide powders," *Bull. Am. Ceram. Soc.*, **65** [8] 1171–1176 (1986).

[139] R.K. Brow and C.G. Pantano, "Compositionally dependent Si 2p binding energy shifts in silicon oxynitride thin films," *J. Am. Ceram. Soc.*, **69** [4] 314–316 (1986).

[140] P.K. Whitman and D.L. Feke, "Colloidal characterization of ultrafine silicon carbide and silicon nitride powders," *Adv. Ceram. Mater.*, **1** 366–370 (1986).

[141] W. Hartmann, F. Van Dijen, R. Metselaar, and C. Siskens, "Slip casting parameters for commercial Si_3N_4 powders," *J. Phys. Colloq.*, **47** [C1] C1-79–C1-83 (1986).

[142] Y. Imamura, K. Ishibashi, and H. Shimodaira, "Characterization of silicon nitride surface;" pp. 828–838 in *10th Annu. Conf. Compos. Adv. Ceram. Mater. Ceram. Eng. Sci. Proc.* John Wiley & Sons, Inc., 1986.

[143] L. Bergstrom and R.J. Pugh, "Interfacial characterization of silicon nitride powders," *J. Am. Ceram. Soc.*, **72** [1] 103–109 (1989).

[144] R.W. Bower, M.S. Ismail, and B.E. Roberds, "Low temperature Si_3N_4 direct bonding," *Appl. Phys. Lett.*, **62** [26] 3485–3487 (1993).

[145] P. Serwer, "Agarose gels: properties and use for electrophoresis," *Electrophoresis*, **4** [6] 375–382 (1983).

[146] Y. Kitasako, T. Nikaido, J. Tagami, K. Ikeda, K. Mitsunari, and S. Nomura, "Surface analysis of human teeth using a pH imaging microscope based on a semiconductor silicon sensor," *Bunseki Kagaku*, **49** [5] 325–327 (2000).

[147] X. Sui, B. Wang, Y. Liu, H. Wu, Q. Li, and Q. Zhang, "Study on microstructure of polysilicic acid in high salinity," *J. Appl. Sci. Eng. Innov.*, **1** [1] 44–48 (2014).

[148] T. Mizutani, H. Nagase, N. Fujiwara, and H. Ogoshi, "Silicic acid polymerization catalyzed by amines and polyamines," *Bull. Chem. Soc. Jpn.*, **71** 2017–2022 (1998).

[149] G. Pezzotti, L. Puppulin, A. La Rosa, *et al.*, "Effect of pH and monovalent cations on the Raman spectrum of water: basics revisited and application to measure concentration gradients at water/solid interface in Si_3N_4 biomaterials," *J. Phys. Chem. C*, **463** 120–136 (2015).

[150] A. Maroudas, "Balance between swelling pressure and collagen tension in normal and degenerate cartilage," *Nature*, **260** [5554] 808–809 (1976).

[151] M.S. McQueney, B.Y. Amegadzie, K. D'Alessio, *et al.*, "Autocatalytic activation of human cathepsin K," *J. Biol. Chem.*, **272** [21] 13955–13960 (1997).

[152] A.J. Barrett, J.F. Woessner, and N.D. Rawlings (eds.), "Cathepsin K;" pp. 1092–1097 in *Handb. Proteolytic Enzym.* Elsevier Academic Press, San Diego, CA, USA, 2012.

[153] Y.T. Konttinen, J. Mandelin, T.-F. Li, *et al.*, "Acidic cysteine endoproteinase cathepsin K in the degeneration of the superficial articular hyaline cartilage in osteoarthritis," *Arthritis Rheum.*, **46** [4] 953–960 (2002).

[154] M. Farr, K. Garvey, A.M. Bold, M.J. Kendall, and P.A. Bacon, "Significance of the hydrogen ion concentration in synovial fluid in rheumatoid arthritis," *Clin. Exp. Rheumatol.*, **3** 99–104 (1985).

[155] D.J. Gorth, S. Puckett, B. Ercan, T.J. Webster, M. Rahaman, and B.S. Bal, "Decreased bacteria activity on Si_3N_4 surfaces compared with PEEK or titanium," *Int. J. Nanomedicine*, **7** 4829–4840 (2012).

[156] P.N. Galgut, "The relevance of pH to gingivitis and periodontitis," *J. Int. Acad. Periodontol.*, **3** [3] 61–67 (2001).

[157] S. Baliga, S. Muglikar, and R. Kale, "Salivary pH: a diagnostic biomarker," *J. Indian Soc. Periodontol.*, **17** [4] 461–465 (2013).

[158] M.G. Zywiel, S.A. Sayeed, A.J. Johnson, T.P. Schmalzried, and M.A. Mont, "State of the art in hard-on-hard bearings: how did we get here and what have we achieved?" *Expert Rev. Med. Devices*, **8** 187–207 (2011).

[159] M. Al-Hajjar, J. Fisher, J.L. Tipper, S. Williams, and L.M. Jennings, "Wear of 36-mm BIOLOX®delta ceramic-on-ceramic bearing in total hip replacements under edge loading conditions," *Proc. Inst. Mech. Eng. H.*, **227** [5] 535–342 (2013).

[160] L. Mattei, F. Di Puccio, B. Piccigallo, and E. Ciulli, "Lubrication and wear modelling of artificial hip joints: a review," *Tribol. Int.*, **44** [5] 532–549 (2011).

[161] H. Reza, S. Hosseinzadeh, A. Eajazi, and A.S. Shahi, "The bearing surfaces in total hip arthroplasty – options, material characteristics and selection;" pp. 163–210 in *Recent Adv. Arthroplast.* Edited by Samo Fokter. 2012.

[162] W.G. Hamilton, J.P. McAuley, D.A. Dennis, J.A. Murphy, T.J. Blumenfeld, and J. Politi, "THA with delta ceramic on ceramic: results of a multicenter investigational device exemption trial," *Clin. Orthop. Relat. Res.*, **468** [2] 358–366 (2010).

[163] Z. Lu and H. McKellop, "Frictional heating of bearing materials tested in a hip joint wear simulator," *Proc. Inst. Mech. Eng. Part H J. Eng. Med.*, **211** [1] 101–108 (1997).

[164] K. Kato and K. Adachi, "Wear of advanced ceramics," *Wear*, **253** [11–12] 1097–1104 (2002).

[165] B. Kerkwijk, L. Winnubst, E.J. Mulder, and H. Verweij, "Processing of homogeneous zirconia-toughened alumina ceramics with high sliding wear resistance," *J. Am. Ceram. Soc.*, **82** [8] 2087–2093 (1999).

[166] K. Adachi, K. Kato, and N. Chen, "Wear map of ceramics," *Wear*, **203–204** 291–301 (1997).

[167] T.E. Fischer, M.P. Anderson, and S. Jahanmir, "Influence of fracture toughness on the wear resistance of yttria-doped zirconium oxide," *J. Am. Ceram. Soc.*, **72** [2] 252–257 (1989).

[168] A.G. Evans and D.B. Marshall, "Wear mechanism in ceramics;" p. 439 in *Fundam. Frict. Wear Mater.* Edited by D.A. Rigney. American Society for Metals, Metals Park, OH, USA, 1981.

[169] A.G. Evans and T.R. Wilshaw, "Quasi-static solid particle damage in brittle solids—I. Observations analysis and implications," *Acta Metall.*, **24** [10] 939–956 (1976).

[170] S.T. Buljan and V.K. Sarin, "The future of silicon nitride cutting tools," *Carbide Tool J.*, **14** 4–7 (1985).

[171] K. Yamada and G. Pezzotti, "Environmental phase stability of ceramics composite for hip prostheses in presence of surface damage," *Key Eng. Mater.*, 361–363 783–786 (2008).

[172] J. Seidel and J.S. Rödel, "Measurement of crack tip toughness in alumina as a function of grain size," *J. Am. Ceram. Soc.*, **80** [2] 433–438 (1997).

[173] T. Fett, *Computation of the Crack Opening Displacements for Vickers Indentation Cracks*. Karlsruhe, Germany, 2002.

[174] J.J. Kruzic, R.L. Satet, M.J. Hoffmann, R.M. Cannon, and R.O. Ritchie, "The utility of R-curves for understanding fracture toughness-strength relations in bridging ceramics," *J. Am. Ceram. Soc.*, **91** [6] 1986–1994 (2008).

[175] X.-K. Zhu and J.A. Joyce, "Review of fracture toughness (G, K, J, CTOD, CTOA) testing and standardization," *Eng. Fract. Mech.*, **85** 1–46 (2012).

[176] A. Neimitz, I. Dzioba, J. Gałkiewicz, and R. Molasy, "A study of stable crack growth using experimental methods, finite elements and fractography," *Eng. Fract. Mech.*, **71** [9–10] 1325–1355 (2004).

[177] P.F. Becher, E.Y. Sun, K.P. Plucknett, *et al.*, "Microstructural design of silicon nitride with improved fracture toughness: I, effects of grain shape and size," *J. Am. Ceram. Soc.*, **81** [11] 2821–2830 (1998).

[178] F.P. Fehlner, *Low Temperature Oxidation: The Role of Vitreous Oxides*. John Wiley and Sons, New York, NY, USA, 1986.

[179] Y. Miura, M. Hasegawa, A. Sudo, G. Pezzotti, and L. Puppulin, "In-vivo degradation of middle-term highly cross-linked and remelted polyethylene cups: modification induced by creep, wear and oxidation," *J. Mech. Behav. Biomed. Mater.*, **51** 13–24 (2015).

[180] G. Pezzotti, R.M. Bock, B.J. McEntire, *et al.*, "Silicon nitride bioceramics induce chemically driven lysis in *Porphyromonas gingivalis*," *Langmuir*, (2016).

[181] G. Pezzotti, Y. Enomoto, W. Zhu, M. Boffelli, E. Marin, and B.J. McEntire, "Surface toughness of silicon nitride bioceramics: I, Raman spectroscopy-assisted micromechanics," *J. Biomed. Mater. Res.*, **54** 328–345 (2015).

[182] B.J. McEntire, Y. Enomoto, W. Zhu, M. Boffelli, E. Marin, and G. Pezzotti, "Surface toughness of silicon nitride bioceramics: II, comparison with commercial oxide materials," *J. Mech. Behav. Biomed. Mater.*, **54** 346–359 (2016).

Index